Introduction to Electrochemical Science and Engineering

Introduction to Electrochemical Science and Engineering

Second Edition

Serguei N. Lvov

CRC Press
Taylor & Francis Group
Boca Raton London New York

CRC Press is an imprint of the
Taylor & Francis Group, an **informa** business

Second edition published 2022
by CRC Press
6000 Broken Sound Parkway NW, Suite 300, Boca Raton, FL 33487-2742

and by CRC Press
2 Park Square, Milton Park, Abingdon, Oxon, OX14 4RN

© 2022 Taylor & Francis Group, LLC

First edition published by CRC Press 2015

CRC Press is an imprint of Taylor & Francis Group, LLC

Library of Congress Cataloging-in-Publication Data
Names: Lvov, Serguei N., author.
Title: Introduction to electrochemical science and engineering / Serguei N. Lvov.
Description: Second edition. | Boca Raton : CRC Press, 2021. | Includes
  bibliographical references and index. | Summary: "This new edition
  updates nearly every chapter, expanding coverage of solild-state
  electrochemistry, hydrogen and fuel cells, energy storage, and
  electrochemical techniques. It includes many case studies related to
  novel electrochemical energy conversion systems and phenomena such as
  batteries, flow cells, fuel cells, electrolyzers, supercapacitors,
  photoelectrochemical systems, bioelectrochemical systems, corrosion in
  high temperature/pressure brines, new pH, and reference electrodes. It
  includes about 360 conceptual and numerical problems and their detailed
  solutions. The book features 8 lab descriptions and instructions and is
  integrated with 9 online videos of lab demonstrations"— Provided by
  publisher.
Identifiers: LCCN 2021031430 (print) | LCCN 2021031431 (ebook) |
  ISBN 9781138196780 (hardback) | ISBN 9781315296852 (ebook)
Subjects: LCSH: Electrochemistry. | Chemical engineering.
Classification: LCC QD553 .L96 2021 (print) | LCC QD553 (ebook) | DDC
  541/.37—dc23
LC record available at https://lccn.loc.gov/2021031430
LC ebook record available at https://lccn.loc.gov/2021031431

ISBN: 978-1-138-19678-0 (hbk)
ISBN: 978-1-032-07300-2 (pbk)
ISBN: 978-1-3152-9685-2 (ebk)

DOI: 10.1201/b21893

Typeset in Times
by codeMantra

Access the Support Material: https://www.routledge.com/9781138196780

Contents

Preface

After publishing the first edition of the book, a new animated video-based course for studying the most common electrochemical techniques was developed. For an instructor who is interested in conducting these laboratories with real instrumentation and chemicals, Appendix C was prepared with detailed description how each of the laboratories can be organized, what data should be collected, and how the data should be treated and presented in a report.

The author recognized that additional examples of the conceptual and numerical problems should be provided to make this book useful not only for the senior undergraduates but also for graduates and professionals. These examples, placed at the end of each chapter, are designed to extend some important aspects of the text.

In addition, a new section on electrochemical impedance spectroscopy (EIS) was prepared and placed in Chapter 7 of this book. The EIS techniques are now widely used and, therefore, should be studied even in the introductory course of electrochemical science and engineering.

Finally, the author tried to correct all typos and uncertainties from the first edition, but is certain that there are still a number of issues to be taken care of for making this book better and, therefore, welcomes any comments and criticism from all readers. The author's email can easily be found at the Penn State website, https://www.psu.edu.

MATLAB® is a registered trademark of The MathWorks, Inc. For product information, please contact:
The MathWorks, Inc.
3 Apple Hill Drive
Natick, MA 01760-2098 USA
Tel: 508-647-7000
Fax: 508-647-7001
E-mail: info@mathworks.com
Web: www.mathworks.com

Acknowledgments

I cannot adequately express appreciation to my wife, Marina, whose patience and permanent support played the key role in sustaining me through the challenge of preparing both the first and second editions of the book. This book would never have been properly completed without a lot of help from my former students, current students, and coworkers. Assistance from my former graduate research associates Mark Fedkin and Justin Beck; graduate students Sarah Hipple, Chunmei Wang, Mark LaBarbera, Soohyun Kim, Rich Schatz, Sanchit Khurana, Debanjan Das, and Ruishu Feng; and undergraduates Eric LaRow, Alexander Morse, and Rosemary Cianni are gratefully acknowledged. Furthermore, I am particularly grateful to my former graduate student Haining Zhao, who has prepared most of the figures. Also, I would like to thank Dr. Andre Anderko and Dr. George Engelhardt of OLI Systems who have given their time to read and comment on all sections of the first edition of the book. Preparation of the second edition of the book would be difficult without significant help from my current students Timothy Duffy, Jiaxi Li, and Clayton Colson. I would like to particularly thank Timothy Duffy for proofreading the text and working on updating the book index. Consultations with my colleagues, particularly with professor, Derek Hall, are gratefully acknowledged. I would also like to thank Dr. George Engelhardt who have given their time to comment on some sections of the updated text. Finally, I thank The Pennsylvania State University for granting me a sabbatical, which I have fruitfully used for preparing the second edition of the book.

Author

Serguei N. Lvov, PhD, is a full professor of energy and mineral engineering and materials science and engineering, and director of the Electrochemical Technologies Program at the EMS Energy Institute of The Pennsylvania State University, University Park, PA, USA. He earned a PhD from the St. Petersburg Institute of Technology, Russia, in 1980 and a DSc in physical chemistry from St. Petersburg State University in 1992. Prior to his tenure at The Pennsylvania State University, he worked at the St. Petersburg School of Mines and the Russian Academy of Science. He has been a visiting scholar at the University of Venice, Italy, the University of Delaware, Newark, USA, and the National Centre for Scientific Research, Vandœuvre-lès-Nancy, France, and was the recipient of a 2019 US Fulbright Scholar Award. His main areas of research are electrochemistry and thermodynamics of aqueous systems under extreme environments such as elevated temperatures and pressures or highly concentrated solutions. He is the author of more than 170 peer-reviewed papers, 6 book chapters, 2 books, and 2 US patents.

Introduction

Power generation demand is constantly growing, but fossil fuels are limited, and we should not entirely count on them in the long run without developing new technologies for efficient, renewable, and clean electric power generation. Many countries throughout the world are starting a number of new initiatives for energy development, and most of them are based on electrochemical energy conversion systems such as fuel and flow cells, electrolyzers, and batteries. At the same time, curricula in engineering education do not give sufficient attention to the electrochemical sciences and commonly emphasize combustion and nuclear engineering processes related to low-cost fossil fuel and nuclear power generation.

The author's experience at The Pennsylvania State University shows that there is growing interest from engineering students in electrochemical energy conversion systems and related technologies. Furthermore, it is evident that interest in these technologies will increase with the continued development of renewable power generation systems.

This introductory-level book aims to fill that gap in the undergraduate engineering education. It introduces the reader to basic principles for a number of growing electrochemical engineering–related technologies, including some of the mentioned earlier. Moreover, this book also provides the reader with a basic understanding of the fundamental concepts of electrochemical science and engineering, such as electrolyte solution, electrochemical cells, electrolyte conductivity, electrode potential, and current-potential relations related to a number of electrochemical systems. Three examples of popular developing technologies directly related to the contents of this book are (1) fuel cell cars, which can significantly reduce CO_2 emission and help to battle global warming; (2) new efficient and inexpensive flow batteries, which can be important for a variety of electric grid applications; and (3) a water electrolysis system combined with solar energy generator, which can be used to produce a clean and renewable fuel—hydrogen. New electrochemical technologies/systems cannot be properly understood without concepts from the electrochemical science and engineering discipline, which are presented in this text.

The educational approach within this book has been verified by the author throughout his 26 years of experience teaching a number of the undergraduate electrochemistry-related courses at The Pennsylvania State University. At this university, these courses are key components in the energy engineering major as well as in the electrochemical engineering undergraduate and electrochemical science and engineering graduate minors that were developed for all engineering and science students.

In addition to undergraduates, this book is beneficial for graduate students, industry professionals, and researchers who have not studied electrochemistry during their college period but need to work in one of the areas related to this subject. Experienced readers can use this textbook to refresh his or her electrochemical background or for teaching the subject. Graduate students and engineers who are looking for a quick introduction to the subject will benefit from the simple structure of this book.

To help foster the understanding of this subject, the reader is provided with more than 360 conceptual and numerical problems distributed over nine quizzes and nine video-based assignments on all nine chapters of the textbook, in addition to exercises given at the end of each chapter. The reader is encouraged to work through the included example problems and exercises, which use reliable reference data from the *CRC Handbook of Chemistry and Physics*. The quizzes in Appendix A provide the readers an opportunity to check their level of understanding for each chapter. Any student can work through all quizzes and assignments by solely relying on the book materials. This feature of the book was found very useful by students who have taken the author's courses.

Another goal of the book is to develop the initial skills for understanding electrochemical experimentation and data analysis without visiting a laboratory. The approach here is to appreciate the capability and applications of a number of electrochemical techniques, which are presented in corresponding videos of laboratory experiments along with video-based assignments located in Appendix B. All laboratories are described in detail in Appendix C. Videos with lab demonstrations are available for the students via YouTube.

This introductory-level textbook attempts to keep a balance between the relative simplicity and accuracy of the materials in all book sections. Therefore, a number of traditional discussions that are common to most electrochemical science and engineering books are excluded. This approach was successfully tested by the author teaching a one-semester course during many years. While there are a few well-known texts addressing electrochemical science and engineering from different angles, the author and his students have found their use to be difficult in undergraduate education. Most of them are listed here in alphabetic order:

V.S. Bagotsky, *Fundamentals of Electrochemistry*, 2nd edn., Wiley, Hoboken, NJ, 2006.

A.J. Bard and L.R. Faulkner, *Electrochemical Methods: Fundamentals and Applications*, Wiley, 2001.

J.O'M. Bockris and A.K.N. Reddy, *Volume 1—Modern Electrochemistry: Ionics*, 2nd edn., Plenum Press, New York, 1998.

J.O'M. Bockris and A.K.N. Reddy, *Volume 2B—Modern Electrochemistry: Electrodics in Chemistry, Engineering, Biology, and Environmental Science*, 2nd edn., Plenum Press, New York, 2000.

J.O'M. Bockris, A.K.N. Reddy, and M. Gamboa-Aldeco, *Volume 2A—Modern Electrochemistry: Fundamentals of Electrodics*, 2nd edn., Plenum Press, New York, 2000.

C.M.A. Brett and A.M.O. Brett, *Electrochemistry: Principles, Methods, and Applications*, Oxford University Press, Oxford, UK, 1993.

T.F. Fuller and J.N. Harb, *Electrochemical Engineering*, Wiley, Hoboken, NJ, 2018.

E. Gileadi, *Electrode Kinetics for Chemical Engineers and Materials Scientists*, Wiley-VCH, New York, 1993.

E. Gileadi, *Physical Electrochemistry: Fundamentals, Techniques and Applications*, Wiley-VCH, Weinheim, Germany, 2011.

E. Gileadi, E. Kirowa-Eisner, and J. Penciner, *Interfacial Electrochemistry, An Experimental Approach*, Addison-Wesley, Reading, MA, 1975.

C.H. Hamann, A. Hamnett, and W. Vielstich, *Electrochemistry*, 2nd edn., Wiley-VCH, Weinheim, Germany, 2007.

C. Lefrou, P. Fabry, and J.-C. Poignet, *Electrochemistry. The Basics, With Examples*, Springer, Heidelberg, Germany, 2012.

J. Newman and N.P. Balsara, *Electrochemical Systems*, 4th edn., Wiley, Hoboken, NJ, 2021.

K. Oldham, J. Myland, and A. Bond, *Electrochemical Science and Technology: Fundamentals and Applications*, Wiley, Chichester, UK, 2012.

N. Perez, *Electrochemistry and Corrosion Science*, Kluwer Academic Publishers, Boston, MA, 2004.

D. Pletcher, *A First Course in Electrode Processes*, RSC Publishing, Cambridge, UK, 2009.

G. Prentice, *Electrochemical Engineering Principles*, Prentice Hall, Upper Saddle River, NJ, 1990.

W. Schmickler, *Interfacial Electrochemistry*, Oxford University Press, New York, 1996.

F. Scholz (Ed.), *Electroanalytical Methods: Guide to Experiment and Applications*, Springer, Berlin, Germany, 2002.

K.J. Vetter, *Electrochemical Kinetics*, Academic Press, New York, 1967.

J. Wang, *Analytical Electrochemistry*, 3rd edn., Wiley-VCH, Hoboken, NJ, 2006.

A.C. West, *Electrochemistry and Electrochemical Engineering. An Introduction*, Columbia University, New York, 2012.

This textbook uses the reference data of *CRC Handbook of Chemistry and Physics* (W.M. Haynes, editor-in-chief, *CRC Handbook of Chemistry and Physics*, CRC Press, 95th edn., 2014–2015), since the new annual additions have not changed the data presented in the "Data Section" of Chapter 10. Most of the quantities, symbols, and units are in agreement with IUPAC recommendations given in its well-known manual (*Quantities, Units and Symbols in Physical Chemistry*, 3rd edn., IUPAC, 2007). Using a consistent source of data, symbols, and units was found to be very important for undergraduate teaching of this subject, with a lot of symbols converging from quite different disciplines such as chemistry, physics, electrical engineering, physical chemistry, and calculus.

As in any other senior-level undergraduate subject, there are valuable prerequisites to improve the study of electrochemical science and engineering. In addition to college-level calculus, physics, and chemistry, readers will benefit from a basic understanding of chemical thermodynamics and mass tranfer, as well as basic concepts from electrical engineering. The author's teaching practice confirms that students who want to succeed in studying electrochemical science and engineering should have a sufficient knowledge in the chemistry of redox reactions, thermodynamics and kinetics of physicochemical processes, and the basics of mass tranfer and physical chemistry of electrolyte solutions.

Most of the material in this book is based on the author's courses taken by senior undergraduate students in The Pennsylvania State University engineering departments such as energy and mineral engineering, materials science and engineering, chemical engineering, civil and environmental engineering, engineering science and mechanics, mechanical engineering and nuclear engineering—these students are gratefully acknowledged for their questions and useful discussions during the author lectures and office hours. The short biographies of the electrochemical science and engineering founders presented in this book were mainly adapted from A.J. Bard, G. Inzelt, and F. Scholz (Eds.), *Electrochemical Dictionary*, Springer, 2008, and Wikipedia. The author acknowledges the opportunity to use all reference material listed in the chapters of this book. Finally, the author will welcome any comments and criticism from all readers.

List of Abbreviations

ac	Alternating current
AFC	Alkaline fuel cell
CAS RN	Chemical Abstracts Service Registry Number
CE	Corrosion rate
CE	Counter electrode
CODATA	Committee on Data for Science and Technology
CV	Cyclic voltammetry
dc	Direct current
DMFC	Direct methanol fuel cell
DP	Decomposition potential
EC	Electrolytic cell
EDL	Electric double layer
EFM	Electrochemical frequency modulation
EIS	Electrochemical impedance spectroscopy
GC	Galvanic cell
IUPAC	International Union of Pure and Applied Chemistry
LSM	Strontium-doped lanthanum manganite
LSV	Linear sweep voltammetry
MCFC	Molten carbonate fuel cell
OCP	Open circuit potential
PAFC	Phosphoric acid fuel cell
PEM	Proton exchange membrane
PEMFC	Proton exchange membrane fuel cell
PTFE	Polytetrafluoroethylene
RDE	Rotating disk electrode
RE	Reference electrode
SHE	Standard hydrogen electrode
SI	International System of Units
SOFC	Solid oxide fuel cell
WE	Working electrode
YSZ	Yttria-stabilized zirconia

List of Symbols, Their Names, and SI Units

LETTERS

A	Cross section (m^2)
A_{DH}	Theoretical constant of Debye–Hückel theory ($(kg\ mol^{-1})^{1/2}$)
$a_{\pm}$	Mean activity (dimensionless)
$a_{\pm, I}$	Mean activity of solution (I) on the left-hand side of the liquid junction (Equation 5.7) (dimensionless)
$a_{\pm, II}$	Mean activity of solution (II) on the right-hand side of the liquid junction (Equation 5.7) (dimensionless)
$\mathring{a}$	Minimum distance between the anion and the cation (m)
$a_{A^-(aq)}$	Activity of anions, $A^-(aq)$ (dimensionless)
$a_{Cl_2(g)}$	Activity of $Cl_2(g)$ (dimensionless)
$a_{Cl^-(aq)}$	Activity of $Cl^-(aq)$ (dimensionless)
$a_{Cl^-(aq,I)}$	Activity of $Cl^-(aq)$ in HCl(aq, I) solution (dimensionless)
$a_{Cl^-(aq,II)}$	Activity of $Cl^-(aq)$ in HCl(aq, II) solution (dimensionless)
$a_{Cu^{2+}(aq)}$	Activity of $Cu^{2+}(aq)$ (dimensionless)
A_{EC}	A semiempirical positive parameter in Equation 8.2 (V)
A_{FC}	A semiempirical positive parameter in Equation 8.1 (V)
$a_{Fe^{2+}(aq)}$	Activity of $Fe^{2+}(aq)$ (dimensionless)
$a_{Fe^{3+}(aq)}$	Activity of $Fe^{3+}(aq)$ (dimensionless)
$a_{H^+(aq)}$	Activity of protons, $H^+(aq)$ (dimensionless)
$a_{H_2(g)}$	Activity of gaseous molecular hydrogen, $H_2(g)$ (dimensionless)
$a_{H_2O(l)}$	Activity of water on the mole fraction concentration scale (dimensionless)
$a_{HA(aq)}$	Activity of molecules, HA(aq), on the molal concentration scale (dimensionless)
a_i	Activity (dimensionless)
$a_i(I)$	Activity of the ith ions in solution I on the left-hand side of the liquid junction (dimensionless)
$a_i(II)$	Activity of the ith ions in solution II on the right-hand side of the liquid junction (dimensionless)
$a_{i, b}$	Activity of the ith component on the molal concentration scale (dimensionless)
$a_{i, x}$	Activity of the ith component on the mole fraction concentration scale (dimensionless)
$a_{OH^-(aq)}$	Activity of hydroxyl ions, $OH^-(aq)$ (dimensionless)
A_T	Intercept of the Tafel equation (V)
$a_{Zn^{2+}(aq)}$	Activity of $Zn^{2+}(aq)$ (dimensionless)
b	Molality (molal concentration) ($mol\ kg^{-1}$)
b_+	Molality (molal concentration) of cation ($mol\ kg^{-1}$)
b_-	Molality (molal concentration) of anion ($mol\ kg^{-1}$)

b_0	Standard molality (= 1.0 mol kg^{-1})
$b_{A^-(aq)}$	Molality of anions, A$^-$(aq) (mol kg^{-1})
B_a'	Anodic Tafel slope in Equation 9.5 (V)
b_{acid}	Total concentration of acid in a buffer solution (mol kg^{-1})
b_{base}	Total concentration of base in a buffer solution (mol kg^{-1})
B_c'	Cathodic Tafel slope in Equation 9.5 (V)
B_{DH}	Theoretical constant of Debye–Hückel theory ((kg mol^{-1})$^{1/2}$ m^{-1})
B_{EC}	A semiempirical positive parameter in Equation 8.2 (V)
B_{FC}	A semiempirical positive parameter in Equation 8.1 (V)
$b_{H^+(aq)}$	Molality of protons, H$^+$(aq) (mol kg^{-1})
$b_{HA(aq)}$	Molality of molecules, HA(aq) (mol kg^{-1})
$b_{HCl(aq)}$	Molality of HCl(aq) (mol kg^{-1})
b_i	Molality (molal concentration) of the ith component (mol kg^{-1})
$b_{OH^-(aq)}$	Molality of hydroxyl ions, OH$^-$(aq) (mol kg^{-1})
b_{salt}	Total concentration of salt in a buffer solution (mol kg^{-1})
B_T	Slope of the Tafel equation using natural logarithm (V)
B_T'	Slope of the Tafel equation using decimal logarithm (V)
C	Empirical parameter in Equation 1.30 (kg mol^{-1})
c	Molarity (molar concentration) (mol m^{-3})
c_a^o	Molar bulk concentration of electrochemically active species at the anode (mol m^{-3})
c_a^s	Molar surface concentration of electrochemically active species at the anode (mol m^{-3})
c_c^o	Molar bulk concentration of electrochemically active species at the cathode (mol m^{-3})
c_c^s	Molar surface concentration of electrochemically active species at the cathode (mol m^{-3})
c_i	Molarity (molar concentration) of the ith component (mol m^{-3})
c_0	Standard molarity (= 1.0 mol L^{-1})
c^o	Molar standard concentration (= 1 mol m^{-3})
c_{Ox}^o	Molar bulk concentration of oxidant (mol m^{-3})
c_{Ox}	Molar concentration of oxidant (mol m^{-3})
CR	Electrochemical corrosion rate (m s^{-1})
c_{Red}	Molar concentration of reductant (mol m^{-3})
c^s	Molar surface concentration of electrochemically active species (mol m^{-3})
c_{Ox}^s	Molar surface concentration of oxidant (mol m^{-3})
$c(x, t)$	Molar concentration of electrochemically active ions as a function of time (t) and distance (x) from electrode (mol m^{-3})
D	Diffusion coefficient of electrochemically active species in Equation 6.17 (m^2 s^{-1})
D_i	Diffusion coefficient of the ith species (m^2 s^{-1})
D_i^0	Limiting ionic diffusion coefficient of the ith ion (m^2 s^{-1})
E	Electric potential difference or simply potential (V)
E	$= E_{eq}$ (in Chapters 4 and 5) (see Section 4.6)
e	Elementary charge (C)
E_m	Potential amplitude (V)

$E(t)$	Potential changing with time t (V)
E_0	$= E_{eq}^0$ (V) (in Chapters 4 and 5)
$E_{AgCl(s)/Ag(s),Cl^-(aq)}^0$	Standard electrode potential of $AgCl(s) + e^- = Ag(s) + Cl^-(aq)$ (V)
$E_{Cl_2(g)/Cl^-(aq)}^0$	Standard electrode potential of $Cl_2(g) + 2e^- = 2Cl^-(aq)$ half-reaction (V)
$E_{Cu^{2+}(aq)/Cu(s)}^0$	Standard electrode potential of $Cu^{2+}(aq) + 2e^- = Cu(s)$ half-reaction (V)
E_{eq}^0	Standard equilibrium (open circuit) potential (V)
$E_{eq,H^+(aq)/H_2(aq)}^0$	Standard potential of the hydrogen electrode (V)
$E_{Fe^{3+}(aq)/Fe^{2+}(aq)}^0$	Standard electrode potential of $Fe^{3+}(aq) + e^- = Fe^{2+}(aq)$ half-reaction (V)
E_{glass}^0	Standard electrode potential of a glass electrode (V)
$E_{Hg_2Cl_2(s)/Hg(s),Cl^-(aq)}^0$	Standard electrode potential of $Hg_2Cl_2(s) + 2e^- = 2Hg(l) + 2Cl^-$ (aq) half-reaction (V)
$E_{HgO(s),H_2O(l)/Hg(l),OH^-(aq)}^0$	Standard electrode potential of $HgO(s) + H_2O(l) + 2e^- = Hg(l) + 2OH^-(aq)$ half-reaction (V)
E_1, E_2, E_3	Electric potential drops at the resistors R_1, R_2, R_3 in Figure 2.7 (V)
E_4	Potential difference provided by a power source in Figure 2.7 (V)
E_D	Decomposition potential (V)
E_{EC}	Potential difference between electrodes (applied potential) in an electrolytic cell (V)
E_{EQ}	Equilibrium potential difference (equilibrium cell potential) (V)
E_{eq}	Equilibrium (open circuit) potential (V)
$E_{eq,H^+(aq)/H_2(g)}$	$= \Delta\phi'$ (V)
E_{FC}	Potential difference between electrodes (cell potential) in a fuel cell (V)
E_{diff}	Liquid junction or diffusion potential (V)
E_{GC}	Potential difference between electrodes (cell potential) in a galvanic cell
E_n	Electric potential drops at the resistors R_n (V)
E_S	Open circuit potential of a pH measurement cell with a standard solution (V)
E_{S1}	Open circuit potential of the first pH measurement cell with a standard solution (V)
E_{S2}	Open circuit potential of the second pH measurement cell with a standard solution (V)
E_T	Total potential difference in Figure 2.8 (V)
E_X	Open circuit potential of a pH measurement cell with a test solution (V)
F	Faraday's constant (C mol^{-1})
f^0	Standard fugacity, which equals standard pressure p_0 (= 1 bar = 0.1 MPa)

$f_{H_2(g)}$	Fugacity of gaseous molecular hydrogen, $H_2(g)$ (Pa)
g	Amount of solute (in gram) per liter of solutions (g L^{-1}) in Table 1.1
I	Ionic strength (dimensionless)
I_b	Ionic strength on the molal concentration scale (mol kg^{-1})
I_c	Ionic strength on the molar concentration scale (mol m^{-3})
I	Electric current (A)
$I(t)$	Current changing with time t (A)
I_m	Current amplitude (A)
i	$(-1)^{1/2}$ square root of negative 1
i	x coordinate unit vector (dimensionless)
I_+	Partial current of cations (A)
I_-	Partial current of anions (A)
I_1, I_2, I_3	Electric currents flowing through resistors R_1, R_2, R_3 shown in Figure 2.5 (A)
I_{EC}	Electric current in an electrolytic cell (A)
I_{FC}	Electric current in a fuel cell (A)
I_{GC}	Electric current in a galvanic cell (A)
I_i	Partial current of the ith ions (A)
I_n	Electric current flowing through resistors R_n (A)
I_T	Total current (A)
j	Current density (A m^{-2})
j	y coordinate unit vector (dimensionless)
j_a	Anodic current density (A m^{-2})
j_c	Cathodic current density (A m^{-2})
j_{corr}	Corrosion current density (A m^{-2})
J_i	Flux vector of the ith species (mol (m^{-2} s^{-1}))
j_∞	Current density at infinite rotation rate of RDE (A m^{-2})
$J_{i,x}$	x component of the flux vector (mol (m^{-2} s^{-1}))
$J_{i,y}$	y component of the flux vector (mol (m^{-2} s^{-1}))
$J_{i,z}$	z component of the flux vector (mol (m^{-2} s^{-1}))
j_{lim}	Limiting current density (A m^{-2})
$j_{lim,a}$	Limiting current density at the anode (A m^{-2})
$j_{lim,c}$	Limiting current density at the cathode (A m^{-2})
j_o	Exchange current density (A m^{-2})
j_o^0	Standard exchange current density (A m^{-2})
j_p	Parasitic current density (A m^{-2})
k	Electric conductivity (S m^{-1})
k	z coordinate unit vector (dimensionless)
k^0	Standard rate constant (mol cm^{-2} s^{-1})
K_a	Equilibrium constant of an acid (dimensionless)
k_B	Boltzmann constant (J K^{-1})
K_b	Product of molalities (dimensionless)
K_{cell}	Cell constant (m^{-1})
K_γ	Product of activity coefficients (dimensionless)
K_w	Dissociation (ionization) constant of water (dimensionless)
L	Electric conductance (S)
l	Distance between electrodes (m)

m	Amount of chemical produced (consumed) due to electrolysis (kg)
M_1	Molar mass of solvent in Table 1.1 (kg mol^{-1})
M_2	Molar mass of solute in Table 1.1 (kg mol^{-1})
M_i	Molecular mass of the ith species (kg mol^{-1})
M_M	Molar mass of corroding metal (kg mol^{-1})
n	Number of electrons in half-reaction (electron number) (dimensionless)
N_A	Avogadro constant (mol^{-1})
n_+	Number density of cations (m^{-3})
n_-	Number density of anions (m^{-3})
n_i	Number density of the ith ions (m^{-3})
n_i	Molar amount of the ith chemical (mol)
Ox	Oxidizing species (oxidant)
P_{EC}	Electric power provided for an electrolytic cell (W)
P_{GC}	Electric power generated in a galvanic cell (W)
p_0	standard pressure (=1 bar)
pH	Negative decimal logarithm of activity of protons, H$^+$(aq) (dimensionless)
pH$_S$	pH of a standard solution (dimensionless)
pH$_{S1}$	pH of the first standard solution (dimensionless)
pH$_{S2}$	pH of the second standard solution (dimensionless)
pH$_X$	pH of a test solution (dimensionless)
pK_a	Negative decimal logarithm of thermodynamic constant of an acid (dimensionless)
pK_{base}	Negative decimal logarithm of thermodynamic constant of a base (dimensionless)
Q	Amount of electric charge (C)
Q	Geometric average (mean) of stoichiometric numbers of an electrolyte (dimensionless)
Q_{FC}	Released heat rate (W m^{-2})
R	Electric resistance (Ω)
R	Molar gas constant (J (mol K)$^{-1}$)
r	Radial coordinate in cylindrical coordinate system (m)
r_{corr}	Area-specific resistance due to corrosion (Ω m^2)
R_e	Resistance of an external resistor (Ω)
r_{EC}	Electrolytic cell area–specific resistance (Ω m^2)
Red	Reducing species (reductant)
r_{FC}	Fuel cell area–specific resistance (Ω m^2)
R_i	Internal resistance (Ω)
r_i	Hydrodynamic radius of the ith ion (m)
T	Thermodynamic temperature (K)
t	Time (s)
t	Celsius temperature (°C)
t_+	Transport number of cation (dimensionless)
t_-	Transport number of anion (dimensionless)
t_i	Transport number of the ith ion (dimensionless)
t_i^0	Limiting transport number of the ith ion (dimensionless)
T_0	Standard temperature (= 298.15 K)

u_+	Mobility of cation (m^2 (Vs)$^{-1}$)
u_-	Mobility of anion (m^2 (Vs)$^{-1}$)
u_i	Electric mobility of the ith ion (m^2 (Vs)$^{-1}$)
u_i^0	Limiting ionic mobility of the ith ion (m^2 (Vs)$^{-1}$)
x	Mole fraction of solute in Table 1.1 (dimensionless)
x_i	Mole fraction of the ith component (dimensionless)
x_w	Mole fraction of water (dimensionless)
z_+	Charge of cation (dimensionless)
z_-	Charge of anion (dimensionless)
z_i	Charge of the ith ion (dimensionless)

GREEKS

α	Degree of dissociation (dimensionless)
β	Symmetry factor (dimensionless)
β_a	Anodic symmetry factor in Equation 9.7 (dimensionless)
β_c	Cathodic symmetry factor in Equation 9.6 (dimensionless)
γ_+	Activity coefficient of cation (dimensionless)
γ_-	Activity coefficient of anion (dimensionless)
$\gamma_\pm$	Mean activity coefficient (dimensionless)
$\gamma_{A^-(aq)}$	Activity coefficient of anions, A$^-$(aq) (dimensionless)
$\gamma_{H^+(aq)}$	Activity coefficient of protons, H$^+$(aq) (dimensionless)
$\gamma_{HA(aq)}$	Activity coefficient of molecules, HA(aq) (dimensionless)
γ_i	Activity coefficient of the ith component (dimensionless)
$\gamma_{i,b}$	Activity coefficient of the ith component on the molal concentration scale (dimensionless)
$\gamma_{i,x}$	Activity coefficient of the ith component on the mole fraction concentration scale (dimensionless)
$\gamma_{w,x}$	Activity coefficient of water on the mole fraction concentration scale (dimensionless)
ΔG	Gibbs energy of a process (J)
$\Delta_f G_\pm^0$	$= \Delta_f G_i^{0,b} \left[J\,mol^{-1} \right]$
$\Delta_f G_{e^-}$	Gibbs energies of the formation of electron ($= 0$, conventionally) (J mol^{-1})
$\Delta_f G_{H^+(aq)}$	Gibbs energy of the formation of aqueous proton, H$^+$(aq) (J mol^{-1})
$\Delta_f G_{H^+(aq)}^0$	Standard Gibbs energy of the formation of aqueous proton, H$^+$(aq) (J mol^{-1})
$\Delta_f G_{H_2(g)}$	Gibbs energy of the formation of gaseous molecular hydrogen, H$_2$(g) (J mol^{-1})
$\Delta G_{H_2(g)}^0$	Standard Gibbs energy of the formation of gaseous molecular hydrogen, H$_2$(g) (J mol^{-1})
$\Delta_f G_i$	Gibbs energy of formation of the ith component (J mol^{-1})
$\Delta_f G_i^0$ (products)	Standard Gibbs energy of the formation of the ith species of products (J mol^{-1})

$\Delta_f G_i^0$ (reactants)	Standard Gibbs energy of the formation of the ith species of reactants (J mol^{-1})
$\Delta_f G_i^{0,b}$	Standard Gibbs energy of the formation of the ith component on the molal concentration scale (J mol^{-1})
$\Delta_f G_i^{\alpha}$	Gibbs energy of formation of the ith species in phase α (J mol^{-1})
$\Delta_f G_i^{\beta}$	Gibbs energy of formation of the ith species in phase β (J mol^{-1})
$\Delta_f G_w$	Gibbs energy of formation of water (J mol^{-1})
$\Delta_f G_w^{0,x}$	Standard Gibbs energy of formation of water on the mole fraction concentration scale (J mol^{-1})
ΔH	Enthalpy of a process (J)
$\Delta\varphi$	Galvanic potential difference (V)
$\Delta\varphi'$	Galvanic potential difference of hydrogen electrode (V)
$\Delta_r C_p^0$	Standard isobaric heat capacity of reaction (J (mol K)$^{-1}$)
$\Delta_r C_{P.T_0}^0$	Standard isobaric heat capacity of reaction at standard temperature T_0 (J (mol K)$^{-1}$)
$\Delta_r C_P$	Isobaric heat capacity of a reaction (J (mol K)$^{-1}$)
$\Delta_r G$	Gibbs energy of a reaction (J mol^{-1})
$\Delta_r G^0$	Standard Gibbs energy of a reaction (J mol^{-1})
$\Delta_r G_{T_0}^0$	Standard Gibbs energy of reaction at standard temperature T_0 (J mol^{-1})
$\Delta_r H_{T_0}^0$	Standard enthalpy of reaction at standard temperature T_0 (J mol^{-1})
$\Delta_r S$	Entropy of a reaction (J (mol K)$^{-1}$)
$\Delta_r S_{T_0}^0$	Standard entropy of reaction at standard temperature T_0 (J (mol K)$^{-1}$)
$\Delta_r V$	Volume (change) of a reaction (m^3 mol^{-1})
ΔS	Entropy of a process (J K^{-1})
δ_N	Nernst diffusion layer (m)
δ_{Pr}	Prandtl layer thickness (m)
ε_o	Electric constant (permittivity of vacuum) (F m^{-1})
ε_r	Solvent relative permittivity (dielectric constant) (dimensionless)
ε_w	Relative permittivity of water (dimensionless)
ξ	Total efficiency of an electrochemical energy conversion system (dimensionless)
ξ_{EC}	Total efficiency of an electrolytic cell (dimensionless)
ξ_{FC}	Total efficiency of a fuel cell (dimensionless)
ξ_i	Current efficiency of an electrochemical energy conversion system (dimensionless)
$\xi_{i, EC}$	Current efficiency of an electrolytic cell (dimensionless)
$\xi_{i, FC}$	Current efficiency of a fuel cell (dimensionless)
ξ_{load}	Efficiency of an electrochemical energy conversion system under load (dimensionless)
$\xi_{load, FC}$	Efficiency of a fuel cell under load (dimensionless)

ξ_{th}	Maximum (thermodynamic) efficiency of an electrochemical energy conversion system (dimensionless)
ξ_{th}^0	Standard maximum (thermodynamic) efficiency (dimensionless)
$\xi_{th,\,EC}$	Maximum (thermodynamic) efficiency of an electrolytic cell (dimensionless)
$\xi_{th,\,FC}$	Maximum (thermodynamic) efficiency of a fuel cell (dimensionless)
ξ_V	Voltage efficiency of an electrochemical energy conversion system (dimensionless)
$\xi_{V,\,EC}$	Voltage efficiency of an electrolytic cell (dimensionless)
$\xi_{V,\,FC}$	Voltage efficiency of a fuel cell (dimensionless)
η	Viscosity of solvent (Pa s)
η	Total overpotential on an electrode (V)
η_{ct}	Charge transfer overpotential on an electrode (V)
η_{EC}	Overpotential of electrolytic cell (V)
η_{GC}	Overpotential of galvanic cell (V)
η_i	The ith contribution to total overpotential on an electrode (V)
η_{mt}	Mass transfer overpotential on an electrode (V)
θ	Nernstian slope ($\approx (\ln 10)$ RT/F) (V)
κ	Inverse radius of the ionic atmosphere (m^{-1})
Λ	Molar conductivity (S m^2 mol^{-1})
Λ^0	Molar conductivity at infinite dilution (limiting conductivity) (S m^2 mol^{-1})
Λ_{I}	Molar conductivities of solution (I) on the left-hand side of the liquid junction (S m^2 mol^{-1})
Λ_{II}	Molar conductivities of solution (II) on the right-hand side of the liquid junction (S m^2 mol^{-1})
Λ_{eq}	Equivalent conductivity (S m^2 mol^{-1})
λ_+	Ionic conductivity of cation (S m^2 mol^{-1})
λ_+^0	Limiting ionic conductivity of cation (S m^2 mol^{-1})
λ_-	Ionic conductivity of anion (S m^2 mol^{-1})
λ_-^0	Limiting ionic conductivity of anion (S m^2 mol^{-1})
λ_i	Ionic conductivity of the ith ion (S m^2 mol^{-1})
λ_i^0	Limiting ionic conductivity of the ith ion (S m^2 mol^{-1})
μ_+	Chemical potential of cation (J mol^{-1})
μ_+^0	Standard chemical potential of cation (J mol^{-1})
μ_-	Chemical potential of anion (J mol^{-1})
μ_-^0	Standard chemical potential of anion (J mol^{-1})
$\mu_\pm$	Chemical potential of a completely dissociated electrolyte on the molal concentration scale (J mol^{-1})
$\mu_\pm^0$	Standard chemical potential of a completely dissociated electrolyte on the molal concentration scale (J mol^{-1})
$\mu_{\mathrm{H^+(aq)}}$	Chemical potential of aqueous proton, H$^+$(aq) (J mol^{-1})
$\mu_{\mathrm{H_2(g)}}$	Chemical potential of gaseous molecular hydrogen, H$_2$(g) (J mol^{-1})
μ_i	Chemical potential of the ith component (J mol^{-1})
μ_i^0	Standard chemical potential of the ith component (J mol^{-1})
$\mu_i^{0,b}$	Standard chemical potential of the ith component on the molal concentration scale (J mol^{-1})

$\mu_i^{0,x}$	Standard chemical potential of the ith component on the mole fraction concentration scale (J mol^{-1})
μ_i^α	Chemical potential on the ith species in phase α (J mol^{-1})
μ_i^β	Chemical potential on the ith species in phase β (J mol^{-1})
$\mu_{i,id}$	Chemical potential of the ith component in the ideal solution
μ_w	Chemical potential of water (J mol^{-1})
$\mu_w^{0,x}$	Standard chemical potential of water on the mole fraction concentration scale (J mol^{-1})
ν	Kinematic viscosity (m^2 s^{-1})
ν	Sum of stoichiometric numbers of cation and anion in an electrolyte (dimensionless)
ν_+	Stoichiometric numbers of cation in an electrolyte (dimensionless)
ν_-	Stoichiometric numbers of anion in an electrolyte (dimensionless)
ν_i	Stoichiometric number of the ith species (dimensionless)
v	Velocity vector (m s^{-1})
v_i	Velocity of the ith ion (m s^{-1})
$v_{i,x}$	x component of the velocity vector (m s^{-1})
$v_{i,y}$	y component of the velocity vector (m s^{-1})
$v_{i,z}$	z component of the velocity vector (m s^{-1})
v_r	Radial component of the solutions velocity (m s^{-1})
v_x	Velocity of solution in one-dimensional system (m s^{-1})
v_z	Axial component of the solution velocity (m s^{-1})
ρ	Density of solvent (kg m^{-3})
ρ_S	Electric resistivity (Ω m)
ρ	Density of solution in Table 1.1 (kg m^{-3})
ρ_W	Density of water (kg m^{-3})
ρ_M	Density of the corroding metal (kg m^{-3})
φ	Electric potential (V)
φ^α	Electric potential in phase α (V)
φ^β	Electric potential in phase β (V)
τ	Transition time (s)
ω	Angular velocity (rad s^{-1})
ω	Mass fraction of solute in Table 1.1 (dimensionless)

The following prefixes are recompensed by IUPAC to denote decimal multiples and submultiples. Note that no space should be between the prefix and the unit symbol.

Multiple	Prefix Name	Prefix Symbol
10^{12}	Tera	T
10^9	Giga	G
10^6	Mega	M
10^3	Kilo	k
10^2	Hecto	h
10^1	Deca	da

Submultiple	Prefix Name	Prefix Symbol
10^{-1}	Deci	d
10^{-2}	Centi	c
10^{-3}	Milli	m
10^{-6}	Micro	μ
10^{-9}	Nano	n
10^{-12}	Pico	p

1 Electrolyte Solutions

1.1 OBJECTIVES

The main objective of this chapter is to introduce the reader to physical chemistry of electrolyte solutions. An electrolyte solution consists of charged species (ions) that make these solutions very useful for electrochemical science and engineering. Concentrations, activities, activity coefficients, and chemical potentials of both solute and solvent are described in detail. The concentration of species in weak electrolytes and pH of aqueous solutions are discussed, and the physical chemistry of buffer solutions is explained.

1.2 FORMATION OF ELECTROLYTE SOLUTIONS

In a dissolution process, heat is consumed or released depending on the internal energy of the crystal or molecule and the energy of solvation (hydration in the case of water as a solvent). Some substances (solutes), forming electrolyte solutions, release heat (e.g., $CaCl_2(s)$, $HCl(l)$) and others can consume heat (e.g., $KCl(s)$, $KNO_3(s)$) when dissolving. Figure 1.1 illustrates how a KCl(aq) solution is created due to interactions between water molecules and KCl(s), a solid salt, which is completely dissociated: $KCl(s) \rightarrow K^+(aq) + Cl^-(aq)$. There are only two species (ions) in the solution, and the concentration of $K^+(aq)$ and $Cl^-(aq)$ is the same as that of KCl(aq). It is important to note that this process goes only one way at ambient conditions. When KCl(s) is placed into water, the salt will be dissolved into ions until a saturation concentration is achieved. The saturation mass fraction [1] of KCl(s) in water at 25°C is 0.2622

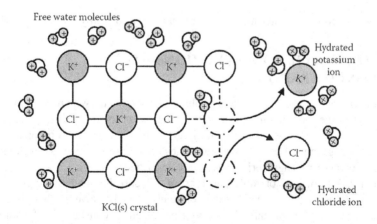

FIGURE 1.1 Formation of KCl(aq) solution due to dissolution reaction: KCl(s) → K+(aq)+Cl−(aq).

DOI: 10.1201/b21893-1

(Table 10.19). However, if the conditions are changed, say, to very high temperature and low density, then it could be a case where dissolved KCl(aq) will mostly be molecules.

Aqueous solutions can approximately be divided into three types: strong electrolytes, weak electrolytes, and nonelectrolytes. Strong electrolytes are completely dissociated (to ions), and weak electrolytes are partially dissociated. In strong electrolytes, there are only ions; in weak electrolytes, there is an equilibrium between partially dissociated molecules and ions. Nonelectrolytes do not produce any ions when dissolved. Examples of nonelectrolyte aqueous solutions are O_2(aq), H_2(aq), and CH_4(aq). At ambient conditions, examples of strong electrolyte aqueous solutions are NaCl(aq), HCl(aq), $CaCl_2$(aq), $LaCl_3$(aq), and NaOH(aq), and examples of weak electrolyte aqueous solutions are CH_3COOH(aq), NH_4OH(aq), and H_2CO_3(aq). In the case of acetic acid, the equilibrium between molecules and ions is as follows:

$$CH_3COOH(aq) = CH_3COO^-(aq) + H^+(aq).$$

There are three species in the solution, and species concentrations are defined by (1) the equilibrium constant, (2) mass balance, and (3) charge balance.

Note that pure water, one of the most important solvents in electrochemical science and engineering, is a weak electrolyte because of its self-ionization:

$$2H_2O(l) = OH^-(aq) + H_3O^+(aq).$$

The hydronium ion, H_3O^+(aq), is formed due to the reaction $H_2O(l) + H^+(aq) = H_3O^+$(aq), the Gibbs energy of which is commonly assumed to be zero because H^+(aq) + H_2O(l) and H_3O^+(aq) are not thermodynamically different. Therefore, the water ionization reaction can also be written as follows:

$$H_2O(l) = OH^-(aq) + H^+(aq).$$

These two reactions of water ionization are thermodynamically indistinguishable.

1.3 ELECTROLYTE CONCENTRATION AND CONCENTRATION SCALES

Concentration is the amount of one of a solution's components with respect to another component or the total amount of the same solution. A solution can be gaseous, liquid, or solid as well as two-component, three-component, or multicomponent. In the following paragraphs, the concentration scales which are most common in electrochemical science and engineering are briefly described.

Molality, b (mol kg^{-1} or mole per 1 kg of solvent), is widely used in physical chemistry and electrochemistry but cannot be used for the solvent. There is no such thing as molality of a solvent. The amount of solvent is constant and for an aqueous solution equals 1 kg or 55.51 mol. The last number is simply 1000 g divided by the molecular mass of water, 18.015 g mol^{-1}. Molality should always be used in equilibrium electrochemistry of aqueous solutions. Note that m symbol can also be used for molality [1] but is not used in this book as it is used for mass.

Molarity, c (mol L^{-1} or mole per 1 L of solution), is used in analytical chemistry, theoretical physical chemistry, and electrochemical kinetics. It cannot be used for a solvent as well. While b and c are numerically very close when the density of solution, ρ, is about 1 g cm^{-3} = 1 kg L^{-1}, they will be different at elevated temperatures as the density of the solvent decreases. This is demonstrated in the "Exercises" section of this chapter. Still, at ambient conditions with dilute solutions, $c \approx b\rho$, and therefore, c is numerically close to b.

Mole fraction, x (dimensionless), is used in theoretical physical chemistry and can be used for either solvent or solute.

Mass fraction (dimensionless) is commonly used and is suitable for either solvent or solute. Volume fraction (dimensionless) is commonly used for gaseous solutions. Percent, % (dimensionless), should actually be called parts per hundred (pph) and is widely used in everyday life. Parts per million, ppm (dimensionless), is usually used for solute, which is in a very small amount. Parts per thousand, ppt (dimensionless), is usually used to define salinity of sea water.

While pph, ppm, ppt, etc., are widely used in the literature and can technically be either on mass or on volume bases, all of them are not recommended by International Union of Pure and Applied Chemistry (IUPAC) [1] and, therefore, are not used in this book.

It is suggested the reader keep in mind the definitions of all concentrations given earlier and, most importantly, be able to convert from one concentration scale to another (see Section 1.4). Some of the listed concentrations depend on temperature and pressure (e.g., molarity and volume fraction), while others do not (e.g., molality and mole fractions). Note the IUPAC [1] recommends using some other names and units of concentrations, but they are not still sufficiently accepted by electrochemists and electrochemical engineers. In this book, molality, molarity, and mole fraction will mainly be used.

1.4 CONVERSION EQUATIONS FOR THE CONCENTRATION OF SOLUTIONS

Relationships between the concentrations described in Section 1.3 can easily be derived by the reader. Otherwise, Table 1.1 should be used.

1.5 CHEMICAL POTENTIAL

Generally, the chemical potential, μ_i, of the ith component in an ideal solution is as follows:

$$\mu_i = \mu_i^0 + RT \ln x_i, \tag{1.1}$$

where
R is the molar gas constant (Table 10.1).
T is the thermodynamic temperature.

TABLE 1.1

Conversion of Concentrations[a]

Given Concentration					Sought concentration
w	x	b	c	g	
—	$\dfrac{xM_2}{xM_2+(1-x)M_1}$	$\dfrac{bM_2}{1000+bM_2}$	$\dfrac{cM_2}{\rho}$	$\dfrac{g}{\rho}$	w
$\dfrac{\dfrac{\omega}{M_2}}{\dfrac{\omega}{M_2}+\dfrac{1-\omega}{M_1}}$	—	$\dfrac{M_1 b}{(M_1 b+1000)}$	$\dfrac{M_1 c}{c(M_1-M_2)+\rho}$	$\dfrac{M_1 g}{g(M_1-M_2)+\rho M_2}$	x
$\dfrac{1000\omega}{M_2(1-\omega)}$	$\dfrac{1000x}{(1-x)M_1}$	—	$\dfrac{1000c}{\rho-(cM_2)}$	$\dfrac{1000g}{M_2(\rho-g)}$	b
$\dfrac{\rho\omega}{M_2}$	$\dfrac{\rho x}{xM_2+(1-x)M_1}$	$\dfrac{\rho b}{1000+M_2 b}$	—	$\dfrac{g}{M_2}$	c
$w\rho$	$\dfrac{\rho x M_2}{xM_2+(1-x)M_1}$	$\dfrac{\rho b M_2}{1000+bM_2}$	cM_2	—	g

[a] The number 1000 (g kg^{-1}) in the table means 1000 g in 1 kg; b—molality (mol kg^{-1}), c—molarity (mol L^{-1}), g—grams of solute per liter of solution (g L^{-1}), x—mole fraction of solute, w—mass fraction of solute, M_1—molar mass of solvent (g mol^{-1}), M_2—molar mass of solute (g mol^{-1}), and ρ—density of solution (g L^{-1}).

x_i is the mole fraction of the ith component.
μ_i^0 is the standard chemical potential.

The chemical potential, μ_i, is a characteristic of the ith component of solution and depends on temperature, pressure, and concentration.

In a real solution, the chemical potential of the ith component is

$$\mu_i = \mu_i^0 + RT \ln a_i = \mu_i^0 + RT \ln \left(x_i \gamma_i \right), \tag{1.2}$$

where a_i is the activity and γ_i is the activity coefficient of the ith component.

Activity was introduced by Gilbert Newton Lewis and plays an important role in thermodynamic description of electrolyte solutions. As can be seen, the chemical potential consists of two contributions: the standard state term and a logarithmic term. The standard state term does not depend on concentration, but the logarithmic term does. Both terms depend on the concentration scale (e.g., molality, molarity, mole fraction) because μ_i^0, x_i, and γ_i in Equation 1.2 all depend on the concentration scale. However, the chemical potential, μ_i, does not depend on the concentration scale.

Gilbert Newton Lewis (1875–1946) was an American physical chemist known for a number of discoveries and contributions in solution thermodynamics and chemistry. He introduced the concept of fugacity, activity, activity coefficient, and the ionic strength.

REFERENCES

A.J. Bard, G. Inzelt, and F. Scholz, *Electrochemical Dictionary*, Springer, Berlin, Germany, 2008.
http://en.wikipedia.org/wiki/Gilbert_N._Lewis

It would be more accurate to rewrite Equation 1.2 using additional symbols as follows:

$$\mu_i = \mu_i^{0,x} + RT \ln \left(x_i \gamma_{i,x} \right) = \mu_i^{0,x} + RT \ln \left(a_{i,x} \right), \tag{1.3}$$

where $\mu_i^{0,x}$ is μ_i when $x_i \to 1$ and $\gamma_{i,x} \to 1$.

The concentration scale of a standard chemical potential and an activity coefficient are specified by additional symbols placed as either the subscript or superscript. For example, the mole fraction scale is specified in Equation 1.3. In this equation, if we want to be precise, $\mu_i^{0,x}$ should be called the standard chemical potential on the mole fraction concentration scale. Equation 1.3 is usually used for solutions of nonelectrolytes, such as $O_2(aq)$, and for solvents, such as water, in electrolyte solutions. Also, this equation can be used for solid solutions such as metal alloys.

For electrolyte solutions, molality (b_i) is commonly used except (1) electrolyte conductivity and (2) electrochemical kinetics, where molarity is commonly used. The chemical potential of the ith solute on the molality scale can be written as follows:

$$\mu_i = \mu_i^{0,b} + RT \ln \left(\gamma_{i,b} b_i / b^0 \right) = \mu_i^{0,b} + RT \ln \left(a_{i,b} \right), \tag{1.4}$$

where $\mu_i^{0,b}$ is $(\mu_i - RT \ln (b_i/b^0))$ when $b_i \to 0$ and $\gamma_b \to 1$. $\mu_i^{0,b}$ in Equation 1.4 is the standard chemical potential on the molal concentration scale, and b^0 is the standard molality, which is equal to 1.0 mol kg^{-1}. Note that Equation 1.4 can also be used for individual ions such as cations or anions. A cation is a positively charged ion, and an anion is a negatively charged ion.

1.6 STANDARD CHEMICAL POTENTIAL AND ACTIVITY COEFFICIENT ON DIFFERENT CONCENTRATION SCALES

The question we would like to answer here is, what is the relationship between $\mu_i^{0,x}$ and $\mu_i^{0,b}$ as well as between $\gamma_{i,x}$ and $\gamma_{i,b}$, taking into account that we know the relationship between x_i and b_i:

$$x_i = \frac{b_i}{(b_i + 55.51)}, \tag{1.5}$$

where 55.51 is the amount of water (in mol) per 1 kg of water, that is, 1000 g kg^{-1} divided by 18.015 g mol^{-1}, which is the molar mass of water. The unit of 55.51 in Equation 1.5 is mol kg^{-1}, which is the same as unit of molality, b_i. Note that IUPAC does not recommend using numbers in equations until the units of the numbers are clearly provided [1]. This is what we have done earlier using a number in Equation 1.5 and immediately clarifying the unit of the number.

Because the chemical potential does not depend on the concentration scale, we can combine Equations 1.3 and 1.4 as follows:

$$\mu_i^{0,x} + RT\ln\left(x_i\gamma_{i,x}\right) = \mu_i^{0,b} + RT\ln\left(\gamma_{i,b}b_i/b^0\right)$$

$$= \mu_i^{0,x} + RT\ln x_i + RT\ln \gamma_{i,x}$$

$$= \mu_i^{0,b} + RT\ln\left(b_i/b^0\right) + RT\ln\left(\gamma_{i,b}\right). \tag{1.6}$$

Substituting Equations 1.5 through 1.6 and assuming an infinitely dilute solution when $x_i \to 0$, $b_i \to 0$, $\gamma_{i,x} \to 1$, $\gamma_{i,b} \to 1$, we can obtain the relationship between $\mu_i^{0,x}$ and $\mu_i^{0,b}$ as follows:

$$\mu_i^{0,x} = \mu_i^{0,b} + RT\ln\left(\frac{55.51}{b^0}\right). \tag{1.7}$$

A relationship between the standard chemical potentials on the molal and molar concentration scales is derived in the "Exercises" section of this chapter.

Considering a similar approach and using Equations 1.5 through 1.7, the reader should be able to easily derive the relationship between $\gamma_{i,x}$ and $\gamma_{i,b}$ ($\gamma_{i,x} = \gamma_{i,b}$ $(1+b_i/55.51)$). A relationship between the activity coefficents on the molal and molar concentration scales is derived in the "Exercises" section of this chapter

Any other relations between the standard chemical potentials or activity coefficients on different concentration scales can be obtained using the same approach. Some of these equations are given in Ref. [2].

Note that the standard chemical potentials have the same dimensions as the chemical potentials, that is, J mol^{-1}, while activities and activity coefficients are dimensionless.

1.7 CHEMICAL POTENTIAL OF SOLVENT AND SOLUTE IN ELECTROLYTE SOLUTION

The chemical potential is probably the most important characteristics of a solution component and was introduced by Josiah Willard Gibbs.

In common aqueous solutions (e.g., sea water), water is the solvent, and dissolved salts are the solutes. The chemical potential of the water solvent, μ_w, is almost always defined using the mole fraction concentration scale:

$$\mu_w = \mu_w^{0,x} + RT\ln\left(x_w\gamma_{w,x}\right). \tag{1.8}$$

For a dilute electrolyte solution, both x_w and $\gamma_{w,x}$ are very close to 1, and therefore, in some applications, we could assume that

$$\mu_w \approx \mu_w^{0,x}. \tag{1.9}$$

As an example, in 0.5 mol kg^{-1} NaCl(aq) where $x_w = 0.9911$, it was experimentally found that $\gamma_{w,x} = 0.9923$ [2], so the product of the water mole fraction and corresponding activity coefficient, $x_w\gamma_{w,x}$, is equal to 0.9835, and $RT\ln(x_w\gamma_{w,x})$ is equal at 298.15 K to -41.32 J mol^{-1}, which can be considered a relatively small value. However, if the solute concentration is high, Equation 1.9 should not be used.

Josiah Willard Gibbs (1839–1903) was an American scientist who made important theoretical contributions to physics, chemistry, and mathematics. He developed the mathematical background of thermodynamics and introduced the concept of chemical potential.

REFERENCES

A.J. Bard, G. Inzelt, and F. Scholz, *Electrochemical Dictionary*, Springer, Berlin, Germany, 2008.
http://en.wikipedia.org/wiki/Josiah_Willard_Gibbs

For an electrolyte solution, Equation 1.4 might be insufficiently convenient because it does not take into account that such solutions consist of ions. Therefore, the chemical potential of a completely dissociated electrolyte, $\mu_\pm$, can be defined as the sum of the ionic contributions of the chemical potentials, μ_+ (cation) and μ_- (anion), taking into account the stoichiometric numbers of ions, v_+(cation) and v_-(anion):

$$\mu_\pm = v_+\mu_+ + v_-\mu_-. \tag{1.10}$$

As examples, for $CaCl_2(aq)$ solution $\mu_\pm = 1\mu_{Ca^{2+}} + 2\mu_{Cl^-}$, for $LaCl_3(aq)$ solution $\mu_\pm = 1\mu_{La^{3+}} + 3\mu_{Cl^-}$, and for $Na_2SO_4(aq)$ $\mu_\pm = 2\mu_{Na^+} + 1\mu_{SO_4^{2-}}$.

The chemical potential of a cation and an anion can be written as follows:

$$\mu_+ = \mu_+^0 + RT\ln\left(\frac{\gamma_+ b_+}{b^0}\right), \quad \mu_- = \mu_-^0 + RT\ln\left(\frac{\gamma_- b_-}{b^0}\right). \tag{1.11}$$

Combining Equations 1.10 and 1.11, the following expression for the chemical potential of an electrolyte can be obtained:

$$\mu_{\pm} = \mu_{\pm}^0 + RT \ln \left(\frac{\gamma_+ b_+}{b^0} \right)^{v+} + RT \ln \left(\frac{\gamma_- b_-}{b^0} \right)^{v-}, \tag{1.12}$$

where $\mu_{\pm}^0 = v_+ \mu_+^0 + v_- \mu_-^0$ is the standard value of $\mu_{\pm}$ when $b_+ \to 0$, $b_- \to 0$, $\gamma_+ \to 1$, and $\gamma_- \to 1$.

As a simplification, we will use in this book μ_+^0 instead of $\mu_+^{0,b}$ as well as γ_+ instead of $\gamma_{+,b}$. These simplifications of symbols are acceptable because the molal concentration scale for the standard chemical potential and the activity coefficients of ions are always used unless otherwise stated.

1.8 ACTIVITY, IONIC STRENGTH, AND ACTIVITY COEFFICIENT

Activity, a_i, and activity coefficient, γ_i, of the ith ion are defined as follows:

$$a_{i,b} = \gamma_{i,b} \left(\frac{b_i}{b^0} \right). \tag{1.13}$$

Activity is a value that should be used in all thermodynamic equations instead of concentration to correctly describe a real solution.

One of the most important characteristics of an electrolyte solution is the ionic strength, I, which is used in the theory of strong electrolytes (e.g., Debye–Hückel theory) and is defined as follows:

$$I = 0.5 \sum \left[z_i^2 \left(\frac{b_i}{b^0} \right) \right], \tag{1.14}$$

where z_i is the charge of the ith ion and the summation is over all solution ions. For example, the ionic strength of 0.5 mol kg^{-1} CaCl$_2$(aq) solution is 0.5 ($2^2 \times 0.5 + 1 \times 2 \times 0.5$) = 1.5. Note that the ionic strength in some textbooks could be dimensional, as given by Equation 1.14 or on the molal concentration scale

$$I_b = 0.5 \sum \left(z_i^2 b_i \right), \tag{1.15}$$

in mol kg^{-1} or could be expressed using the molar (mol L^{-1}) concentration scale:

$$I_c = 0.5 \sum \left(z_i^2 c_i \right). \tag{1.16}$$

Let us now define the physical meaning of the activity and activity coefficient. The relationship between the activity and activity coefficient is such that when $I \to 0$, $a_{i,b} \to b_i$, $\gamma_{i,b} \to 1$. Therefore, the activity coefficient shows the difference between the chemical potential of the ith species in real and ideal solutions with the same concentration. Also, we could say that the activity is a *corrected concentration*, which should be used in all thermodynamic equations to keep real solution equations the same as for ideal solutions.

1.9   ACTIVITY COEFFICIENT OF AN ELECTROLYTE AND AN ION

The activity coefficient of an ion cannot be experimentally measured (as any other thermodynamic property of a single ion) but can be theoretically calculated for a dilute solution when I_b is less than about 0.05 mol kg^{-1}. However, the activity coefficient of an electrolyte, $\gamma_{\pm}$, can be experimentally measured, and such values are given in Tables 10.16 and 10.17. $\gamma_{\pm}$ is called the mean activity coefficient and is related to the activity individual coefficient of the cation, γ_{+}, and anion, γ_{-}, as follows:

$$\gamma_{\pm} = \left(\gamma_{+}^{\nu_{+}}\gamma_{-}^{\nu_{-}}\right)^{1/\nu},\tag{1.17}$$

where $\nu = \nu_{+}+\nu_{-}$.

Therefore, the mean activity coefficient of an electrolyte is simply the geometric average of the individual activity coefficients of the electrolyte cation and anion.

It is interesting to note the different limitations between theoretically and experimentally determining activity coefficients for dilute and concentrated solutions. Theoretical calculations are possible, and quite accurate, at low concentrations where any experimental measurements are usually difficult to perform. However, at high concentrations, it is possible to experimentally measure the mean activity coefficients, whereas accurate theoretical calculations from first principles are not currently possible.

Let us find out where the relationship (1.17) is coming from taking into account that in a completely dissociated electrolyte $b_{+} = \nu_{+} b_i$ and $b_{-} = \nu_{-} b_i$, where b_i is the concentration of a strong (completely dissociated to ions) electrolyte:

$$\mu_{\pm} = \left(\nu_{+}\mu_{+}^{0} + \nu_{-}\mu_{-}^{0}\right) + RT\ln\left(\gamma_{+}b_{+}/b^{0}\right)^{\nu_{+}} + RT\ln\left(\gamma_{-}b_{-}/b^{0}\right)^{\nu_{-}}$$

$$= \mu_{\pm}^{0} + RT\ln\left(\gamma_{+}^{\nu_{+}}\gamma_{-}^{\nu_{-}}\right) + RT\ln\left(b_{+}/b^{0}\right)^{\nu_{+}} + RT\ln\left(b_{-}/b^{0}\right)^{\nu_{-}}$$

$$= \mu_{\pm}^{0} + RT\ln\left(\gamma_{\pm}\right)^{\nu} + RT\ln\left(b_i/b^{0}\right)^{\nu} + RT\ln\left(Q\right)^{\nu}$$

$$= \mu_{\pm}^{0} + RT\ln\left(a_{\pm}\right)^{\nu},\tag{1.18}$$

where $Q = \left(\nu_{+}^{\nu_{+}}\nu_{-}^{\nu_{-}}\right)^{1/\nu}$.

Now, comparing Equations 1.4 and 1.18 and taking into account that $\mu_{\pm}$ and μ_i are the same values as well as $\mu_{\pm}^{0} = \mu_i^{0,b}$, the activity of a strong electrolyte, $a_{i,b}$, and the activity coefficient $\gamma_{i,b}$ can be defined as follows:

$$a_{i,b} = \left(a_{\pm}\right)^{\nu} = \left[\gamma_{\pm}\left(\frac{b_i}{b^{0}}\right)Q\right]^{\nu}.\tag{1.19}$$

$$\gamma_{i,b} = \left(\gamma_{\pm}Q\right)^{\nu}\left(\frac{b_i}{b^{0}}\right)^{\nu-1}.\tag{1.20}$$

Based on Equation 1.17 if we assume that $\gamma_+ = \gamma_-$, then $\gamma_\pm = \gamma_+ = \gamma_-$ for any strong electrolyte (e.g., 1–1 electrolytes such as KCl, 2–1 electrolytes such as $CaCl_2$, 1–2 electrolytes such as Na_2SO_4, and 3–1 electrolytes such a $LaCl_3$, etc.). Note that $\gamma_\pm = \gamma_+ = \gamma_-$ is a poor but, in most cases, unavoidable approximation except, probably, KCl(aq), where we could assume that the activity coefficients of K^+(aq) and Cl^-(aq) are almost the same because the cation and anion have almost the same size (as shown in Figure 1.1), electron structure and, therefore, the same level of hydration. Likewise, the transport numbers (see Chapter 3) of these two ions are very similar. Because many properties of these two ions are similar, we can believe that the thermodynamic properties of these two ions are similar too. Note that KCl(aq) has a special importance in electrochemical science and engineering where the electrochemical and transport properties of a single ion can experimentally be detected.

1.10 CHEMICAL POTENTIAL AND GIBBS ENERGY OF FORMATION

There is a fundamental problem in thermodynamics: we do not know the absolute values of some thermodynamic functions. For example, the absolute values of such properties like internal energy, enthalpy, and Gibbs energy of a chemical are not known. Absolute values of the standard chemical potentials ($\mu_i^{0,x}$, $\mu_i^{0,b}$, and $\mu_w^{0,x}$) are not known too, and we use instead the standard Gibbs energies of formation ($\Delta_f G_i^{0,x}$, $\Delta_f G_i^{0,b}$, and $\Delta_f G_w^{0,x}$), which can be obtained experimentally. There is also an inconvenience with the Gibbs energies of formation, because some reference books do not specify the concentration scale for these values. We should always be alert when using the standard Gibbs energies of formation from a reference book and make sure that the concentration scale is well defined.

The equations for the Gibbs energy of formation of water and the ith electrolyte can be written as follows:

$$\Delta_f G_w = \Delta_f G_w^{0,x} + RT\ln\left(x_w\gamma_{w,x}\right), \tag{1.21}$$

$$\Delta_f G_i = \Delta_f G_i^{0,b} + RT\ln\left(\frac{\gamma_{i,b}b_i}{b^0}\right)$$

$$= \Delta_f G_\pm^0 + RT\ln\left(\gamma_\pm\right)^\nu + RT\ln\left(\frac{b_i}{b^0}\right)^\nu + RT\ln(Q)^\nu, \tag{1.22}$$

where $\Delta_f G_i^{0,b} = \Delta_f G_\pm^0 = \nu_+\Delta_f G_+^0 + \nu_-\Delta_f G_-^0$.

As an example of using the Gibbs energies of formation, let us consider the saturated NaCl(aq) solution at 25°C and 1 bar shown in Figure 1.2 and take into account that the activity of water, $a_{w,x} = x_w\gamma_{w,x}$, can be calculated as fugacity of water vapor (see Figure 1.2) above solution, f_w, divided by fugacity of water vapor above pure water, f_o, at the same temperature (Raoult's law) [3].

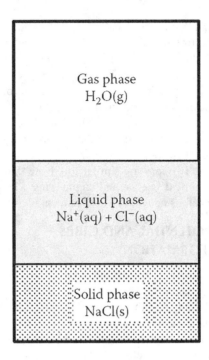

FIGURE 1.2   The NaCl(s)–NaCl(aq)–H₂O(g) system.

Note that fugacity of a pure substance can be calculated using the corresponding equation of state [3] and a gas component fugacity can be calculated similarly to activity, that is, as a product of the component partial pressure and the fugacity coefficient. Details of such calculations can be found in any chemical thermodynamics and some physical chemistry books [3].

Using data from Ref. [4], $\Delta_f G_w$ and $\Delta_f G_i$ can be calculated for the NaCl(aq) saturated solution. Employing a short extrapolation for calculating $\gamma_\pm$ of the saturated NaCl(aq) solution, the following values can be obtained: $\Delta_f G_w = -237.83$ kJ mol⁻¹ and $\Delta_f G_i = -384.38$ kJ mol⁻¹, which can be checked for their validity because this is a three-phase system and all phases are in equilibrium. From Table 10.3, the Gibbs energy of formation of solid NaCl, $\Delta_f G_{NaCl(s)} = -384.10$ kJ mol⁻¹, and the Gibbs energy of formation of the gaseous water, which is above the solution saturated with NaCl(s), can be calculated as -237.84 kJ mol⁻¹, which both are very similar to $\Delta_f G_i$ and $\Delta_f G_w$ calculated previously. Details of calculations described in this paragraph are as follows:

1. Concentration of the saturated NaCl(aq) solutions is 6.153 mol kg⁻¹ (Table 10.21).
2. The mean activity coefficient of NaCl(aq) in the saturated solutions can be found as 0.953 by the extrapolation of the data from (Table 10.17).
3. $\Delta_f G^0_{Na^+} = -261.9$ kJ mol⁻¹ and $\Delta_f G^0_{Cl^-} = -131.2$ kJ mol⁻¹ (Table 10.5), and therefore, $\Delta_f G_i^{0,b} = \Delta_f G^0_\pm = (-261.9 - 131.2) \times 10^3 = -393,100$ J mol⁻¹.
4. $RT \ln (\gamma_\pm)^\nu = 2 \times 8.314 \times 298.15 \times \ln (0.953) = -239$ J mol⁻¹.

5. $RT \ln (b_i/b^0)^\nu = 2 \times 8.314 \times 298.15 \times \ln (6.153/1) = 9008$ J mol^{-1}.

6. $Q = \left(\nu_+^{\nu+} \nu_-^{\nu-} \right)^{1/\nu} = 1, \quad RT \ln(Q)^\nu = 0.$

7. $\Delta_f G_i = \Delta_f G_\pm^0 + RT \ln(\gamma_\pm)^\nu + RT \ln \left(\dfrac{b_i}{b^0} \right)^\nu + RT \ln(Q)^\nu$

$$= -393{,}100 - 239 + 9008 + 0 = -384{,}331 \, \mathrm{J\,mol}^{-1}.$$

8. From Table 10.3, $\Delta_f G_{\mathrm{NaCl(s)}} = -384.10$ kJ mol^{-1}, which is close to the value obtained in (7): -384.3 kJ mol^{-1}.

9. $p_w = 2.401$ kPa (Table 10.25), $p_o = 3.1699$ kPa (Table 10.23).

10. $a_{w,x} = \dfrac{f_w}{f_o} = x_w \gamma_{w,x} \approx x_w = \dfrac{p_w}{p_o} = \dfrac{2.401}{3.1699} = 0.7574.$

11. $\Delta_f G_w^{0,x} = -237{,}141 \, \mathrm{J\,mol}^{-1}$ (Tables 10.3 and 10.4).

12. $\Delta_f G_{w(l)} = \Delta_f G_w^{0,x} + RT \ln \left(x_w \gamma_{w,x} \right)$

$$= -237{,}141 + 8.314 \times 298.15 \times \ln(0.7574) = -237.83 \, \mathrm{kJ\,mol}^{-1}.$$

13. $\Delta_f G_{w(g)} = \Delta_f G_{w,g}^{0,p} + RT \ln \left(f_{w(g)} / f^0 \right)$

$$= \Delta_f G_{w,g}^{0,p} + RT \ln \left(p_{w(g)} / p^0 \right)$$

$$= -228.584 + 8.314 \times 298.15 \ln \left(2.401 \times 10^{-2} \right)$$

$$= -237.83 \, \mathrm{kJ\,mol}^{-1}, \text{ which is the same as obtained in Ref. [12].}$$

A simple recipe on calculating the fugacity, f_i, when given a partial pressure, p_i, can be found in Ref. [5].

In the end of this section, we need to note that based on a convention in chemical thermodynamics, the standard Gibbs energy of formation of any element in its reference state (phase) is zero at all temperatures at 1 bar of pressure. For example, the standard Gibbs energy of formation of metal copper, gaseous hydrogen, and solid graphite is equal to zero.

1.11 THE DEBYE–HÜCKEL THEORY OF DILUTE ELECTROLYTE SOLUTIONS

The Debye–Hückel theory provides a method to calculate the activity coefficient of a single ion, γ_i (γ_+ or γ_-), which describes interactions of the ith ion with all other ions in the solution (Figure 1.3) and shows deviation of the chemical potential of the ith ion, μ_i (μ_+ or μ_-), from its ideal value, $\mu_{i,\,id}$ ($\mu_{+,\,id}$ or $\mu_{-,\,id}$):

$$\ln \gamma_i = \frac{(\mu_i - \mu_{i,id})}{(RT)} = \frac{\left[\mu_i - \mu_i^0 - RT \ln \left(b_i / b^0 \right) \right]}{(RT)}. \tag{1.23}$$

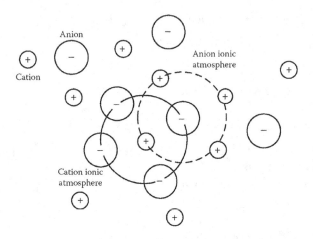

FIGURE 1.3   Ionic atmosphere in a solution.

In 1923, Peter Debye and Erick Hückel obtained the following equation for $(\mu_i - \mu_{id})$ [2,3]:

$$\mu_i - \mu_{i,id} = -\frac{z_i^2 e^2 \kappa}{8\pi \varepsilon_r \varepsilon_o \left(1 + \kappa \mathring{a}\right)}, \tag{1.24}$$

where
   κ is the inverse radius of the ionic atmosphere (see Figure 1.3).
   $\mathring{a}$ is the minimum distance between centers of the anion and cation.
   ε_r is the solvent relative permittivity (dielectric constant).
   ε_o is the electric constant (permittivity of vacuum), which is a fundamental constant (Table 10.1).
   z_i is charge of the ith ion.
   e is the elementary charge, which is a fundamental constant and equals 1.6022×10^{-19} C (Table 10.1).

Peter Joseph William Debye (1884–1966) was a Dutch-American physicist and physical chemist who was awarded Nobel Prize in Chemistry in 1936.

He has a number of significant scientific contributions and one of them was the development of a theory of strong electrolyte solutions. The theory was created together with his assistant Erich Hückel and is known as Debye–Hückel theory.

REFERENCES

A.J. Bard, G. Inzelt, and F. Scholz, *Electrochemical Dictionary*, Springer, Berlin, Germany, 2008.
http://en.wikipedia.org/wiki/Peter_Debye

Erich Armand Arthur Joseph Hückel (1896–1980) was a German physical chemist who collaborating with Peter Debye developed a theory of strong electrolyte solutions known as Debye–Hückel theory.

REFERENCES

A.J. Bard, G. Inzelt, and F. Scholz, *Electrochemical Dictionary*, Springer, Berlin, Germany, 2008.
http://en.wikipedia.org/wiki/Erich_H%C3%BCckel

The radius of the ionic atmosphere shown as large cycles in Figure 1.3 and κ (in m^{-1}) is defined as follows:

$$\kappa^2 = \frac{2e^2 N_A \rho b^0}{\varepsilon_r \varepsilon_o k_B T} I_b, \tag{1.25}$$

where

ρ is the solvent density.

T is the thermodynamic temperature.

N_A is the Avogadro constant (Table 10.1).

1.12 CALCULATION OF ACTIVITY COEFFICIENT USING THE DEBYE–HÜCKEL THEORY

Combining Equations 1.23 through 1.25, an expression for the activity coefficient of an individual ion can be obtained:

$$\ln \gamma_{i,b} = -\frac{A_{DH} z_i^2 I_b^{1/2}}{\left(1 + B_{DH} \mathring{a} I_b^{1/2}\right)}$$

$$\ln \gamma_+ = -\frac{A_{DH} z_+^2 I_b^{1/2}}{\left(1 + B_{DH} \mathring{a} I_b^{1/2}\right)}, \tag{1.26}$$

$$\ln \gamma_- = -\frac{A_{DH} z_-^2 I_b^{1/2}}{\left(1 + B_{DH} \mathring{a} I_b^{1/2}\right)}$$

where A_{DH} and B_{DH} depend on temperature, density of solvent, and its static relative permittivity (dielectric constant) available from Table 10.22.

Using the definition of the mean activity coefficient (Equation 1.17), the following equation can be obtained to calculate $\gamma_\pm$:

$$\ln \gamma_\pm = -\frac{A_{DH} |z_+ z_-| I_b^{1/2}}{\left(1 + B_{DH} \mathring{a} I_b^{1/2}\right)}. \tag{1.27}$$

Note that Equations 1.26 and 1.27 represent second approximation of the Debye–Hückel theory.

If the ionic strength is below 0.001 mol kg^{-1}, the Debye–Hückel equations for an individual ion can be simplified as follows:

$$\ln \gamma_{i,b} = -A_{DH} z_i^2 I_b^{1/2}$$

$$\ln \gamma_+ = -A_{DH} z_+^2 I_b^{1/2}, \tag{1.28}$$

$$\ln \gamma_- = -A_{DH} z_-^2 I_b^{1/2}$$

and for the electrolyte as follows:

$$\ln \gamma_\pm = -A_{DH} \left| z_+ z_- \right| I_b^{\frac{1}{2}}. \tag{1.29}$$

Note that Equations 1.28 and 1.29 represent the Debye–Hückel limiting law or fist approximation of the theory.

Using theoretical equation for A_{DH} and B_{DH} from Ref. [2] and properties of water given in (Table 10.22), one can calculate $A_{DH} = 1.172$ (kg mol^{-1})$^{\frac{1}{2}}$ and $B_{DH} = 3.28 \times 10^9$ (kg mol^{-1})$^{\frac{1}{2}}$ m^{-1} = 0.328 (kg mol^{-1})$^{\frac{1}{2}}$ (Å)$^{-1}$ at a temperature of 25°C and a pressure of 1 bar. $\mathring{a}$ in Equations 1.26 and 1.27 is a model parameter and can be taken as 4.5 Å, which corresponds to common diameter of an aqueous ion, including its hydration shell. When the solution consists of more than two ions, $\mathring{a}$ becomes a purely empirical parameter and is usually chosen somewhere between 3 and 5 Å.

1.13   CALCULATED AND OBSERVED ACTIVITY COEFFICIENTS

The comparison between calculated and observed $\gamma_\pm$ values shown in Table 1.2 and Figure 1.4 clearly demonstrates that the Debye–Hückel second approximation,

TABLE 1.2
Mean Activity Coefficients (Table 10.17 and Ref. 2)

				$\gamma\pm$	
b (mol kg^{-1})	I_b (mol kg^{-1})	Equation 1.29	Equation 1.27[a]	HCl(aq)	KCl(aq)
		1–1 Electrolytes			
0.001	0.001	0.964	0.965	0.965	0.965
0.002	0.002	0.949	0.952	0.952	0.951
0.005	0.005	0.921	0.928	0.929	0.927
0.010	0.010	0.889	0.903	0.905	0.901
0.020	0.020	0.847	0.872	0.876	0.869
0.050	0.050	0.770	0.822	0.832	0.816
	2–1 and 1–2 Electrolytes			CaCl$_2$(aq)	Na$_2$SO$_4$(aq)
0.001	0.003	0.880	0.888	0.888	0.886
0.002	0.006	0.834	0.850	0.851	0.846
0.005	0.015	0.750	0.785	0.787	0.777
0.010	0.030	0.666	0.725	0.727	0.712
0.020	0.060	0.563	0.658	0.664	0.637
0.050	0.150	0.403	0.564	0.577	0.529
	3–1 Electrolytes			LaCl$_3$(aq)	
0.001	0.006	0.762	0.784	0.790	
0.002	0.012	0.680	0.719	0.729	
0.005	0.030	0.544	0.617	0.636	
0.010	0.060	0.423	0.533	0.560	
0.020	0.120	0.296	0.449	0.483	
0.050	0.300	0.146	0.348	0.388	

[a] $\mathring{a} = 4.5$ Å.

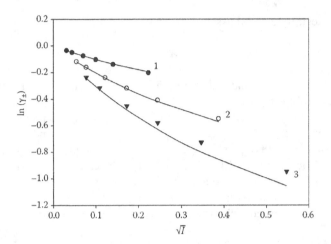

FIGURE 1.4   Mean activity coefficients of (1) KCl, (2) $CaCl_2$, and (3) $LaCl_3$ vs. square root of the dimensionless ionic strength. Data points are from Table 10.17 and Ref. 3, and solid lines are obtained using Equation 1.27 with $\mathring{a} = 4.5$ Å.

Equation 1.27, can be used at $I < 0.05$ ($b = 0.05$ mol kg^{-1}) for 1–1, at $I < 0.06$ ($b = 0.02$ mol kg^{-1}) for 2–1 and 1–2, and at $I < 0.006$ ($b = 0.001$ mol kg^{-1}) for 1–3 strong electrolytes. Also, Table 1.2 shows that for 1–1, 1–2, and 2–1 electrolytes, the range of applicability of the limiting law, Equation 1.29, is shifted to $I < 0.002$.

Obviously, Equation 1.27 can be used for a larger concentration range than Equation 1.29, and the data presented in Table 1.2 and Figure 1.4 (Table 10.17 and Ref. 3) confirm this statement.

1.14   MEAN ACTIVITY COEFFICIENT IN CONCENTRATED AQUEOUS SOLUTIONS

Figure 1.5 shows that there is strong concentration dependence of the mean activity coefficient, $\gamma_\pm$, and at high concentrations, the activity coefficient can be either well below or above 1. The strong concentration dependence of $\gamma_\pm$ is caused by a variety of interactions between ions and, also, due to the association/dissociation processes between ions and molecules. The theory of concentrated electrolyte solutions is still under development because the processes of the ion association/dissociation and hydration significantly complicate the development of such a theory. As a result, some semi-empirical approaches are well accepted and have been usually used thus far [2]. An example of empirically modifying the Debye–Hückel equation by adding an additional term to the second approximation of the theory is given as follows:

$$\ln \gamma_\pm = -\frac{A_{DH}|z_+z_-|I_b^{\frac{1}{2}}}{\left(1 + B_{DH}\mathring{a}_o I_b^{\frac{1}{2}}\right)} + CI_b. \tag{1.30}$$

$$\ln \gamma_{i,b} = -\frac{A_{DH}z_i^2 I_b^{\frac{1}{2}}}{\left(1 + B_{DH}\mathring{a}_o I_b^{\frac{1}{2}}\right)} + CI_b. \tag{1.31}$$

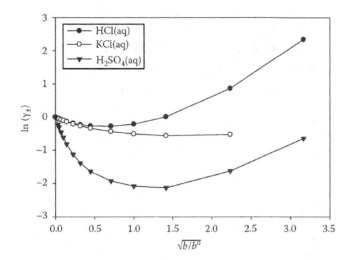

FIGURE 1.5 Dependence of the mean activity coefficient of HCl(aq), KCl(aq), and H_2SO_4(aq) from the square root of the dimensionless molality. The data are taken from Table 10.17.

Equations 1.30 and 1.31 that can be called the Debye–Hückel third approxima-tion work pretty well for single electrolyte aqueous solutions but cannot be easily used when two or more electrolytes have high concentrations. In a multicompo-nent aqueous solution, Equations 1.30 and 1.31 can be used only if one of the electrolytes is dominating and concentrations of all others are much smaller. A well-known example of such a solution is sea water (or underground brine) where NaCl is a dominating electrolyte. As can be seen from Table 1.3, NaCl is highly dominating in sea water, and NaCl molality (around 0.48 mol kg^{-1}) is about an order magnitude larger than another major electrolyte, $MgCl_2$ (around 0.052 mol kg^{-1}). In such a solution, using the experimental mean activity coefficients of NaCl(aq) (Table 10.17) to calculate parameter C in Equations 1.30 and 1.31 would be a reasonably accurate approach. In Section 1.20, there is an exercise showing a simple way to calculate parameter C of Equation 1.30 for KCL(aq) solution for a concentration range from 0 to 0.2 mol kg^{-1}.

1.15 SPECIATION IN WEAK ELECTROLYTES

The solution speciation here means a list of the ions and molecules as well as their concentrations. A weak electrolyte, for example, HA (weak acid) or BOH (weak base), when dissolved in a solvent (water), consists of both neutral molecules and ions due to the following reactions:

$$HA(aq) = H^+(aq) + A^-(aq)(\text{ionization}) \text{ or } HA(aq) + H_2O(l)$$

$$= H_3O^+(aq) + A^-(aq)(\text{hydrolysis}).$$

TABLE 1.3

Composition of Sea Water at a Salinity of 35 ppt with an Ionic Strength of 0.6675 (Table 10.27)

Aqueous Ions	b (mol kg^{-1})
Cl$^-$	0.56200
Br$^-$	0.00087
F$^-$	0.00007
SO$_4^{2-}$	0.01140
HCO$_3^-$	0.00143
NaSO$_4^-$	0.01080
KSO$_4^-$	0.00012
Na$^+$	0.47200
K$^+$	0.01039
Mg^{2+}	0.04830
Ca^{2+}	0.00143
Sr^{2+}	0.00009
MgHCO$_3^+$	0.00036
MgSO$_4^0$	0.00561
CaSO$_4^0$	0.00115
NaHCO$_3^0$	0.00020
H$_3$BO$_3^0$	0.00037
Ionic strength	0.6675

$$BOH(aq) = B^+(aq) + OH^-(aq)(\text{ionization}) \text{ or } B(aq) + H_2O(l)$$

$$= HB^+(aq) + OH^-(aq)(\text{hydrolysis}).$$

For example, in case of acetic acid, $CH_3COOH(aq)$:

$$CH_3COOH(aq) = H^+(aq) + CH_3COO^-(aq)(\text{ionization}),$$

or

$$CH_3COOH(aq) + H_2O(l) = H_3O^+(aq) + CH_3COO^-(aq)(\text{hydrolysis}),$$

and in case of ammonium hydroxide, $NH_4OH(aq)$:

$$NH_4OH(aq) = NH_4^+(aq) + OH^-(aq)(\text{ionization}) \text{ or } NH_3(aq) + H_2O(l)$$

$$= NH_4^+(aq) + OH^-(aq)(\text{hydrolysis}).$$

Note that in these reactions of hydrolysis

$$NH_3(aq) + H_2O(l) = NH_4OH(aq) \text{ and } H^+(aq) + H_2O(l) = H_3O^+(aq),$$

the Gibbs energy of these reactions is assumed to be zero, and therefore, the concentration of $NH_3(aq)$ and $NH_4OH(aq)$ in the ammonium hydroxide solution is the same as the concentration of $H^+(aq)$ and $H_3O^+(aq)$ in acetic acid solution.

The total concentration of the weak electrolyte (e.g., with a weak acid) is a sum of the concentration of ions and molecules, $b_{HA} = b_{HA(aq)} + b_{H+(aq)} = b_{HA(aq)} + b_{A-(aq)}$ (mass balance equation), and the solution should be electrically neutral, that is, $b_{H+(aq)} = b_{A-(aq)}$ (charge balance equation), in addition to the relationship between all concentrations in accordance with the law of mass action:

$$K_a = \frac{a_{H+(aq)} a_{A-(aq)}}{a_{HA(aq)}} = \left[\frac{b_{H+(aq)} b_{A-(aq)}}{b_{HA(aq)} b^o} \right] \left[\frac{\gamma_{H+(aq)} \gamma_{A-(aq)}}{\gamma_{HA(aq)}} \right] = [K_b][K_\gamma]. \quad (1.32)$$

A numerical exercise of using the mas balance, charge balance and two mass action equations (four equations with four variables) is given in Section 1.20 for an important technological and environmental system of carbonic acid in water, $H_2CO_3(q)$. In electrochemical studies, we should always keep in mind that the atmospheric carbon dioxide reacting with water and producing carbonic acid significantly changes its acidity and usually should be removed by purging argon or nitrogen through the solution.

To simplify Equation 1.32, the degree of dissociation, $\alpha = b_{H+(aq)}/b_{HA} = b_{A-(aq)}/b_{HA}$, $(1-\alpha) = b_{HA(aq)}/b_{HA}$, can be introduced, and the dissociation constant, K_a, can be presented using only α and b_{HA}:

$$K_a = \alpha^2 b_{HA} K_\gamma / \left[(1-\alpha) b^o \right]. \quad (1.33)$$

If all activity coefficients equal 1, then Equation 1.33 becomes

$$K_a = \alpha^2 b_{HA} / \left[(1-\alpha) b^o \right], \quad (1.34)$$

which is the well-known dilution law of Friedrich Wilhelm Ostwald.

Equations 1.33 and 1.34 are widely used because α is an experimentally measurable value (e.g., using conductivity measurement), and, therefore, K_a can be experimentally estimated, or vice versa; if K_a is known, the concentration of all species can be found. Note that any thermodynamic constant, like K_a, is a dimensionless value, and, therefore, logarithm of the thermodynamic constant can be freely taken. Negative decimal logarithm of K_a is pK_a; that is, $pK_a = -\log_{10} K_a = -\log K_a$. pK_a values for a number of weak electrolytes can be found in (Tables 10.14 and 10.15).

There is a way to further simplify Equation 1.34 in a case when α is significantly smaller than 1, and this is a quite common case when the electrolyte is really very weak. In this approximation, Equation 1.34 is simplified as follows:

$$K_a \approx \frac{\alpha^2 b_{HA}}{b^0},$$

Friedrich Wilhelm Ostwald (1853–1932) was a Baltic German physical chemist, who was awarded the Nobel Prize in Chemistry in 1909 for his work on a number of chemical problems including chemical equilibria and kinetics. He was one of the first scientists to develop a theory of electrolyte solutions.

REFERENCES

A.J. Bard, G. Inzelt, and F. Scholz, *Electrochemical Dictionary*, Springer, Berlin, Germany, 2008.
http://en.wikipedia.org/wiki/Wilhelm_Ostwald

and the degree of dissociation, α, can immediately be calculated at any total concentration of acid, b_{HA}, using the equilibrium constant K_a:

$$\alpha = \left(\frac{K_a b^0}{b_{HA}} \right)^{1/2}.$$

If the approximation $\alpha \ll 1$ cannot be used, then a quadratic Equation 1.34 can analytically be solved and the positive solution (α cannot be negative) of the quadratic equation should be chosen. If alpha is estimated, the concentration of all three species of the weak electrolyte can be easily calculated.

1.16   pH OF AQUEOUS SOLUTIONS

pH is an important characteristic of aqueous solution and is directly related to the activity of protons, $a_{H+(aq)}$:

$$\text{pH} = -\log_{10}\left[a_{H^+(aq)} \right]. \tag{1.35}$$

If $a_{H+(aq)}$ is known, the activity of $OH^-(aq)$, $a_{OH^-(aq)}$, can be calculated using the dissociation (ionization) constant of water, K_w (Table 10.6):

$$K_w = \frac{a_{H^+(aq)} a_{OH^-(aq)}}{a_{H_2O(l)}}, \tag{1.36}$$

where $a_{H_2O(l)}(= a_w)$ is the activity of water on the mole fraction concentration scale. The activity of water molecules, $a_{H_2O(l)} = x_{H_2O(l)}\gamma_{H_2O(l),x}$, in pure water is very close to 1 but can deviate from 1 in a high concentrated solution. Also, in pure water, the activity coefficients of $H^+(aq)$ and $OH^-(aq)$ are very close to 1 (this can be verified using the Debye–Hückel theory), and therefore, for pure water, Equation 1.36 can be simplified as follows:

$$K_w \approx \frac{b_{H^+(aq)} b_{OH^-(aq)}}{\left(b^0\right)^2}. \tag{1.37}$$

The right part of Equation 1.37 is called the ion product of water.

Note that K_w, as any other equilibrium constant, is independent of concentration but depends on temperature and pressure, while the ion product depends on the activity coefficients and concentration of water in the solution. The most reliable K_w values over a wide range of temperatures and pressures can be found in Table 10.6. Note that it is useful to keep in mind that at 25°C and 1 bar $K_w/(b^0)^2 = 10^{-14}$, so the concentrations of $H^+(aq)$ and $OH^-(aq)$ are 10^{-7} mol kg^{-1} and pH of pure water is 7. If a strong electrolyte containing either $H^+(aq)$ or $OH^-(aq)$ is added to pure water, pH will be drastically changed, while pH will be slightly changed if a strong electrolyte like NaCl is added due to the change of the activity coefficients of $H^+(aq)$, $OH^-(aq)$, and $H_2O(l)$. The reader may want to practice in calculating the concentration of $H^+(aq)$ and $OH^-(aq)$ in the saturated NaCl(aq) using the data given in Section 1.10.

While it is usually thought that pH of an aqueous solution can easily be measured using two electrodes and a pH meter, in fact, pH can easily be defined only in dilute solutions. There are two main problems in defining pH: (1) pH is a thermodynamic characteristic of an individual ion and therefore cannot be experimentally found without some assumptions that are outside of thermodynamics, and (2) available electrochemical probes cannot work in many environments of practical interest (e.g., very concentrated brine) and, therefore, should be specially developed for particular applications. Generally, to measure pH of a test solution, pH_x, a calibration of reliable electrochemical probe using a number of standard solutions should be carried out with an assumption that pH of the standard solutions, pH_s, is well defined. The standard solutions should preferably be buffer solutions (see Section 1.17) so when a small amount of acid or base is added, pH of the standard solution is not significantly changed.

1.17 pH OF BUFFER SOLUTIONS

An aqueous solution consisting of a coupled weak acid or weak base and salt with, respectively, same anion or cation is considered a buffer solution. Such a solution is resistant to pH changes upon small addition of acid or base and can be useful when constant pH during an electrochemical experiment is required. An example of the acidic buffer solution is a mixture of $CH_3COOH(aq)$ and $CH_3COONa(aq)$, and an example of the basic buffer solution is a mixture of $NH_4OH(aq)$ and $NH_4Cl(aq)$. Note that most of the salts are strong electrolytes, and $CH_3COONa(aq)$ and $NH_4Cl(aq)$ are not exemptions. They are both fully dissociated.

In the case of the $[CH_3COOH + CH_3COONa](aq)$ buffer solution, we can use Equation 1.32 as follows:

$$pH = pK_a + \log\left(\frac{a_{A-(aq)}}{a_{HA(aq)}}\right) = pK_a + \log\left[\frac{b_{A-(aq)}\gamma_{A-(aq)}}{\left(b_{HA(aq)}\gamma_{HA(aq)}\right)}\right], \qquad (1.38)$$

where $pK_a = -\log K_a$, $b_{A-(aq)}$, and $b_{HA(aq)}$ are, respectively, the molal concentration of anion of the coupled salt and molecules of the weak acid, and $\gamma_{A-(aq)}$ and $\gamma_{HA(aq)}$ are the activity coefficients of the corresponding species on the molal concentration scale. Taking into account that (1) $b_{A-(aq)}$ is practically equal to the total concentration of the coupled salt, b_{salt}; (2) $b_{HA(aq)}$ is practically equal to the total concentration of the coupled weak acid, b_{acid}; (3) $\gamma_{HA(aq)}$ is close to 1; and (4) the individual activity coefficient of the salt anion is approximately equal to the mean activity coefficient of the salt, $\gamma_{\pm}$, pH of the acidic buffer solution can be calculated as follows:

$$pH \approx pK_a + \log\left[\frac{b_{salt}\gamma_{\pm}}{b_{acid}}\right]. \qquad (1.39)$$

Making similar derivations, the equations for pH of the basic buffer solution can also be obtained:

$$pH \approx pK_{base} + \log\left[\frac{b_{salt}\gamma_{\pm}}{b_{base}}\right]. \qquad (1.40)$$

Because we cannot exactly calculate the activity coefficient of an individual ion in high concentrated solutions and, therefore, Equations 1.39 and 1.40 cannot give a precise value, pH of buffer solutions should experimentally be measured rather than calculated. Such measurements have been carried out and are available in Table 1.4. The experimental approach to obtain pH of an aqueous solution will be considered in Chapter 5.

1.18   SUGGESTED pH OF STANDARD SOLUTIONS

Standard buffer solutions have been developed in a way so that each of them should have (1) inexpensive chemicals, (2) simple recipe, and (3) reproducibility and stability in time.

Standard buffer solutions presented in Table 1.4 cover a wide range of pH from 1 to 13 so that a pH sensor can be calibrated and used at any pH in this range.

It should be mentioned that the common pH values of aqueous solutions at 25°C and 1 bar are between 1 and 13, while pH of a high concentrated acid can be <1 and even negative while pH of a highly concentrated base can be >13 and even >14

TABLE 1.4
pH of Ten Buffer Solutions at 25°C[a] [4]

A		B		C		D		E		F		G		H		I		J	
pH	x	pH	x	pH	x	pH	x	pH	x	pH	x	pH	x	pH	x	pH	x	pH	x
1.00	67.0	2.20	49.5	4.10	1.3	5.80	3.6	7.00	46.6	8.00	20.5	9.20	0.9	9.60	5.0	10.90	3.3	12.00	6.0
1.10	52.8	2.30	45.8	4.20	3.0	5.90	4.6	7.10	45.7	8.10	19.7	9.30	3.6	9.70	6.2	11.00	4.1	12.10	8.0
1.20	42.5	2.40	42.2	4.30	4.7	6.00	5.6	7.20	44.7	8.20	18.8	9.40	6.2	9.80	7.6	11.10	5.1	12.20	10.2
1.30	33.6	2.50	38.8	4.40	6.6	6.10	6.8	7.30	43.4	8.30	17.7	9.50	8.8	9.90	9.1	11.20	6.3	12.30	12.8
1.40	26.6	2.60	35.4	4.50	8.7	6.20	8.1	7.40	42.0	8.40	16.6	9.60	11.1	10.00	10.7	11.30	7.6	12.40	16.2
1.50	20.7	2.70	32.1	4.60	11.1	6.30	9.7	7.50	40.3	8.50	15.2	9.70	13.1	10.10	12.2	11.40	9.1	12.50	20.4
1.60	16.2	2.80	28.9	4.70	13.6	6.40	11.6	7.60	38.5	8.60	13.5	9.80	15.0	10.20	13.8	11.50	11.1	12.60	25.6
1.70	13.0	2.90	25.7	4.80	16.5	6.50	13.9	7.70	36.6	8.70	11.6	9.90	16.7	10.30	15.2	11.60	13.5	12.70	32.2
1.80	10.2	3.00	22.3	4.90	19.4	6.60	16.4	7.80	34.5	8.80	9.6	10.00	18.3	10.40	16.5	11.70	16.2	12.80	41.2
1.90	8.1	3.10	18.8	5.00	22.6	6.70	19.3	7.90	32.0	8.90	7.1	10.10	19.5	10.50	17.8	11.80	19.4	12.90	53.0
2.00	6.5	3.20	15.7	5.10	25.5	6.80	22.4	8.00	29.2	9.00	4.6	10.20	20.5	10.60	19.1	11.90	23.0	13.00	66.0
2.10	5.10	3.30	12.9	5.20	28.8	6.90	25.9	8.10	26.2	9.10	2.0	10.30	21.3	10.70	20.2	12.00	26.9		
2.20	3.9	3.40	10.4	5.30	31.6	7.00	29.1	8.20	22.9			10.40	22.1	10.80	21.2				
		3.50	8.2	5.40	34.1	7.10	32.1	8.30	19.9			10.50	22.7	10.90	22.0				
		3.60	6.3	5.50	36.6	7.20	34.7	8.40	17.2			10.60	23.3	11.00	22.7				
		3.70	4.5	5.60	38.8	7.30	37.0	8.50	14.7			10.70	23.8						
		3.80	2.9	5.70	40.6	7.40	39.1	8.60	12.2			10.80	24.25						

(Continued)

TABLE 1.4 (CONTINUED)
pH of Ten Buffer Solutions at 25°C[a] [4]

A		B		C		D		E		F		G		H		I		J	
pH	x	pH	x	pH	x	pH	x	pH	x	pH	x	pH	x	pH	x	pH	x	pH	x
		4.00	0.1	5.90	43.7	7.60	42.4	8.80	8.5										
						7.70	43.5	8.90	7.0										
						7.80	44.5	9.00	5.7										
						7.90	45.3												
						8.00	46.1												

[a] Compositions of buffer solutions given in the table: A—25 mL of 0.2 molar KCl + x mL of 0.2 molar HCl; B—50 mL of 0.1 molar potassium hydrogen phthalate + x mL of 0.1 molar HCl; C—50 mL of 0.1 molar potassium hydrogen phthalate + x mL of 0.1 molar NaOH; D—50 mL of 0.1 molar potassium dihydrogen phosphate + x mL of 0.1 molar NaOH; E—50 mL of 0.1 molar tris(hydroxymethyl)aminomethane + x mL of 0.1 molar HCl; F—50 mL of 0.025 molar borax + x mL of 0.1 molar HCl; G—50 mL of 0.025 molar borax + x mL of 0.1 molar NaOH; H—50 mL of 0.05 molar sodium bicarbonate + x mL of 0.1 molar NaOH; I—50 mL of 0.05 molar disodium hydrogen phosphate + x mL of 0.1 molar NaOH; J—25 mL of 0.2 molar KCl + x mL of 0.2 molar NaOH. In all buffer solutions, the final volume of mixture should be 100 mL by adding water.

due to a contribution from a large value of the activity coefficient. For example, pH of 5 mol kg^{-1} HCl(aq) can be as low as -1.1 if the mean activity coefficient HCl(aq) (Table 10.17) is used for the approximate estimation of pH $\approx -\log_{10}(5\times2.38) = -1.08$.

One of the common problems to be solved is to calculate the concentration of protons, H$^+$(aq), if the pH value is given or obtained from an experiment. In Section 1.20, such an exercise is provided.

1.19 SUMMARY

- Aqueous electrolyte solutions are formed due to interactions between ions of a solid, for example, KCl(s) or a liquid, for example, HCl(l), with water molecules forming individual ions such as K$^+$(aq), H$^+$(aq), and Cl$^-$(aq).
- There are strong and weak electrolytes as well as nonelectrolytes. Strong electrolytes are completely dissociated to ions, weak electrolytes are partially dissociated, and nonelectrolytes are not dissociated.
- Generally, a variety of concentration scales are used, and it is important to be able to convert from one concentration scale to another. Molality, molarity, and mole fraction are mainly used in this book.
- The chemical potential consists of two contributions: the standard state term and a logarithmic term. Both terms depend on the concentration scale. However, the chemical potential does not depend on the concentration scale.
- Because the absolute value of the standard chemical potential is not known, we use instead the standard Gibbs energy of formation. The standard Gibbs energy of formation depends on the concentration scale and therefore might be different for the same species in different reference books.
- The activity coefficient depends on the concentration scale, too. Therefore, the standard Gibbs energy of formation and the activity coefficient for a species should be on the same concentration scale.
- The activity coefficient of an ion (γ_+ or γ_-) cannot be experimentally measured but can be theoretically calculated for a dilute solution. However, the activity coefficient of an electrolyte, $\gamma_\pm$, can be measured in a wide range of concentrations except for very dilute solutions where precision of the measurements can be insufficient.
- For any electrolyte, if $\gamma_+ = \gamma_-$, then $\gamma_\pm = \gamma_+ = \gamma_-$. This assumption can safely be used for KCl(aq) even at relatively high concentrations. For most of the electrolytes, the use of the assumption that $\gamma_+ = \gamma_-$ is quite questionable but can be unavoidable if calculations with individual activity coefficients are needed.
- To calculate the ionic activity coefficient, the Debye–Hückel equation can be used up to about $I_b = 0.05$ mol kg^{-1} at ambient conditions and up to $I_b = 0.1$ mol kg^{-1} at elevated temperatures (e.g., 250°C).
- The theory of concentrated electrolyte solutions is under development. It is because of the processes of ion association/dissociation and hydration, which significantly complicate the development of such a theory. As a result, a semi-empirical approach is well accepted and usually used to calculate the activity coefficients.

- When measuring pH, the activity of $H^+(aq)$ (or $OH^-(aq)$) ions can experimentally be obtained. A calibration approach using a number of buffer solutions should be used in such measurements.
- An aqueous solution, consisting of a coupled weak acid and salt or weak base and salt, that resists pH changes upon small addition of acid or base represents an example of the buffer solution.
- To measure pH, a set of standard (buffer) solutions should be used to calibrate pH sensor. The compositions of buffer solutions are well known, and they cover a wide range of pH from 1 to 13.

1.20 EXERCISES

PROBLEM 1

The mass fraction of KCl in aqueous solution is 0.1. Calculate mole fraction, molality, and molarity of the solute at 25 and 90°C assuming that the density of the solution changes with temperature similarly to the density change of pure water.

Solution:

Equations for conversion of concentration can be found in Table 1.1. To convert mass fraction, w, to mole fraction, x, the following equation can be used:

$$x = \frac{\dfrac{w}{M_2}}{\dfrac{w}{M_2} + \dfrac{1-w}{M_1}},$$

where M_1 and M_2 are, respectively, the molar mass of solvent and solute. From Table 10.29, $M_1 = 18.015$ g mol^{-1} and $M_2 = 74.551$ g mol^{-1}. Then, $x = 0.026148$ and is independent of temperature.

To convert mass fraction, w, to molality, b, the following equation can be used:

$$b = \frac{1000w}{M_2(1-w)},$$

where the number 1000 is from converting g into kg. Then, $b = 1.4904$ mol kg^{-1} and is independent of temperature.

To convert mass fraction, w, to molarity, c, the following equation can be used:

$$c = \frac{1000\rho w}{M_2},$$

where ρ is density of solution in g cm^{-3}. At 20°C, $\rho_{20} = 1.0633$ g cm^{-3} (Table 10.18). To calculate ρ at 25°C we will assume that the temperature dependence of the solution density is the same as that of water (Table 10.22). This assumption gives us at 25°C $\rho = 1.0621$ g cm^{-3} and then $c_{25} = 1.4246$ mol L^{-1}. The solution density at 90°C cannot be found in Chapter 10, so we will make the same assumption that temperature dependence of solution density is the same as that of pure water (Table 10.22). Then, at 90°C $\rho = 1.0283$ g cm^{-3} and $c = 1.3793$ mol L^{-1}, which shows that molarity and molality are quite different at an elevated temperature.

PROBLEM 2

What is the difference between the chemical potentials of the ith solute on the molal, $\mu_i^{0,b}$, and molar, $\mu_i^{0,c}$, concentration scales?

Solution:

$$\mu_i = \mu_i^{0,c} + RT \ln\left(c_i \gamma_{i,c}/c_i^0\right) = \mu_i^{0,b} + RT \ln\left(b_i \gamma_{i,b}/b_i^0\right),$$

$$\mu_i^{0,c} - \mu_i^{0,b} = RT \ln\left[b_i c_i^0/\left(b_i^0 c_i\right)\right] + RT \ln\left(\gamma_{i,b}/\gamma_{i,c}\right),$$

when $b_i \to 0$, $c_i \to 0$, $c_i/b_i \to \rho_s$, $\gamma_{i,b} \to 1$, $\gamma_{i,c} \to 1$, then

$$\mu_i^{0,c} - \mu_i^{0,b} = RT \ln\left[c_i^0/\left(b_i^0 \rho_s\right)\right].$$

PROBLEM 3

What is the difference between the activity coefficient of the ith solute on the molal, $\gamma_{i,b}$, and molar, $\gamma_{i,c}$, concentration scales?

Solution:

Using the solution of Problem 2

$$\mu_i^{0,c} - \mu_i^{0,b} = RT \ln\left[b_i c_i^0/\left(b_i^0 c_i\right)\right] + RT \ln\left(\gamma_{i,b}/\gamma_{i,c}\right) = RT \ln\left[c_i^0/\left(b_i^0 \rho_s\right)\right],$$

the relationship between $\gamma_{i,b}$ and $\gamma_{i,c}$ is as follows:

$$\gamma_{i,c} = \gamma_{i,b} b_i \rho_s/c_i.$$

PROBLEM 4

Using the mean activity coefficients of KCl(aq) (see Table 10.17), calculate parameter C of Equation 1.30 suitable for a concentration range from 0 to 0.2 mol kg^{-1}.

Solution:

Empirical parameter C in Equation 1.30

$$\ln\gamma_\pm = -\frac{A_{DH}|z_+z_-|I_b^{0.5}}{1+B_{DH}\mathring{a}I_b^{0.5}} + CI_b,$$

can be calculated for all concentrations of KCl(aq) and is given in Table 10.17 as follows:

b_{KCl} (mol kg⁻¹)	$\gamma_\pm$ (exp)	C (kg mol⁻¹)	$\gamma_\pm$ (calc) with $C = -0.084$ kg mol⁻¹
0.001	0.965	−0.2180	0.965
0.002	0.951	−0.5366	0.952
0.005	0.927	−0.1521	0.927
0.01	0.901	−0.2124	0.902
0.02	0.869	−0.1645	0.870
0.05	0.816	−0.1261	0.818
0.1	0.768	−0.1128	0.770
0.2	0.717	−0.0848	0.717

In these calculations $A_{DH} = 1.172$ (kg mol⁻¹)⁰·⁵, $B_{DH} = 3.28\times10^9$ (kg mol⁻¹)⁰·⁵ m⁻¹, $\mathring{a} = 4.5\times10^{-10}$ m.

If parameter C is taken as −0.0848 kg mol⁻¹ at molality of KCl of 0.2 mol kg⁻¹, the calculated mean activity coefficents are almost the same as the experimental ones. Therefore, $C = -0.0848$ kg mol⁻¹ is appropriate to use in the third approximation of the Debye–Hückel theory.

PROBLEM 5

If the total concentration of carbon in an aqueous solution of H_2CO_3 is 0.1 mol kg⁻¹, the following reactions take place:

$$\text{Reaction 1}: H_2CO_3(aq) = HCO_3^-(aq) + H^+(aq),$$

$$\text{Reaction 2}: HCO_3^-(aq) = CO_3^{2-}(aq) + H^+(aq),$$

calculate molal concentration of $H_2CO_3(aq)$, $HCO_3^-(aq)$, $CO_3^{2-}(aq)$, and $H^+(aq)$ at temperature of 25°C and pressure of 1 bar assuming that the activity coefficients of all species equal 1.

Solution:

Four equations should be constructed to solve for the four unknown concentrations. The first two are the mass action equations (assuming the activity coefficients of all species equal 1):

$$K_1 = \frac{b_{H^+} b_{HCO_3^-}}{b_{H_2CO_3}},$$

$$K_2 = \frac{b_{H^+} b_{CO_3^{2-}}}{b_{HCO_3^-}},$$

where K_1 and K_2 are the thermodynamic equilibrium constants available in Table 10.14 for Reactions 1 and 2. The third equation shows the mass balance of carbon:

$$b_t = b_{H_2CO_3} + b_{HCO_3^-} + b_{CO_3^{2-}} = 0.1 \, \text{mol kg}^{-1}.$$

and the fourth equation shows the charge balance:

$$b_{H^+} = b_{HCO_3^-} + 2 \times b_{CO_3^{2-}}.$$

Numerically solving the four equations given above, the following concentrations are obtained:

$$b_{H^+} = 2.111 \times 10^{-4} \, \text{mol kg}^{-1},$$

$$b_{H_2CO_3} = 9.998 \times 10^{-2} \, \text{mol kg}^{-1},$$

$$b_{HCO_3^-} = 2.111 \times 10^{-4} \, \text{mol kg}^{-1},$$

$$b_{CO_3^{2-}} = 4.677 \times 10^{-11} \, \text{mol kg}^{-1}.$$

PROBLEM 6

What is the molal concentration of HCl(aq) for a solution with pH $= 0.091$?

Solution:

To calculate the acid concentration of the HCl(aq) solution, the activity coefficient of H^+(aq) should be known:

$$pH = -\log_{10}\left(b_{H^+} \gamma_{H^+}\right).$$

While the concentration of H^+(aq) at ambient conditions is the same as concentration of HCl(aq) (strong electrolyte), the ionic activity coefficient γ_{H^+} cannot be experimentally measured and therefore is unknown. As an approximation, we can assume that γ_{H^+} equals to the mean activity coefficient of HCl(aq), $\gamma_\pm$, which is experimentally measurable and given in Table 10.17 for a wide range of concentrations. The table below shows pH calculations using data from Table 10.17, and it is clear that the solution's HCl concentration is equal to 1 mol kg^{-1} as shown in bold in the table below. Note that as can be seen from the table below, pH values can be negative when the concentration and activity coefficient are high enough.

$b_{HCL(aq)} = b_{H^+}$	$\gamma_\pm$	$pH = -\log_{10}\left(b_{H^+}\gamma_{H^+}\right) \approx -\log_{10}\left(b_{HCL(aq)}\gamma_\pm\right)$
0.001	0.965	3.015
0.002	0.952	2.720
0.005	0.929	2.333
0.01	0.905	2.043
0.02	0.876	1.756
0.05	0.832	1.381
0.1	0.797	1.098
0.2	0.786	0.803
0.5	0.759	0.421
1	**0.811**	**0.091**
2	1.009	−0.304
5	2.38	−1.075
10	10.4	−2.017

REFERENCES

1. E.R. Cohen et al., *Quantities, Units, and Symbols in Physical Chemistry*, 3rd ed., RSC Publishing, Cambridge, UK, 2007.
2. R.A. Robinson and R.H. Stokes, *Electrolyte Solutions: Second Revised Edition*, Dover, Mineola, NY, 2002.
3. P. Atkins and J. de Paula, *Atkins' Physical Chemistry*, 11th ed., Freeman, New York, 2018.
4. J. Rumble (Editor-in-Chief), *Handbook of Chemistry and Physics*, 101st ed., CRC Press, Boca Raton, FL, 2020.
5. I.N. Levine, *Physical Chemistry*, 6th ed., McGrow-Hill, Boston, MA, 2009.

2 Electrochemical Cells

2.1 OBJECTIVES

The main objective of this chapter is to introduce a reader to the design and components of typical electrochemical cells. Some other topics, mainly related to the background of electrochemical science and engineering, will also be covered. Faraday's law of electrolysis and the current–cell potential dependence will be introduced in the chapter. A description of the Daniell cell operating in equilibrium, electrolytic, and galvanic modes will be given. A new approach in presenting the electrochemical diagrams is proposed. A short review on the most important quantities, their names, and units and the key laws of electric circuits are briefly discussed.

2.2 BASIC AND DERIVED SI UNITS

The treatment of the electrochemical data requires using a set of units to be well defined and self-consistent. In this book, the International System of Units, abbreviated as SI, will be mainly used. The SI core is the seven base units given in Table 2.1.

Definitions of the base units are given in Refs. [1,2]. Here, we are presenting an example of one of them, ampere (A), which is a key unit in electrochemical science and engineering.

> The ampere is that constant current which, if maintained in two straight parallel conductors of infinite length, of negligible circular cross-section, and placed 1 m apart in vacuum, would produce between these conductors a force equal to 2×10^{-7} newton per meter of length.

While the definition is rather complicated, it allows experimentally defining ampere by measuring a force. All SI base units are supposed to be experimentally defined.

TABLE 2.1
Base SI Units [1,2]

Base Quantity	Unit Name	Unit Symbol
Length	meter	m
Mass	kilogram	kg
Time	second	s
Electric current	ampere	A
Thermodynamic temperature	kelvin	K
Amount of substance	mole	mol
Luminous intensity	candela	cd

DOI: 10.1201/b21893-2

There are also a number of derived SI units that may be expressed in terms of the base units using the mathematical symbols of multiplication and division. Some of the derived units can have special names and symbols, and they may themselves be used in combination with those for base and other derived units to express the units of other quantities. Examples of derived units for some physical quantities are given in Ref. [1]. From that list, some of them related to the topic of the book are given in Table 2.2.

For additional convenience, some of the derived units have been given special names and symbols [1]. A few examples can be found in Table 2.3. These names and symbols may themselves be used to express other derived units and can be expressed using the base units.

Still, such quantities as electric current (in A), electric charge (in $C = A\ s$), electric potential difference (in $V = W\ A^{-1} = m^2 kg\ s^{-3}\ A^{-1}$), electric power (in $W = A\ V = m^2 kg\ s^{-3}$), and electric resistance (in $\Omega = V\ A^{-1} = m^2 kg\ s^{-3}\ A^{-2}$) are the most important and will extensively be used in this book. There will be other units introduced later.

TABLE 2.2
Examples of Derived SI Units [1]

Physical Quantity	Unit Name	Unit Symbol
Area	square meter	m^2
Volume	cubic meter	m^3
Density	kilogram per cubic meter	$kg\ m^{-3}$
Specific volume	cubic meter per kilogram	$m^3 kg^{-1}$
Current density	ampere per square meter	$A\ m^{-2}$
Concentration	mole per cubic meter	$mol\ m^{-3}$

TABLE 2.3
Examples of Derived SI Units with Special Names and Symbols [1]

Physical Quantity	Special Name	Symbol	Units	Base SI Units
Force	newton	N	—	$m\ kg\ s^{-2}$
Pressure	pascal	Pa	$N\ m^{-2}$	$m^{-1} kg\ s^{-2}$
Energy	joule	J	$N\ m$	$m^2 kg\ s^{-2}$
Power	watt	W	$J\ s^{-1}$	$m^2 kg\ s^{-3}$
Electric charge	coulomb	C	—	As
Electric potential difference	volt	V	WA^{-1}	$m^2 kg\ s^{-3}\ A^{-1}$
Capacitance	farad	F	CV^{-1}	$m^2 kg^{-1} s^4 A^2$
Electric resistance	ohm	Ω	VA^{-1}	$m^2 kg\ s^{-3}\ A^{-2}$
Electric conductance	siemens	S	AV^{-1}	$m^{-2} kg^{-1} s^3\ A^2$
Celsius temperature	degree	°C	—	K

2.3   ELECTROCHEMICAL CELL (SYSTEM)

There are a variety of electrochemical systems around us, and it is assumed that there will be more systems developed in the future to efficiently convert and store energy. Traditionally, an electrochemical system can also be called electrochemical cell. A typical electrochemical cell consists of at least (1) electrodes, (2) electrolyte, and (3) external circuit. The cell shown in Figure 2.1 is the electrochemical cell that produces chemicals, Cu(s) and $Cl_2(g)$, using electric energy from an electric power supply (e.g., a battery). This electrochemical system is called an electrolytic cell (EC), and the process taken place in the cell is electrolysis.

The electrochemical reactions that take place at the electrodes shown in Figure 2.1 are as follows:

$$Cu^{2+}(aq) + 2e^-(Pt) \rightarrow Cu(s),$$

which is a reduction half-reaction to electrochemically reduce $Cu^{2+}(aq)$ to Cu(s) and

$$2Cl^-(aq) \rightarrow Cl_2(g) + 2e^-(Pt),$$

which is an oxidation half-reaction to electrochemically oxidize $Cl^-(aq)$ to $Cl_2(g)$. In the above half-reactions, the species accepting electrons is called oxidizing agent or oxidant and the species loosing electrons is reducing agent or reductant.

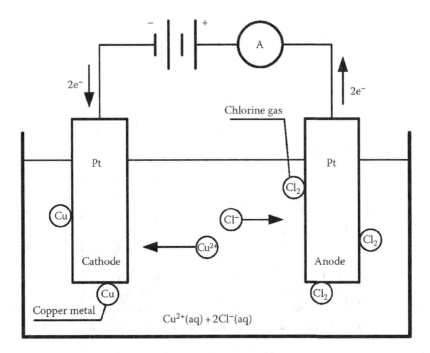

FIGURE 2.1   Cu/Cl$_2$ EC, which produces chemicals, Cu(s) and Cl$_2$(g), using an electric power supply, for example, battery.

The total reaction occurring in the electrochemical cell can be obtained by adding these two half-reactions:

$$Cu^{2+}(aq) + 2Cl^-(aq) \rightarrow Cu(s) + Cl_2(g).$$

Note that the mass balance and the charge balance should always be provided in any half-reaction and total electrochemical reaction. Also, the number of electrons in the reduction and oxidation half-reactions should be the same.

The left-hand electrode in Figure 2.1 is called the cathode (cations go to the electrode), which is a negatively charged electrode in the EC. The right-hand electrode is the positively charged anode, and anions go to the electrode. The polarity of the electrodes in an EC is defined by the polarity of the power supply, for example, the battery shown in Figure 2.1.

2.4   ELECTROLYTIC AND GALVANIC CELLS

While Figure 2.1 shows an EC, Figure 2.2 demonstrates a galvanic cell (GC). The GC spontaneously converts chemical energy into electric energy as long as the cell chemicals, $H_2(g)$ and $Cl_2(g)$, are available.

In the GC shown in Figure 2.2, the following electrochemical reactions take place:

$$Cl_2(g) + 2e^-(Pt) \rightarrow 2Cl^-(aq),$$

is the cathodic (reduction) half-reaction,

$$H_2(g) \rightarrow 2H^+(aq) + 2e^-(Pt),$$

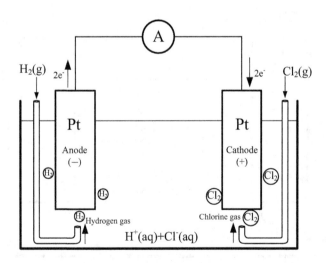

FIGURE 2.2 H_2/Cl_2 galvanic (fuel) cell, which consumes $H_2(g)$ and $Cl_2(g)$ and spontaneously converts the chemical energy into electric energy.

is the anodic (oxidation) half-reaction, and

$$H_2(g) + Cl_2(g) \rightarrow 2Cl^-(aq) + 2H^+(aq),$$

is the total GC reaction, which occurs spontaneously as long as the required chemicals for the reactions, $H_2(g)$ and $Cl_2(g)$, are provided. Note that, in all the three reactions, the mass balance and charge balance requirements are fulfilled.

Any EC or GC has at least two electrodes: the cathode and the anode. A reduction reaction takes place at the cathode, and an oxidation reaction occurs at the anode. Generally, either the cathode or the anode could be positively (or negatively) charged. In a GC (e.g., Figure 2.2), the cathode is more positive than the anode. In an EC (e.g., Figure 2.1), the anode is positively charged with respect to the cathode.

The GC shown in Figure 2.2 can also be called a fuel cell because hydrogen is a fuel and electric energy can be produced when an oxidant (Cl_2) is used in combination with the fuel in the electrochemical energy conversion system. Certainly, a less expensive and nonpoisonous oxidant should be used to make the fuel cell practical.

2.5 ELECTROCHEMICAL DIAGRAMS (TRADITIONAL)

It takes some time and effort to prepare schematics of the electrochemical cells shown in Figures 2.1 and 2.2. Therefore, electrochemists have developed an approach to present electrochemical cells in a form of relatively simple diagrams. In an electrochemical diagram, all chemicals, phases, and phase boundaries should be shown. While it is not common practice, it is not a bad idea to make sure that you have the same material (metal) for both terminals of any electrochemical cell, and this habit can make a theoretical analysis of the cell more consistent. The main practical reason for using the same metal for both terminals is to avoid forming an additional potential difference at the interfaces between different metals if they are at different temperatures.

Following are the electrochemical diagrams corresponding to the two previously considered electrochemical cells. The Cu/Cl$_2$ EC (Figure 2.1) diagram is

$$Cathode(-)Pt(s)|Cu(s)|CuCl_2(aq)|Cl_2(g)|Pt(s)(+)Anode,$$

and the H$_2$/Cl$_2$ GC (Figure 2.2) diagram is

$$Anode(-)Pt(s)|H_2(g)|HCl(aq)|Cl_2(g)|Pt(s)(+)Cathode.$$

Note that a vertical line, |, in the above diagrams shows a phase boundary (e.g., between a metal and solution).

Because the location of the cathode and the anode as well as their polarities (positive and negative) is usually well known, this information may not be shown in the electrochemical diagrams.

While the electrochemical diagrams of the Cu/Cl$_2$ EC and H$_2$/Cl$_2$ GC given previously are very common, they have a serious problem of not showing a three-phase boundary, which commonly occurs in electrochemical cells. A new type of electrochemical diagram is proposed in this book and described here.

2.6 NEW ELECTROCHEMICAL DIAGRAMS

Let us consider the anodic reaction of the H_2/Cl_2 GC shown in Figure 2.2 as an example of the difficulties using the traditional diagram. The hydrogen oxidation reaction taking place at the anode is

$$H_2(g) \rightarrow 2H^+(aq) + 2e^-(Pt),$$

and is described in a cell electrochemical diagram as follows:

$$Pt(s)\mid H_2(g)\mid HCl(aq).$$

The hydrogen oxidation half-reaction consists of three species (hydrogen molecule, proton, and electron) located in three phases (gas, aqueous solution, and platinum, respectively), which are supposed to be in intimate contact for the reaction to take place. However, the three-phase contact is not shown in the preceding diagram and, in principle, cannot be shown in any traditional electrochemical diagram. The traditional electrochemical diagram suggests that at the anode, there are two-phase boundaries, while there should be a three-phase boundary corresponding to the half-reaction. To show the three-phase boundaries in the electrochemical cells, we propose the following diagram for the H_2/Cl_2 GC:

$$H_2(g) \qquad Cl_2(g)$$

$$Anode\,(-)\,e^-\,(Pt) \times HCl(aq) \times e^-\,(Pt)(+)\,Cathode,$$

which can be used to show (1) all species (including electrons) participating in the electrochemical half-reactions and (2) the intersection of three phases corresponding to the half-reactions.

The new electrochemical diagram introduces a new symbol, $\times$, which allows us to show three- and even four-phase boundaries. In the new diagram, it is clearly seen that at both the anode and the cathode, three phases are in an intimate contact while the half-reactions show the electrochemically active species corresponding to these phases. In addition, the new diagram allows demonstrating all electrically conductive phases in a single horizontal row, while in the traditional diagram, the conductive phases are separated by a nonconductive phase, such as $H_2(g)$ or $Cl_2(g)$. Note that it is useful before developing an electrochemical diagram to find out the electrochemically active and inactive species as well as electric conductive and nonconductive phases.

2.7 FARADAY'S LAW OF ELECTROLYSIS

The words "electrochemical" or "electrochemistry" suggest that there is a close relationship between "electricity" and "chemistry." One such relationship was discovered by Michael Faraday in 1833.

Michael Faraday (1791–1867) was an English scientist who contributed to a number of fields in physics and chemistry. One of his major discoveries was related to the laws of electrolysis. He invented a system of oxidation numbers and popularized electrochemical terms such as anode, cathode, electrode, and ion.

REFERENCES

A.J. Bard, G. Inzelt, and F. Scholz, *Electrochemical Dictionary*, Springer, Berlin, Germany, 2008.
http://en.wikipedia.org/wiki/Michael_Faraday.

Faraday's law of electrolysis is used to directly connect the amount of electric charge, Q (in C), passed in an electrochemical cell (electrolytic or galvanic) and the amount of the ith chemical, m_i (in kg), produced (consumed) at the cell electrodes:

$$m_i = \frac{Q v_i M_i}{nF}, \tag{2.1}$$

where
v_i is the stoichiometric number of the ith species.
M_i is its molecular mass in kg mol^{-1}.
n is the number of electrons in the half-reaction.
F is Faraday's constant (96,485 C mol^{-1}) (Table 10.1).

If current, I, in the cell is constant, then $Q = I \times t$. However, in general, $Q = \int I \, dt$, where the integration should be done for the whole time period, t, of the electrolytic or galvanic process.

Faraday's law (2.1) simply shows that the amount of a substance produced or consumed at an electrode of an electrochemical cell is proportional to the amount of charge passed through the electrode, assuming that there are not any parasitic (sometimes unknown electrochemical processes in the cell) reactions.

Faraday's constant is the key fundamental value used in electrochemical science and engineering and can be calculated using two other fundamental constants: the Avogadro constant, N_A, and the elementary charge, e, so $F = N_A e$. This expression shows the physical meaning of Faraday's constant, which is the total charge of 1 mol of species with a unit charge, positive or negative. Examples of such species are electrons (e^-), protons $H^+(aq)$, $OH^-(aq)$, and $Cu^+(aq)$.

2.8 FUNDAMENTAL CONSTANTS

Table 2.4 gives the 2006 self-consistent set of values of a few fundamental constants recommended by the Committee on Data for Science and Technology (CODATA) to be frequently used in this book. An extended list of the fundamental constants can be found in Table 10.1.

The number of fundamental constants that a reader remembers (up to 4–5 digits) shows how frequently he/she does the numerical calculations. In this book, four fundamental constants—R, N_A, e, and F—will be used extensively and, therefore, are recommended for remembering.

2.9 WATER ELECTROLYSIS

We will use water electrolysis, which is a process to produce hydrogen and oxygen using electric power, to demonstrate how Faraday's law should be used. Water electrolysis is a well-known process which has been in existence for a long time. The main purpose of water electrolysis is to produce hydrogen gas. Hydrogen gas is a chemical widely used in modern society for the production of ammonia, metals, and plastics, for petroleum processing, and can also be used as a fuel. In principle, the electrochemical splitting of water does not require any other chemicals. However, the

TABLE 2.4
Values of Some Fundamental Constants (Table 10.1)

Symbol	Numerical Value	Unit
e	$1.602176565(35) \times 10^{-19}$	C
N_A	$6.02214129(27) \times 10^{23}$	mol^{-1}
F	$96,485.3365(21)$	$C \, mol^{-1}$
R	$8.314472(15)$	$J \, mol^{-1} \, K^{-1}$

Note: Uncertainty of the constants is shown by digits in brackets.

electric resistivity of pure water is very high due to the low concentration of ions. To decrease resistivity between electrodes, either a strong acid or a strong base can be added. A basic aqueous solution is usually less corrosive than acidic, and therefore, KOH(aq) solution is used in commercial electrolyzers to decrease the solution resistivity. The total electrochemical reaction, which takes place in a water electrolyzer, is as follows:

$$H_2O(l) \rightarrow 0.5O_2(g) + H_2(g),$$

which can be split into the cathodic and anodic half-reactions as shown in Figure 2.3.

Note that in the process shown in Figure 2.3, $K^+(aq)$ is an electrochemically inactive species and, therefore, does not participate in any electrochemical reaction.

For applying Faraday's law, Equation 2.1, for water electrolysis shown in Figure 2.3, we should use $n = 2$, $F = 96,485\,C\,mol^{-1}$ and, then, the stoichiometric numbers, v_i, and molecular masses, M_i, for each participating in the electrolysis chemical as follows:

Cathode:

$$H_2O(l)(\text{consumed}): v_i = 2,\ M_i = 18.015 \times 10^{-3}\,kg\,mol^{-1}.$$

$$H_2(g)(\text{produced}): v_i = 1,\ M_i = 2.016 \times 10^{-3}\,kg\,mol^{-1}.$$

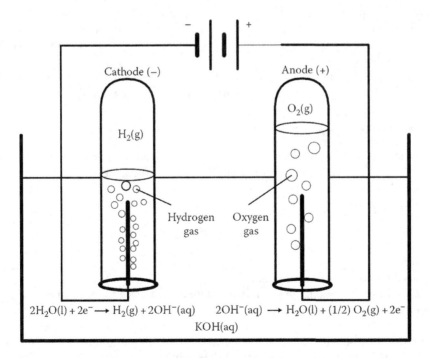

FIGURE 2.3 Electrolysis in KOH(aq) solution.

Anode:

$$H_2O(l)(\text{produced}): v_i = 1,\ M_i = 18.015 \times 10^{-3}\ \text{kg}\,\text{mol}^{-1}.$$

$$O_2(g)(\text{produced}): v_i = 0.5,\ M_i = 31.999 \times 10^{-3}\ \text{kg}\,\text{mol}^{-1}.$$

Overall cell reaction:

$$H_2O(l)(\text{consumed}): v_i = 1,\ M_i = 18.015 \times 10^{-3}\ \text{kg}\,\text{mol}^{-1}.$$

$$H_2(g)(\text{produced}): v_i = 1,\ M_i = 2.016 \times 10^{-3}\ \text{kg}\,\text{mol}^{-1}.$$

$$O_2(g)(\text{produced}): v_i = 0.5,\ M_i = 31.999 \times 10^{-3}\ \text{kg}\,\text{mol}^{-1}.$$

Note that a water electrolyzer can also be used to find out the amount of charge passed by measuring the volume of hydrogen or oxygen produced due to the electrolysis. The ideal gas law can be used to calculate the mass using volume, and the ratio of the amount of hydrogen and oxygen (in mole or volume units) can be defined without any calculations by recognizing the stoichiometric numbers in the electrochemical reaction. Clearly, in water electrolysis, the volume of produced hydrogen will be two times larger than the volume of oxygen as shown in Figure 2.3.

In Section 2.6, we introduced the new electrochemical diagram which is useful when more than two phases are in contact and all of them have electrochemically active species participating in electrochemical reactions. Both the anodic and cathodic half-reactions shown in Figure 2.3 are such reactions. An exercise in Section 2.17 shows how the new diagram can be constructed for the electrolysis reaction discussed in this section.

Other exercises in Section 2.17 provide examples of estimating the amounts of chemicals produced due to electrolysis when current is not constant.

2.10 OHM'S LAW

Electric resistance, R, of either electron or ion (or mixed) conductive material is an important characteristic in electrochemical science and engineering and is defined by Ohm's law:

$$R = \frac{E}{I}, \tag{2.2}$$

where
I is the electric direct current, dc (in A).
E is the corresponding to dc electric potential difference (in V), commonly and incorrectly called voltage.

Note that for a pure resistance, the same equation is valid in case of alternating current $I(t)$ and corresponding potential $E(t)$: $R = E(t)/I(t)$.

The SI unit of electric resistance is Ω, which is called ohm. The resistance depends on the sample geometry and material itself:

$$R = \frac{\rho l}{A}, \tag{2.3}$$

where

ρ is the material's electric resistivity in Ω m (commonly Ω cm).
l is the distance between electrodes.
A is the cross section of material between the electrodes (Figure 2.4).

If the sample shown in Figure 2.4 is to be tested, two electrodes would be connected to two terminals of an electrometer to measure E and I or just R (resistance). If resistivity of a metal rod (wire) is to be measured, the appropriate length and diameter should be chosen to make sure that the wire resistance is substantially larger than the resistance of the electrometer wires.

The resistivity, ρ, in Equation 2.3 is an important characteristic of material and should not depend on the sample geometry. As an example, in Table 2.5, there are values of resistivity for a number of widely used materials that are common in electrochemical studies. An exercise of employing Equation 2.3 for an electrochemical system called flow battery (see Figure 3.6) is given in Section 2.17.

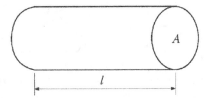

FIGURE 2.4 A sample of an electrically conductive material with length l and cross section A to measure its resistance.

TABLE 2.5
Examples of Electric Resistivity of Materials That Are Commonly Used in Electrochemical Studies

Material	Temperature (°C)	Resistivity (Ω m)
Copper	25	1.7×10^{-8}
Platinum	25	1.1×10^{-7}
Mercury	25	1.0×10^{-6}
Graphite	25	10^{-5} (approximately)
Nation	25	0.1 (approximately)
Yttria-stabilized zirconia (YSZ)	1000	0.2 (approximately)
Millipore deionized water (H_2O)	25	1.8×10^5
Glass	25	10^{10}–10^{14}

2.11   KIRCHHOFF'S LAW

Based on Kirchhoff's law, the algebraic sum of the currents entering a node is equal to the algebraic sum of the currents leaving the node, that is, $\Sigma I = 0$.

Considering a node shown in Figure 2.5, we can obtain a relationship between the three currents shown in the figure:

$$I_1 = I_2 + I_3. \tag{2.4}$$

Another illustration of Kirchhoff's law is shown in Figure 2.6, where the total current I_T flows through n parallel connected resistors R_n, so the relationship between I_T and the currents $I_1, I_2, \ldots, I_n$ is as follows:

$$I_T = I_1 + I_2 + \cdots + I_n, \tag{2.5}$$

with the obvious relation between total resistance, R, and resistors of the circuit R_n:
$$R^{-1} = (R_1)^{-1} + (R_2)^{-1} + \cdots + (R_n)^{-1}.$$

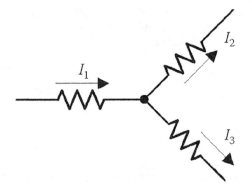

FIGURE 2.5   An electric node with three currents I_1, I_2, and I_3 for illustrating Kirchhoff's law.

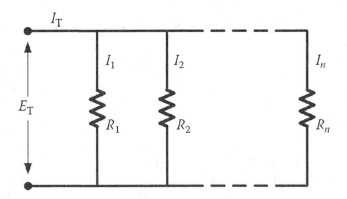

FIGURE 2.6   An electric circuit with parallel resistors for illustrating Kirchhoff's law.

Kirchhoff's law can also be used for the potential difference and is formulated as follows. The algebraic sum of the potential difference (voltage) over the circuit elements around any closed circuit loop must be zero, that is, $\Sigma E = 0$.

For the electric circuit shown in Figure 2.7, which consists of three series-connected resistors (R_1, R_2, and R_3) and a power source providing a potential difference E_4, Kirchhoff's law can be written as follows:

$$E_1 + E_2 + E_3 + E_4 = 0, \tag{2.6}$$

where E_1, E_2, and E_3 are the electric potential drops at the resistors R_1, R_2, and R_3, respectively. Obviously, the relationship between total resistance in this circuit, R, and resistors R_1, R_2, and R_3 is as follows: $R = R_1 + R_2 + R_3$.

A common use of Kirchhoff's law using electric potentials is demonstrated in Figure 2.8, which shows that the applied total potential difference E_T is a sum of the potential drops E_1, E_2, ..., E_n:

$$E_T = E_1 + E_2 + \cdots + E_n = I(R_1 + R_2 + \cdots + R_n) = IR, \tag{2.7}$$

where R is the total resistance of the circuit shown in Figure 2.8.

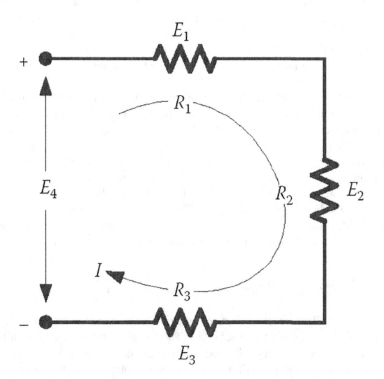

FIGURE 2.7 An electric circuit with three resistors (R_1, R_2, and R_3) and a power source E_4 providing a potential difference E_4. An illustration of Kirchhoff's law.

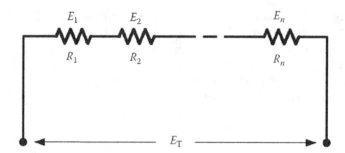

FIGURE 2.8   An electric circuit with the total potential difference E_T equals a sum of the potential drops $E_1, E_2, ..., E_n$.

2.12   ELECTROCHEMICAL CELLS WITH TRANSFER

The previously discussed electrochemical cells shown in Figures 2.1 and 2.2 are called cells without transfer because both ions of the electrolyte are electrochemically active and participate in the respective electrochemical reactions, which is, generally, a rare case. Usually, there are two electrolytes and only one of the ions of each electrolyte is electrochemically active. Hence, the cathode and the anode should be divided by a separator that must be an ionic conductor. Note that some of the electrochemical cells without transfer should also have a separator to make sure that chemicals of the cathode and the anode are not directly reacting to one another. For example, in Figure 2.2, if there is not any separator between the cathode and the anode, Cl_2 and H_2 can directly react, and this is what should not be occurring in an electrochemical system.

A classical example of the electrochemical cell with transfer is the Daniell cell, which is shown in Figure 2.9, and it is in equilibrium mode. This mode corresponds to a measurement when an electrometer with a very high internal resistance of around 10^{15} Ω is used. As a result, there is very small current, of around 10^{-15} A, going through the circuit, and the electrodes and both half-reactions are very close to equilibrium. E_{EQ}, shown in Figure 2.9, is the equilibrium potential difference (equilibrium cell potential), which can be calculated theoretically using thermodynamic data as will be discussed in Chapter 4. The half-cell compartments are separated by a salt bridge (separator), and composition of the bridge should be shown in both traditional and new electrochemical diagrams. For example, the Daniell traditional diagram

$$(-)Cu(s)|Zn(s)|ZnSO_4(aq)|KCl(aq)|CuSO_4(aq)|Cu(s)(+),$$

shows the salt bridge, which consists of aqueous KCl solution and electrically connects $ZnSO_4$(aq) and $CuSO_4$(aq) solutions. The bridge can be made using a highly concentrated KCl(aq) solution with two porous plugs at the ends (Figure 2.9). A reason for using such a solution will be described in Chapter 3. Note that in the case of the Daniell cell, the new kind of electrochemical diagram is not required because there are not any three-phase boundaries in the cell. It is because the metal electrode

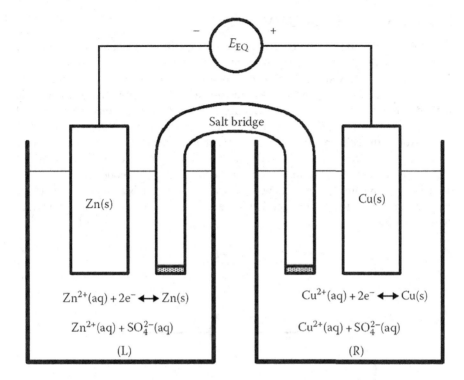

FIGURE 2.9 The Daniell cell in equilibrium mode. One should avoid using *cathode* and *anode* in such a mode.

serves both the metal atoms (e.g., Zn(s)) and electrons (e.g., $e^-(Zn)$). In other words, either the left-hand or right-hand electrochemical half-reaction shown in Figure 2.9 consists of three species, but two of them are in the same phase.

It is important to emphasize here the difference between cells without transfer and cells with transfer. A cell with transfer has two additional potential differences between the salt bridge and the electrolytes at each end of the bridge. These potentials can be minimized and almost eliminated in a number of ways. The additional potentials are referred to as the diffusion or liquid junction potentials, which will be discussed in Chapter 3.

Almost any study of an electrochemical system should be started with the equilibrium mode. In such a mode, the electrodes should not be called the cathode or the anode, and both half-reactions should be shown as reduction reactions. This is because, by common convention, the reference data on the (standard) electrode potentials are given for the reduction reactions. The convention will be discussed in Chapter 4. However, one electrode of an electrochemical cell should be more positive than another one, and the polarity can be experimentally found using the high-resistance electrometer. Also, the polarity of the electrodes in some equilibrium cells can theoretically be calculated using thermodynamic data. This will also be discussed in Chapter 4.

2.13 DANIELL CELL IN GALVANIC MODE

Let us electrically connect the Daniell cell electrodes through a load (e.g., an external resistor with a resistance R_e) using a circuit shown in Figure 2.10.

What can be observed are the electrons spontaneously flowing from the negatively charged electrode (anode) through the resistor to the positively charged electrode (cathode) as long as the GC chemicals are in contact with the electrodes and all necessary chemicals are available. By common convention, the electric current, I_{GC}, flows through the cell from the cathode to the anode when the electrons are traveling in the opposite direction from the anode to the cathode. The electric potential difference (or cell potential) measured between the cathode and the anode, E_{GC}, should be, from a thermodynamic point of view, a positive value, which is always less than E_{EQ}. An important value to be estimated in the electrochemical measurements of a GC is the internal resistance, R_i, defined by the following equation:

$$R_i = \frac{\left(E_{EQ} - E_{GC}\right)}{I_{GC}}, \qquad (2.8)$$

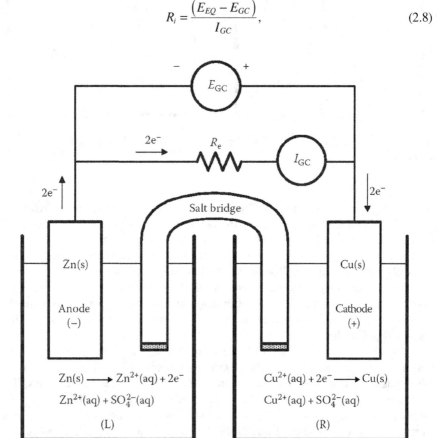

FIGURE 2.10 The Daniell cell in galvanic mode: the cathode is positively charged, and the anode is negatively charged. A reduction half-reaction takes place at the cathode, and oxidation half-reaction is at the anode.

which is a result of combining Kirchhoff's and Ohm's laws that are discussed earlier.

The internal resistance includes the resistances of the two electrolyte solutions, the salt bridge and all interfaces between the metal electrodes. Based on Kirchhoff's law, the relationship between the external resistance, R_e, and the internal resistance, R_i, is as follows:

$$E_{EQ} = I_{GC}(R_e + R_i), \qquad (2.9)$$

so, combining Equations 2.8 and 2.9 we can obtain that

$$E_{GC} = I_{GC}R_e, \qquad (2.10)$$

which is simply Ohm's law.

The electrochemical diagram of the Daniell cell shown in Figure 2.10 is as follows:

$$\text{Anode}(-)\text{Cu(s)}|\text{Zn(s)}|\text{ZnSO}_4\,(\text{aq})\|\text{KCl}(\text{aq})\|\text{CuSO}_4\,(\text{aq})\,|\,\text{Cu(s)}(+)\text{Cathode},$$

where the positive cathode (reduction reaction takes place) and the negative anode (oxidation reaction takes place) are explicitly shown. Moreover, the Cu electrode is again positive, similar to what was observed in the equilibrium mode.

Note that running the Daniell cell in the galvanic mode, Faraday's law of electrolysis can be applied with an experimental possibility to better understand and verify the law.

2.14 DANIELL CELL IN ELECTROLYTIC MODE

If we want to run the Daniell cell as an electrolytic cell (EC), in a way opposite to the galvanic mode, and carry out electrolysis, we should apply a potential E_{EC} using a battery or any other direct current (dc) power supply. The magnitude of E_{EC} should be larger than that of the equilibrium potential of the EC, that is, $|E_{EC}| > |E_{EQ}|$, and this causes the current to flow in the opposite direction of the current in a GC, as shown in Figure 2.11.

In EC, we should use the decomposition potential (E_D) rather than equilibrium potential (E_{EQ}). The decomposition potential is the minimum theoretical potential at which electrolysis takes place, so E_D is directly related to the equilibrium potential of the GC as follows:

$$E_D = -E_{EQ}. \qquad (2.11)$$

Because the Gibbs energy of any electrolytic reaction is positive (electrolytic reaction is not spontaneous), the decomposition potential is always negative, while the equilibrium potential of the GC is always positive. The relationship between the potential difference of the EC or simply the applied potential (E_{EC}), the decomposition potential (E_D), current in the EC (I_{EC}), and the internal resistance (R_i) is as follows:

$$E_{EC} = E_D + I_{EC}R_i, \qquad (2.12)$$

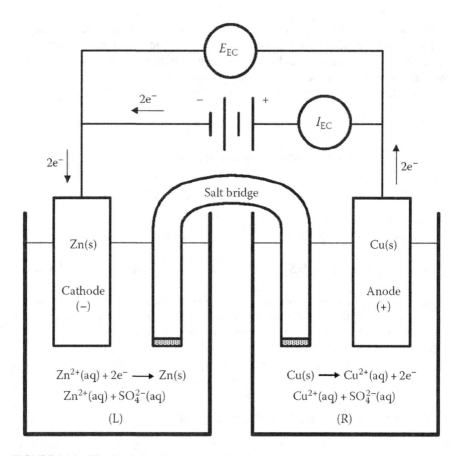

FIGURE 2.11 The Daniell cell in electrolytic mode: the cathode is negatively charged and the anode is positive with respect to the cathode.

where again Kirchhoff's and Ohm's laws were applied.

The electrochemical diagram of the Daniell cell operating in an electrolytic mode, shown in Figure 2.11, is as follows:

$$\text{Cathode}(-)\,Cu(s)\,\big|\,Zn(s)\,\big|\,ZnSO_4(aq)\,\big\|\,KCl(aq)\,\big\|\,CuSO_4(aq)\,|\,Cu(s)(+)\,\text{Anode}.$$

As can be seen from Figures 2.9 through 2.11 and corresponding electrochemical diagrams, the Daniell cell does not change the polarity of the electrodes when operating in either equilibrium or galvanic or electrolytic mode. The polarity of the Cu electrode is always positive and polarity of Zn electrode is always negative, whereas the cathode and the anode do change positions when the cell is switched between electrolytic and galvanic modes. When a cell is regularly switched from the electrolytic mode to the galvanic mode, the electrodes should be named just positive and negative, not the cathode and the anode. For example, a rechargeable battery does not have any cathode and anode as commonly but inconveniently presented in some literature and websites.

2.15   CELL POTENTIAL-CURRENT DEPENDENCE

Now, the previously described relationships between the cell potentials, the electric currents, and resistances R_i and R_e are shown in Figure 2.12, which is consistent with the rules of electrochemical science and engineering and also with thermodynamics. The idea of a single diagram for a cell to run in all three modes was presented in Ref. [3]. In this book, we have further described the diagram applicability for equilibrium and galvanic and electrolytic modes and provided some important details of using the diagram.

The electric power generated in the GC, P_{GC}, or provided for electrolysis in the EC, P_{EC}, is defined based on Joule's law:

$$P_{GC} = I_{GC}E_{GC}. \tag{2.13}$$

$$P_{EC} = I_{EC}E_{EC}. \tag{2.14}$$

Note that in the equations given above, E_{EQ}, E_{GC}, I_{GC} are always positive and E_D, E_{EC}, I_{EC} are always negative. The positive values of P_{GC} and P_{EC} mean that electric power is generated in the galvanic mode by the electrochemical cell and by an electric power supply in the electrolytic mode. It is important to note that the magnitude of E_{GC} should always be less than that of E_{EQ}, and the magnitude of E_{EC} should always be greater than that of E_D. There is neither generation nor production of power in any equilibrium cell.

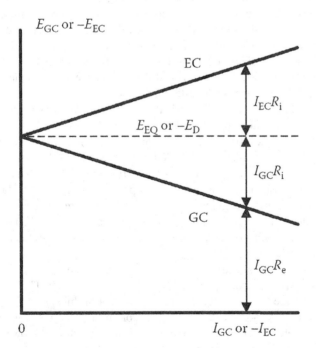

FIGURE 2.12   Cell potential–current dependences (simplified and shown as straight lines) in equilibrium cells, GCs, ECs.

2.16   SUMMARY

- In this book, the SI units (or allowed by IUPAC units [2]) are mainly used. The reader should be familiar with a variety of units (chemical, thermodynamic, electric, etc.) and be able to make necessary unit conversions. Current density (in A cm^{-2}) is important in electrochemistry. While current (in A) is measured, the current density can be calculated assuming that the electrode surface is known, which is not always the case.
- A typical electrochemical cell consists of (1) electrodes, (2) electrolyte, and (3) external circuit connected to the electrodes.
- An electrochemical cell can be either equilibrium or electrolytic or galvanic. In the equilibrium cell, there are no electrochemical reactions going on in the cell, and therefore, there is not any anode and cathode, whereas, in any electrolytic or galvanic cell, the electrodes can be called the cathode and the anode. A reduction reaction takes place at the cathode, and an oxidation reaction occurs at the anode.
- The GC spontaneously converts chemical energy into electric energy. In an EC, electric energy of a battery (a dc power supply) is used to produce chemicals. A fuel cell is a GC in which fuel is used as one of the cell chemicals.
- The cathode (and the anode) can be either positive or negative. In a GC, the cathode is positive, and in an EC, the anode is positive. If an electrochemical cell can work in both modes (galvanic and electrolytic), each of the electrodes can be either the cathode or the anode. However, the electrode polarity remains the same. In a rechargeable battery, the electrodes should be called positive electrode and negative electrode.
- Electrochemical diagrams show the materials (chemicals) and phases of electrochemical cell. The traditional diagrams have some challenges, so a new type of diagrams is suggested in this book.
- Faraday's law connects the chemical and electric processes taking place in an electrochemical system.
- In this book, the 2006 self-consistent set of values of the fundamental constants recommended by the CODATA are used. While it may sound unusual, it is suggested to remember four fundamental constants (R, N_A, e, and F) to be used in solving a variety of problems offered in the book.
- The reader should recognize the difference between resistance and resistivity (conductance and conductivity). It is important to know the relationship between the conductor resistance (conductance), resistivity (conductivity), and the conductor geometry.
- Ohm's, Kirchhoff's, and Joule's laws are used to connect the cell potential, current, resistances, and power of an electrochemical cell. The Daniell cell is used as an example to demonstrate these relationships.
- The cell potential–current dependences for equilibrium cell, GC, and ECs are considered, and recommendations on making the self-consistent and thermodynamically correct diagrams are provided.

2.17   EXERCISES

PROBLEM 1

Construct the traditional and new electrochemical diagrams for the electrolytic cell shown in Figure 2.3 to electrolyze water using KOH(aq) solution. Describe why the traditional diagram might be inconvenient for understanding the cell operation.

Solution:

The traditional electrochemical diagram of electrolysis shown in Figure 2.3 can be presented as

$$\text{Cathode}(-)\text{Pt}(s)\big|H_2(g)\big|\text{KOH}(aq)\big|O_2(g)\big|\text{Pt}(s)(+)\text{Anode},$$

and the new electrochemical diagram can be drawn as

$$H_2(g) \qquad O_2(g)$$
$$\text{Cathode}(-)e^-(\text{Pt}) \times \text{KOH}(aq) \times e^-(\text{Pt})(+)\text{Anode}.$$

In the new diagram the cross symbols, $\times$, shows there are three phases in an intimate contact at both the anode and cathode. At the cathode solid Pt, $H_2(g)$ and KOH(aq) are forming the three-phase boundary, and at the anode, the three-phase boundary is formed by the same aqueous KOH(aq), $O_2(g)$, and another solid Pt. These three-phase boundaries cannot be shown in the traditional diagram. In addition, the new diagram shows that current can go throw the electrolytic cell due to ionic conductivity in KOH(aq) and electronic conductivity in Pt, while the traditional diagram shows that the current path is interrupted by two gas phases, $H_2(g)$ and $O_2(g)$.

PROBLEM 2

Using Faraday's law, calculate the mass of deposited Cu(s) due to electrolysis from $CuSO_4$(aq) solution. The electrolysis was conducted for 10 minutes with a current of 1 A and then for an additional 15 minutes with a current of 0.4 A.

Solution:

First, we calculate the amount of electric charge, Q, in the electrolysis, which was conducted in two steps with a constant current in each of them. If the time in the first and the second steps are, respectively, $t_1 = 10$ minutes and $t_2 = 15$ minutes, then

$$Q = \int I\ dt = I_1 t_1 + I_2 t_2 = 1 \times 10 \times 60 + 0.4 \times 15 \times 60 = 960 \text{ C},$$

where I_1 and I_2 are the currents in the first and second electrolysis steps, respectively.

The deposited Cu(s) due to electrolysis from $CuSO_4$(aq) solution can be presented by a half-reaction as $Cu^{2+}(aq) + 2e^- \rightarrow Cu(s)$. Using Faraday's law, Equation 2.1, with

$v_{Cu^{2+}} = 1$, $M_{Cu} = 0.06355\,kg\,mol^{-1}$ (Table 10.2), and $n = 2$, the amount of deposited Cu(s) can be calculated as

$$m_{Cu} = \frac{Qv_{Cu^{2+}}M_{Cu}}{nF} = 960 \times 1 \times \frac{0.063546}{2 \times 96{,}485} = 0.0003161\,kg = 0.3161\,g.$$

PROBLEM 3

Calculate the volumes of $Cl_2(g)$ and $H_2(g)$ produced by electrolysis of HCl(aq) solution at 25°C and 1 bar during 1 hour if electrolysis current was linearly decreased from 2 to 1 A. How will the produced gas volumes be changed if electrolysis is carried out at an elevated pressure of 2 bar and an elevated temperature of 50°C?

Solution:

The electrolysis of HCl(aq) can be presented by the following reaction:

$$2HCl(aq) \rightarrow H_2(g) + Cl_2(g),$$

with $H_2(g)$ production at the cathode

$$2H^+(aq) + 2e^- \rightarrow H_2(g),$$

and $Cl_2(g)$ production at the anode

$$2Cl^-(aq) \rightarrow Cl_2(g) + 2e^-.$$

Let us, first, calculate the amount of charge used for the electrolysis:

$$Q = \int I\,dt = \left(\frac{2+1}{2}\right) \times 1 \times 3600 = 5400\,C.$$

In the above equation, the integration was carried out taking into account that the current is linearly decreased from 2 to 1 A.

Now we can calculate the molar amounts of produced $Cl_2(g)$ and $H_2(g)$ by applying Equation 2.1 (without the molar mass term), where $v_{Cl_2} = v_{H_2} = 1$ and $n = 2$:

$$n_{Cl_2} = n_{H_2} = \frac{Qv_{Cl_2}}{nF} = \frac{5400 \times 1}{2 \times 96{,}485} = 0.002798\,mol.$$

The ideal gas law $PV = n_i RT$ can then be used to calculate the volume produced at 25°C (298.15 K) and 1 bar:

$$V_{Cl_2} = V_{H_2} = \frac{n_{Cl_2}RT}{P} = \frac{0.002798 \times (8.314 \times 10^{-2}) \times 298.15}{1} = 0.6937\,L.$$

The same equation can be used to calculate the volumes at 2 bar and 50°C (323.15 K):

$$V_{Cl_2} = V_{H_2} = \frac{n_{Cl_2} RT}{P} = \frac{0.002798 \times \left(8.314 \times 10^{-2}\right) \times 323.15}{2} = 0.3759 \text{L}.$$

PROBLEM 4

Calculate conductance of a Nafion membrane (a polymeric ion conductor) installed between the positive and negative electrodes of a vanadium flow battery working at ambient temperature of 25°C. The cell has a 5 × 3 cm rectangular shape, and the thickness of the membrane is 0.2 mm.

Solution:

The conductance of a material, L, in S can be obtained using Equation 2.3

$$L = \frac{1}{R} = \frac{A}{\rho l},$$

where ρ is the material resistivity (Ω m), l is the length between the two electrodes (m), and A is the cross-sectional area of the material (m^2).

The resistivity of Nafion can be found in Table 2.5 and equals 0.1 Ω m, so the conductance of the Nafion membrane given in this problem is

$$L = \frac{A}{\rho l} = \frac{\left(15 \times 10^{-4}\right)}{0.1 \times \left(0.2 \times 10^{-3}\right)} = 75 \text{S}.$$

PROBLEM 5

Five resistors are connected as shown in the figure below. Find the equivalent resistance between A and B.

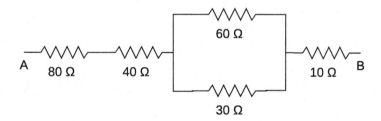

Solution:

Here, we will apply the Ohm's and Kirchhoff's laws presented in Sections 2.10 and 2.11. Taking into account that three of the resistors (80, 40, and 10 Ω) of the circuit are in series and two others (60 and 30 Ω) are in parallel, the equivalent resistance (R) between A and B can be calculated as follows:

$$R = 80 + 40 + 10 + \left[(60 + 30)/(60 \times 30)\right] = 130.05 \Omega.$$

REFERENCES

1. J. Rumble (Editor-in-Chief), *Handbook of Chemistry and Physics*, 101st ed., CRC Press, Boca Raton, FL, 2020.
2. E.R. Cohen et al., *Quantities, Units, and Symbols in Physical Chemistry*, 3rd ed., RSC Publishing, Cambridge, UK, 2007.
3. C.H. Hamann, A. Hamnett, and W. Vielstich, *Electrochemistry*, 2nd ed., Wiley-VCH, Weinheim, Germany, 2007.

3 Electric Conductivity

3.1 OBJECTIVES

The main objective of this chapter is to introduce students to an important electrochemical phenomenon, conductance (resistance), as well as a significant property of a material, conductivity (resistivity). The electrolyte conductivity origin, measurements, and interpretation are considered. The difference between conductances due to direct current (dc) and alternating current (ac) is discussed. Some examples with conductivity of aqueous solutions are given. It is explained that the efficiency of electrochemical energy conversion systems can significantly depend on the conductance of its components.

3.2 TYPES OF ELECTRIC CONDUCTIVITY

Electric conductivity can be provided by any charged species assuming that there is a mechanism of transferring this species or transferring just a charge due to an electric potential difference between electrodes.

In a metal (Figure 3.1), electric conductivity is provided by negatively charged electrons, which can easily dissociate from the lattice atoms and therefore can travel through them due to the electric field established in the electron conductor. The charge of an electron is well known and equal to -1.60218×10^{-19} C, which is a negative value of a fundamental constant, e, named elementary charge.

A variety of metal and metal alloy wires and electrodes that are employed in electrochemical cells are typical electron conductors.

A semiconductor has two types of charged species: electrons and holes. Both species provide electric conductivity in the semiconductor. Figure 3.2 schematically

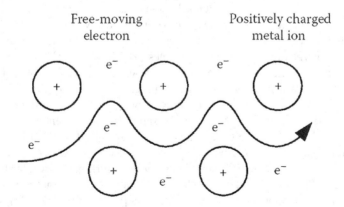

FIGURE 3.1 Electron conductivity in metals. There are free electrons moving along the positively charged lattice ions.

DOI: 10.1201/b21893-3

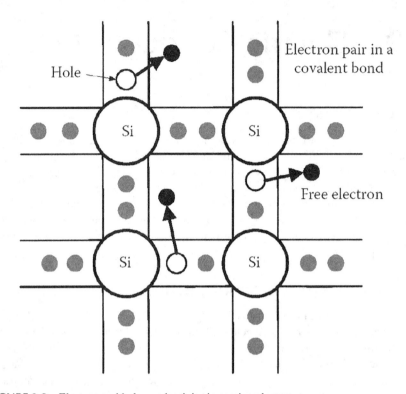

FIGURE 3.2   Electron and hole conductivity in semiconductors.

shows how these two charged species are created in a silicon crystal. In the silicon lattice, an electron can be knocked loose from its position and create an electron deficiency called a hole. The hole charge is $+1.60218 \times 10^{-19}$ C.

A semiconductor can be used as an electrode in an electrochemical cell, but such cells will not be covered in this book.

In electrolyte solutions and molten salts, the electric conductivity is provided by charged atomic species (ions), which can be either positive or negative. The magnitude of the ion charge can be doubled, tripled, etc., of the elementary charge. For example, the charge of Ca^{2+}(aq) ion is +2 times 1.60218×10^{-19} C, and the charge of PO_4^{3-} (aq) ion is −3 times 1.60218×10^{-19} C. When an electric potential difference is set up between two electrodes, negatively charged ions can move toward the positively charged electrode, and positively charged ions can move toward the negatively charged electrode (Figure 3.3). Note that the significant movement of the ions in an electrochemical cell using dc of a battery can occur only if some electrochemical reactions take place at the electrodes.

In Figure 3.3, the reaction of reduction of protons forming molecular hydrogen, $2H^+$(aq) $+ 2e^-$(Pt) $\rightarrow H_2$(g), can take place at the negatively charged electrode, and the reaction of oxidation of chloride ions, $2Cl^-$(aq) $\rightarrow Cl_2$(g) $+ 2e^-$(Pt), can take place at the positively charged electrode with the total electrolytic cell reaction $2HCl$(aq) $\rightarrow H_2$(g) $+ Cl_2$(g). However, these half-reactions will occur at a relatively

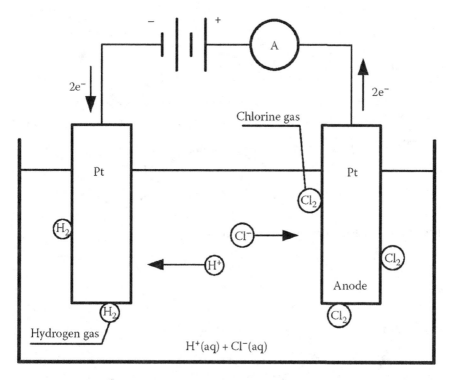

FIGURE 3.3   Electrolytic conductivity in HCl(aq) solution.

high potential difference, so another anodic reaction requiring a smaller potential difference, $H_2O(l) \rightarrow 0.5O_2(g) + 2H^+(aq) + 2e^-(Pt)$, can take place at a lower potential difference.

Electrically conductive polymers can also provide electric conductivity, which can be either the electron conductor, the ion conductor or both. We will briefly discuss here only the ion-conducting polymers, which are called ionomers. In Figure 3.4, the structure of a well-known ionomer in electrochemical science and engineering, Nafion, is shown. In Nafion, perfluorovinylether groups terminated with negatively charged sulfonate groups are incorporated onto a poly tetrafluoroethylene (Teflon) backbone. Nafion has a nanoporous structure (Figure 3.4), which allows movement of protons (or other cations) but does not conduct anions or electrons.

At high temperature, some ceramic materials can conduct electricity via transit of charged species from one site to another via point defects called vacancies in the crystal lattice. Yttria-stabilized zirconia (YSZ) is a zirconium oxide (ZrO_2), zirconia–based ceramic, in which the particular crystal structure is made ion-conducting by an addition of yttrium oxide, Y_2O_3 (yttria). YSZ is actually a solid solution (Figure 3.5), which has a number of properties different from the pure components, zirconia and yttria. For example, pure zirconia is not an ion conductor, while YSZ has quite high O^{2-} conductivity at temperatures above 700°C. When yttria is added to zirconia, some of the Zr^{4+} ions in the zirconia lattice are replaced with Y^{3+} ions. Due to different

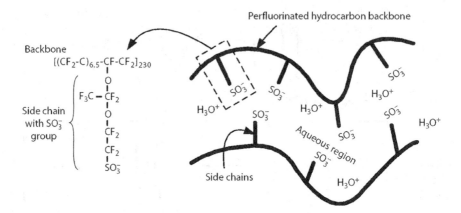

FIGURE 3.4   An ion conductive polymer (Nafion).

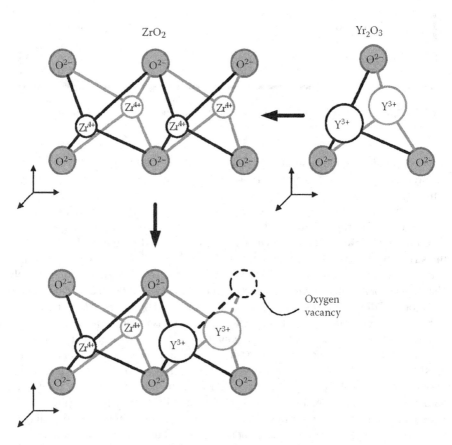

FIGURE 3.5   Development of oxygen vacancies in YSZ to provide high ionic conductivity at temperatures of 700°C–900°C.

charges of the ions, oxygen vacancies are produced when three O^{2-} ions replace four O^{2-} ions. The presence of oxygen vacancies allows YSZ to conduct electric current via transport of O^{2-} ions. The ability to conduct O^{2-} ions allows YSZ to be used in high-temperature electrochemical cells such as solid oxide fuel cells (SOFC) operating at temperatures at 700°C–900°C. Note that a concentration of yttria above 18 mol.% should be used to avoid a phase transition in YSZ between tetragonal and cubic lattices.

In this book, we will discuss the electric conductivity of electrolyte solutions, ionomers, ion-conducting ceramics, and metals but will skip semiconductors.

3.3 SIGNIFICANCE OF ION CONDUCTIVE MATERIALS FOR ELECTROCHEMICAL ENGINEERING

In Figures 3.6–3.8, three electrochemical energy conversion systems are schematically presented. In all three systems, an ion conductive material is used to physically separate two electrodes while providing electric conductance between them. The ability to conduct electric current through the membrane/separator directly relates to the efficiency of the electrochemical cell. The membrane/separator should have the highest possible ionic conductivity while maintaining a low permeability to all chemical species except the conducting ions. The ratio of membrane conductivity

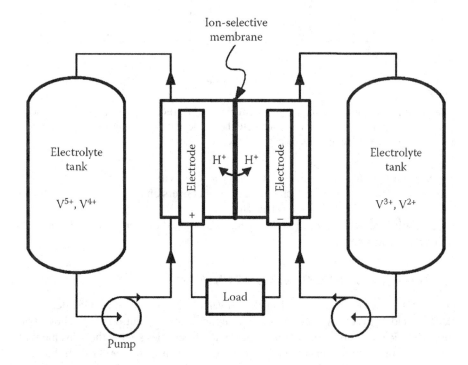

FIGURE 3.6 Vanadium flow battery. An ion conductive membrane is used to separate the positive and negative electrodes of the flow cell and also to provide electric conductance between the electrodes.

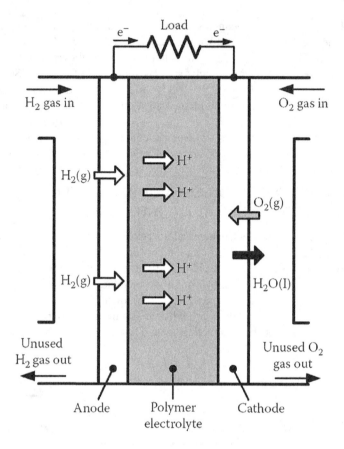

FIGURE 3.7 Proton exchange membrane fuel cell (PEMFC). A proton conductive membrane (polymer electrolyte) is used to separate the anode and the cathode of the fuel cell and also to provide electric conductance between the electrodes.

to permeability can be called the membrane's selectivity and can be used to compare membranes for any electrochemical energy conversion cell with a goal to find a membrane with highest possible selectivity.

A detailed description of the electrochemical cells shown in Figures 3.7 and 3.8 is given in Chapter 8.

3.4 ORIGIN OF IONIC CONDUCTIVITY

Figure 3.9 shows a simple model of the movement of two hydrated ions due to an electric field between two electrodes. The figure assumes, but does not show, that some chemical reactions should take place at the electrodes so that the movement of ions is possible. The $Na^+(aq)$ ion has more molecules in its hydration shell due to a smaller diameter but has the same charge as $Cl^-(aq)$.

In this consideration, the main characteristic of the solvent is viscosity, η. The ions can move while under the influence of an electric field, E/l (in V m^{-1}), and a steady

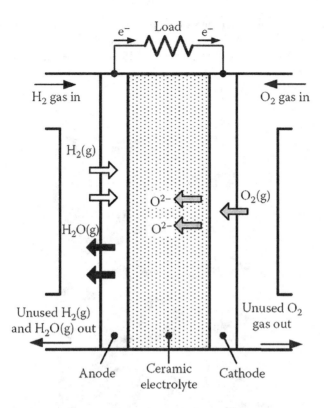

FIGURE 3.8 Solid oxide fuel cell. An O^{2-} ion conductive membrane is used to separate the anode and the cathode of the fuel cell and provide electric conductance between the electrodes.

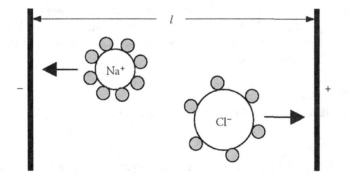

FIGURE 3.9 Movement of the hydrated Na^+(aq) and Cl^-(aq) ions in a solvent.

velocity of the ion, v_i, is established due to the balance between the electric field force ($|z_i| eE$) and the frictional force ($6\pi\eta r_i v_i$). If the forces are balanced, then [1]

$$|z_i|e\left(\frac{E}{l}\right) = 6\pi\eta r_i v_i,$$ (3.1)

where r_i is the hydrodynamic radius of the ith ion, which takes into account the surrounding ion molecules that are moving with the ion. Equation 3.1 is actually due to George Gabriel Stokes and was proposed in 1845 as a simple model showing a possible relation between the applied electric field and ion velocity depending on some properties of ion and of solvent.

George Gabriel Stokes (1819–1903) was a scientist and politician born in Ireland. He made seminal contributions to fluid dynamics and mathematical physics including the well-known the Navier–Stokes equation describing the motion of fluid substances.

REFERENCE

http://en.wikipedia.org/wiki/Sir_George_Stokes,_1st_Baronet

The ion's velocity divided by the electric field represents an important characteristic of the ion that is named mobility, u_i, and can easily be derived from Equation 3.1 as follows:

$$u_i = \frac{|z_i|e}{(6\pi\eta r_i)},\qquad(3.2)$$

showing that the ion mobility, u_i, depends on the properties of the ion (z_i and r_i) and a property of solvent, viscosity (η).

Note that the electric field and the ion velocity are vectors while the ion mobility is a scalar quantity with unit m^2 $(V\ s)^{-1}$. However, traditionally in most of the studies, the mobility is reported with the unit of cm^2 $(V\ s)^{-1}$. Note that Equation 3.2 is not a fundamental law but represents an idealized model where a sphere with a fixed radius is moving in a medium with macroscopic viscosity and, therefore, the applicability of Equation 3.2 for real solutions is quite limited. However, this equation helps to see the physical meaning of the ion mobility.

Here, we introduce the electrolyte conductivity, k, which is the ability of an ion conductive material, liquid or solid, to sustain the passage of electric current. Conductivity depends on the mobility, u_i, of the electrolyte ions, their charges, z_i, and their number densities n_i (in m^{-3}). If an electrolyte consists of a cation and an anion and their mobilities, charges, and number densities are, respectively, $u+$ and $u-$, $z+$ and $z-$, and $n+$ and $n-$, then k can be given as follows:

$$k = e\left(n_+u_+z_+ + n_-u_-|z_-|\right). \tag{3.3}$$

The conductivity unit is $\Omega^{-1}\ m^{-1} = S\ m^{-1}$ (or $S\ cm^{-1}$). The quantity that is inverse of conductivity is resistivity, $\rho = 1/k$, with unit $\Omega\ m$ (or $\Omega\ cm$). If there are more than two ions in the solution, Equation 3.3 should be generalized as follows:

$$k = e\Sigma\left(n_iu_i|z_i|\right), \tag{3.4}$$

so all ions of the solution should be considered, and the summation should be taken over all ions, including those that are produced due to the water self-ionization, $H_2O(l) = H^+(aq) + OH^-(aq)$ or dissociation of molecules to ions.

3.5 CONDUCTANCE AND CONDUCTIVITY OF AN ELECTROLYTE

Measuring the resistance, R, of the electrolyte sample shown in Figure 3.10, combining Equations 2.3 and 3.3, and taking into account that the (electric) conductance $L = 1/R = I/E$, we can express the conductance of a solution in terms of independently measurable values:

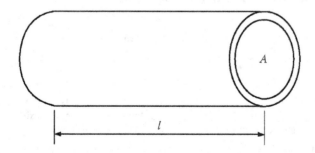

FIGURE 3.10 A glass tube with an electrolyte solution inside. The length of the tube is l, and its cross section is A. Two electrodes should be attached at each end of the tube to provide an electric field moving the ions.

$$L = \left(\frac{A}{l}\right)k = \left(\frac{A}{l}\right)e\left(n_+u_+z_+ + n_-u_-|z_-|\right) = \left(\frac{A}{l}\right)\left(\frac{e}{F}\right)\left(n_+\lambda_+ + n_-\lambda_-\right), \quad (3.5)$$

where $\lambda_+ = F\,u_+\,z_+$ and $\lambda_- = F\,u_-\,|z_-|$ are introduced as the ionic conductivities of the cation and the anion, respectively. Equation 3.5 is applicable only if two ions are in the solution, which is quite a rare case. If more than two ions are considered, then conductance is defined using Equation 3.4 as follows:

$$L = \left(\frac{A}{l}\right)e\Sigma\left(n_iu_i|z_i|\right) = \left(\frac{A}{l}\right)\left(\frac{e}{F}\right)\Sigma\left(n_i\lambda_i\right). \quad (3.6)$$

As we can see from Equations 3.5 and 3.6, conductance is proportional to the number densities (concentrations) of the ions, n_i, and their ionic conductivities λ_i; and the proportionality factor is (A/l). Therefore, at constant composition of the electrolyte, if the conductor cross section is larger and/or the distance between electrodes is shorter, then the conductance is larger and resistance is smaller. To finalize the description of the relationships between conductance, conductivity, resistance, and resistivity, the following equations might be useful:

$$k = \frac{1}{\rho} = L\frac{l}{A} = \frac{l}{AR}. \quad (3.7)$$

In electrochemical science and engineering, we are interested in both measuring and calculating the conductivity of a variety of ion conductors. The experimentally measured conductivities for a variety of materials can be found in Ref. [2]. As can be seen from Table 3.1, where some examples for commonly used in electrochemical science and engineering materials are given, the conductivity can vary by many orders of magnitude.

In calculating an electrolyte conductivity and its conductance in an electrochemical cell, the following problems can arise: (1) difficulties in estimating the

TABLE 3.1
Conductivities of Different Substances

System	t (°C)	κ (S m^{-1})
Copper	25	5.9×10^7
Platinum	25	9.1×10^6
Mercury	25	1.0×10^6
Graphite	25	10^3 (approximately)
Nafion	25	10 (approximately)
YSZ	1000	5 (approximately)
0.01 mol kg^{-1} KCl(aq)	25	0.1408
Millipore deionized water (H$_2$O)	25	5.5×10^{-6}
Glass	25	10^{-10} to 10^{-14}

concentration of ions due to the association/dissociation reactions, (2) absence of the ionic conductivities for some ions, and (3) the cell geometry (l and A) that is usually not well defined. Still, the conductivity of very pure water can theoretically be calculated, and is absolutely reliable as demonstrated below.

3.6   ELECTRIC CONDUCTIVITY OF PURE WATER

Let us use Equation 3.3 to calculate water conductivity at a temperature of 25°C and a pressure of 1 bar using the CRC Handbook reference data [2]. As was described earlier, we know that water dissociates to two ions H^+(aq) and OH^-(aq), and concentration of the ions can easily be calculated. There are ten major steps in such calculation:

1. The ionization (dissociation) constant of water, K_w, is about 10^{-14} (Table 10.6).
2. $K_w \approx b_+ b_-/(b^0)^2$ (Chapter 1).
3. $b_+/b^0 = b_-/b^0 = (K_w)^{0.5}$; $b_+ = b_- = 10^{-7}$ mol kg^{-1}.
4. $c_+ = c_- \approx b_+ \rho = c_- \rho$ (Chapter 1).
5. Density of pure water is 0.99705 g cm^{-3} (Table 10.22).
6. $c_+ = c_- = 0.99705 \times 10^{-7}$ mol L^{-1}.
7. $n_+ = n_- = c_+ N_A = c_- N_A = (0.99705 \times 10^{-7})(6.022 \times 10^{23}) = 6.004 \times 10^{16}$ L^{-1} $= 6.004 \times 10^{19}$ m^{-3}.
8. $u_+ = \lambda_+/(F z_+)$ and $u_- = \lambda_-/(F|z_-|)$.
9. $\lambda_+ \approx \lambda_+^0 = 349.65 \times 10^{-4}$ m^2 S mol^{-1}; $\lambda_- \approx \lambda_-^0 = 198 \times 10^{-4}$ m^2 S mol^{-1} (Table 10.12).
10. $k = 1.602 \times 10^{-19} \times 6.004 \times 10^{19} \times 10^{-4} (349.65 + 198)/96,485 = 5.459 \times 10^{-6}$ S m^{-1}.

The value obtained is close to the experimental conductivity given in Table 3.1 for the conductivity of Millipore water, which is produced by a commercial system for laboratories from tap water. If a salt is added to water, the solution conductivity is a sum of the conductivity of the water ions and salt ions, and Equation 3.4 should be used to calculate the total conductivity of the aqueous solution. It will be shown in Section 3.19 how to calculate the conductivity of water contaminated by a weak electrolyte.

3.7   CONDUCTANCE BY DIRECT AND ALTERNATING CURRENTS

There is an electric double layer (EDL) at any electrode/electrolyte interface. Hermann Ludwig Ferdinand von Helmholtz was the first who introduced EDL in 1879. In the first approximation, the EDL can be considered as a capacitor, which can be charged and discharged if an external potential is applied. In Figure 3.11, the metal electrode is assumed to be attached to the negative terminal of a battery so positively charged ions approach the electrode surface.

Hermann Ludwig Ferdinand von Helmholtz (1821–1894) was a German scientist who made significant contributions to a number of areas in chemistry and physics. In physics, he contributed to the development of the conservation of energy theory and made fundamental input to the formulation of chemical thermodynamics. He is well known for a formulation of the electric double layer and concepts of the electrode/electrolyte interface.

REFERENCES

A.J. Bard, G. Inzelt, and F. Scholz, *Electrochemical Dictionary*, Springer, Berlin, Germany, 2008.
http://en.wikipedia.org/wiki/Hermann_von_Helmholtz

Clearly, direct current (dc) cannot go through the electrode/electrolyte interface until a potential equal to the decomposition potential is applied (Figure 3.12). Therefore, in the case of dc current, charge is not transferred unless an electrochemical reaction takes place at the electrode. The dc at any potential above the decomposition potential, E_D, is due to both the solution conductivity and electrolysis processes at the electrodes. Therefore, the electrolyte conductivity cannot be separately measured using dc. However, if an alternating current (ac) is applied, a linear current–cell potential dependence is observed as shown in Figure 3.12 at least at relatively low cell potential values.

Let us assume that an electric conductor has both the ionic conductivity and electron conductivity. Then, an ac test will allow us to measure the conductor resistance due to both the ions and electrons. However, with a dc test, only electron conductance

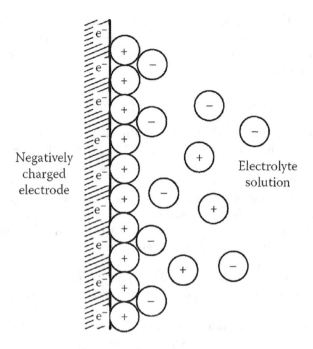

FIGURE 3.11 Interface between a negatively charged electrode and an electrolyte solution.

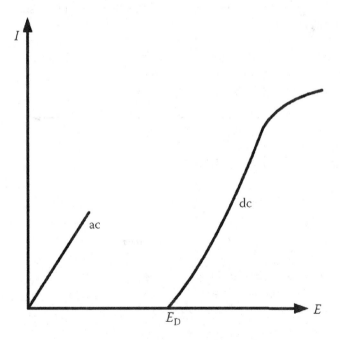

FIGURE 3.12 Magnitude of the current, I, in an electrolytic cell as a function of the applied ac and dc potential, E. E_D is the magnitude of the decomposition potential.

can be measured, if a potential below the decomposition potential is applied. The separation between the ionic and electron conductance can also be carried out using a sophisticated electrochemical technique called electrochemical impedance spectroscopy (EIS), which is described in Chapter 7.

3.8 ELECTRODE/ELECTROLYTE INTERFACE

The electrode/electrolyte interface presents a hurdle for transporting the charged species across the interface when dc potential is applied. However, the conductance can be observed by measuring the ac response. When ac potential is applied (Figure 3.13a), the EDL is charged and discharged in a similar manner as an EDL capacitor (Figure 3.13b), and the electric current can flow without any necessity for the actual transfer of charge across the electrode/electrolyte interface.

The impedance (see Section 7.10) of an electrochemical cell is defined on the basis of passing ac when the sinusoidal perturbation of the electrode potential is applied. The applied alternating potential, $E(t)$, has a form $E_m\cos(2\pi ft)$, where f is the ac frequency and E_m is the potential amplitude (Figure 3.13a). The applied potential is changing in time, t, as shown in Figure 3.13a, and corresponding alternating current can be measured.

This approach is used to measure resistance/conductance of electrolyte solutions in a conductivity cell schematically shown in Figure 3.14a. When ac is used in the electrochemical conductivity cell, consisting of two electrodes and an electrolyte between them, the cell can theoretically be represented by an equivalent electric circuit shown in Figure 3.14b. In this figure, R_s is the resistance of the solution located between the electrodes, $R_{ct(-)}$ and $R_{ct(+)}$ are the faradaic resistances of the negatively

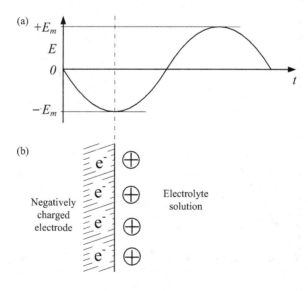

FIGURE 3.13 Time dependence of the ac potential (a) and corresponding EDL charging similar to a capacitor (b).

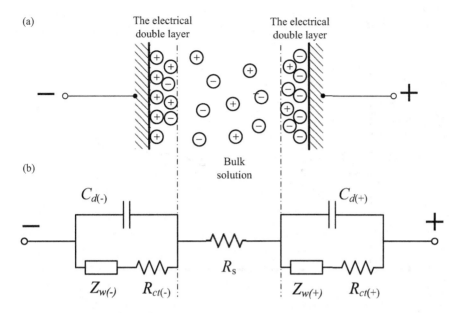

FIGURE 3.14 Schematically shown are the ac electrochemical conductivity cell with two electrodes and an electrolyte between them (a). The cell can be represented by an equivalent electric circuit (b).

and positively charged electrodes, and $C_{d(-)}$ and $C_{d(+)}$ are the interfacial EDL capacitances of the negatively and positively charged electrodes, respectively. $Z_{w(-)}$ and $Z_{w(+)}$ are called Warburg elements, which are impedances related to diffusion of the electrochemically active species in the EDL region. They are shown as rectangulars due to a possibility to include both the ohmic and capacitance components. One of the useful applications of the EIS is extracting the solution resistance from the total impedance of the conductivity cell [1] and will be described in Section 7.10.

3.9 EXTRACTING THE ELECTROLYTE RESISTANCE

The theory of EIS, which is briefly described in Section 7.10, shows that if the alternating potential, $E(t)$, has a form $E = E_m\cos(2\pi ft)$, the electrolyte resistance, R_s, can approximately be found by extrapolation when the ac frequency $f \to \infty$ ($1/f^{0.5} \to 0$), that is, the cell resistance $R \to R_s$ when $f \to \infty$. Therefore, at a very high frequency, all contributions to the impedance except R_s can be ignored. A simple approach of extracting R_s is shown in Figure 3.15, where the experimentally obtained cell resistance is plotted as a function of ($1/f^{0.5} \to 0$) and R_s is obtained by extrapolation if more or less linear dependence is observed. Unfortunately, it is not a common case, and more sophisticated EIS techniques should be used to correctly estimate R_s. Note that one of the problems in using this simple method is significant scattering of observed R values in a region of high frequencies. As shown in Figure 3.15, this scattering can reduce the precision and reliability of the liner extrapolation.

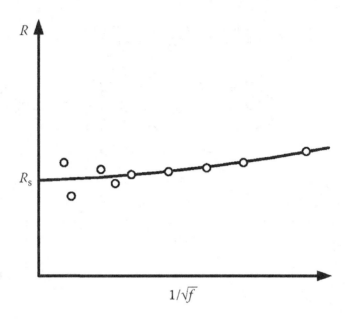

FIGURE 3.15 An approach to extract the electrolyte resistance, R_s, by extrapolating the measured cell resistance R vs. $1/f^{0.5}$ to $f = \infty$.

To avoid electrochemical decomposition of the electrolyte, the potential amplitude magnitude, $|E_m|$, should be less than $|E_D|$. Also, it is beneficial for the conductivity measurements when the EDL capacitances, $C_{d(-)}$ and $C_{d(+)}$, are large values. To increase $C_{d(-)}$ and $C_{d(+)}$, the surface of the electrode should be as large as possible. In addition, in the first approximation, the Warburg elements, $Z_{w(-)}$ and $Z_{w(+)}$, can be removed from the equivalent electric circuit given in Figure 3.14.

More sophisticated method to extract the solution resistance from the total impedance of a conductivity cell will be described in Section 7.10.

3.10   CONDUCTIVITY MEASUREMENTS

The schematic of a simple immersion conductivity cell for rapid measurements of solution conductivity is shown in Figure 3.16, and the schematic of an experimental setup for such measurements is shown in Figure 3.17. The cell consists of two tetragonal platinized (electrochemically covered by fine Pt powder called platinum black) platinum electrodes positioned in parallel at a distance of a few millimeters. The platinized platinum has a true surface area much larger than the geometrical surface area of the electrode and, therefore, increases the EDL capacitances, $C_{d(-)}$ and $C_{d(+)}$. The setup consists of a digital bridge, a conductivity cell, and a thermometer. The electrolyte conductivity strongly depends on temperature, so temperature control is very important. Any conductivity measurement is started with obtaining the conductivity cell constant $K_{cell} = l/A$ because neither the distance between electrodes, l, nor the cross section, A, can be accurately measured due to the uncertainty of the trajectory of the ion movement between the electrodes.

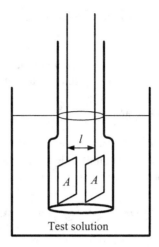

FIGURE 3.16 A simple immersion conductivity cell.

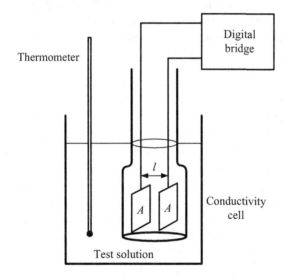

FIGURE 3.17 Schematic of experimental conductivity measurement setup.

The cell constant can be measured if the conductivity of a standard solution, k, is well known and the conductance of the solution, L, can be precisely measured. One of the common standard solutions is KCl(aq). At a concentration of 0.01 mol L^{-1}, its conductivity equals 0.140823 S m^{-1} at 25°C and 0.127303 S m^{-1} at 20°C (both at an ambient pressure of 1 bar) (Table 10.9). The cell constant can now be found by measuring the standard solution conductance, L:

$$K_{cell} = \frac{k}{L}. \tag{3.8}$$

The common conductivity measurement procedure includes the following main requirements:

1. The test solutions should be free of CO_2 and other impurities.
2. $H^+(aq)$ and $OH^-(aq)$ ions should be taken into account in precise estimations of conductivity, particularly in diluted solutions.
3. Measurements should be carried out at a precisely measured and constant temperature.
4. A standard solution to estimate the cell constant, K_{cell}, should be available.
5. An electronic system (e.g., a digital bridge) to measure conductance at different frequencies should be available.

3.11 MOLAR AND EQUIVALENT CONDUCTIVITY

The concentration dependence of the ionic conductivity is quite complicated. If there were not any obstacles, it would be a linear dependence of k vs. molar concentration, c. As can be seen in Figure 3.18, generally, there is not such a linear dependence at high concentrations. It was found that $\Lambda = k/c$, which is called molar conductivity, is more convenient to use for electrolyte solutions. In SI, the units of Λ are S m² mol⁻¹, so k is in S m⁻¹ and c is in mol m⁻³. However, in almost all studies, k is given in S cm⁻¹, and the common molarity unit is mol L⁻¹ = mol dm⁻³. This small contradiction between the SI requirements and research traditions should be kept in mind. In this book, we will follow the habit of using S cm⁻¹ for k and mol L⁻¹ for c.

Note that in the past, the equivalent conductivity $\Lambda_{eq} = \Lambda/(v_\pm z_\pm)$ was widely used, where $v_\pm z_\pm$ is defined as

$$v_\pm z_\pm = v_+ z_+ = v_- |z_-|.\tag{3.9}$$

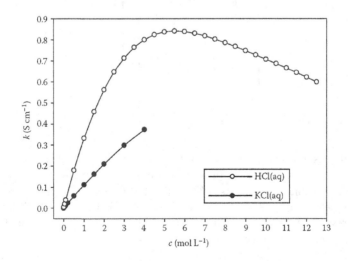

FIGURE 3.18 Conductivity of two aqueous solutions as a function of molar concentration (Table 10.10).

For example, for Na_2SO_4, $v_+ = 2$, $z_+ = 1$, $v_- = 1$, $z_- = -2$, $v_{\pm}z_{\pm} = 2$. Clearly, $\Lambda = \Lambda_{eq}$ for 1–1 electrolytes (e.g., NaCl). Connected to the equivalent conductivity, the equivalent molar concentration, $N = c/(v_{\pm}z_{\pm})$, called normality, was used (and is still used) in the past. Presently, N is not recommended by IUPAC [3]. The molar conductivity of an electrolyte is directly connected to the mobilities and conductivities of individual ions as

$$\Lambda = v_{\pm}z_{\pm}F(u_+ + u_-) = v_+\lambda_+ + v_-\lambda_-, \tag{3.10}$$

while the equivalent conductivity is related to the ionic mobilities and ionic conductivities as

$$\Lambda_{eq} = F(u_+ + u_-) = \frac{\lambda_+}{z_+} + \frac{\lambda_-}{|z_-|}. \tag{3.11}$$

The literature is confusing with regard to the molar and equivalent conductivities. The preceding equations should help to sort this mess out.

3.12   KOHLRAUSCH'S LAW AND LIMITING CONDUCTIVITY

While Λ should not depend on concentration as strongly as k does, the molar conductivity decreases as the concentration increases. Friedrich Wilhelm Georg Kohlrausch, in 1900, experimentally found that for dilute strong electrolytes, there is a linear dependence of Λ from $c^{1/2}$:

$$\Lambda = \Lambda^0 - K(c)^{1/2}, \tag{3.12}$$

which represents Kohlrausch's law. In Equation 3.12, K is a constant and Λ^0 is the molar conductivity at infinite dilution (limiting conductivity), a state when there are not any interactions between ions. Figure 3.19 shows the linear dependence of $\Lambda(c^{1/2})$

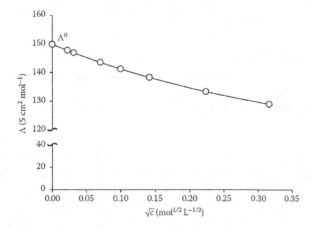

FIGURE 3.19   Molar conductivity of KCl(aq) as a function $c^{1/2}$ in a region of diluted solutions up to infinitely diluted solution with $\sqrt{c} = 0$ (Table 10.11).

for KCl(aq), which is a good example of a strong electrolyte. While K is the empirical constant, there is a theory of conductivity of the dilute strong electrolytes, the Debye–Hückel–Onsager limiting law [1], which can be used to theoretically calculate K for a limited concentration range. This concentration range for three electrolytes, KCl(aq), HCL(aq), and $ZnSO_4$(aq), was in investigated in on the problems in Section 3.19. Also, that problem allows calculating K values at temperature of 25°C and 1 bar.

Friedrich Wilhelm Georg Kohlrausch (1840–1910) was a German physicist who investigated the conductive properties of electrolytes and contributed to the knowledge of their behavior.

REFERENCES

A.J. Bard, G. Inzelt, and F. Scholz, *Electrochemical Dictionary*, Springer, Berlin, Germany, 2008.
https://en.wikipedia.org/wiki/Friedrich_Kohlrausch_(physicist)

Due to the association of ions into molecules or ion pairs as the electrolyte concentration increases, there is not such linear dependence of $\Lambda(c^{1/2})$. In this case, because the dependence of Λ from $c^{1/2}$ is not linear, Λ^0 cannot be estimated by a linear extrapolation using Kohlrausch's law. The dependence $\Lambda(c^{1/2})$ for a weak electrolyte, acetic acid, is shown in Figure 3.20 along with data for two strong electrolytes, HCl(aq) and KCl(aq). As is shown in Section 3.19 for 1-1 strong electrolytes (e.g. KCl(aq) or HCl(aq), Equation 3.12 is valid only at concentrations up to 0.01 mol L^{-1}.

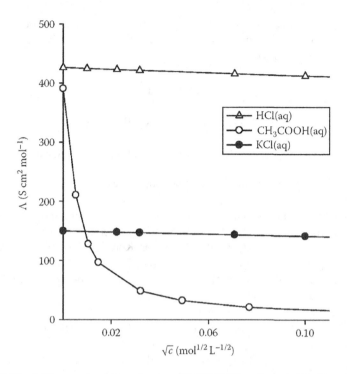

FIGURE 3.20 Molar conductivity of strong (Table 10.11) and weak [1] electrolytes as a function $\sqrt{c}$.

3.13 ADDITIVITY OF ELECTROLYTE LIMITING CONDUCTIVITY

The limiting molar conductivity of an electrolyte, Λ^0, is a sum of the limiting ionic conductivities, λ_i^0, taking into account the number of ions, v_i, in the electrolyte formula, $K_{v+}A_{v+}$:

$$\Lambda^0 = v_+\lambda_+^0 + v_-\lambda_-^0, \tag{3.13}$$

where λ_+^0 and λ_-^0 are the limiting ionic conductivities of the cation and the anion, respectively. For example, the limiting molar conductivity of $CaCl_2$ is equal to λ_+^0 (Ca^{2+}) + $2\lambda_-^0$ (Cl^-). If there is more than one electrolyte in the solution, all ions should be taken into account using the summation: $\Lambda^0 = \sum v_i\lambda_i^0$.

The limiting ionic conductivity is a transport property of an ion, which does not depend on concentration but depends on temperature and pressure (or density of the solvent). Contrary to any thermodynamic property of an individual ion, the limiting ionic conductivities, as transport properties, of individual ions can experimentally be obtained and are tabulated in (Table 10.12) for a number of anions and cations. Therefore, the molar conductivity of an electrolyte at infinite dilution, Λ^0, is an additive value and can be calculated using the limiting ionic conductivities. This possibility is particularly useful for the weak electrolytes, when Λ^0 cannot be experimentally obtained using Kohlrausch's law by an extrapolation to the infinite dilution. For example, the limiting molar conductivity of acetic acid, $CH_3COOH(aq)$, can be calculated as follows:

$$\lambda_+^0 = 349.65\,S\,cm^2\,mol^{-1}\,[\text{Table 10.12}],$$

$$\lambda_-^0 = 40.9\,S\,cm^2\,mol^{-1}\,[\text{Table 10.12}], v_+ = 1,$$

$$v_- = 1, \Lambda^0\left(CH_3COOH, aq\right) = 349.65 + 40.9 = 390.55\,S\,cm^2\,mol^{-1}.$$

If this value is compared with the molar conductivities of $CH_3COOH(aq)$, shown in Figure 3.20, it can be seen that Λ^0 is about two times larger than the value in Figure 3.20 for the smallest experimentally determined molarity of 0.00002801 mol L⁻¹. Therefore, there is no way the limiting molar conductivity of a weak electrolyte can be obtained using an extrapolation based on Kohlrausch's law. However, Λ^0 of $CH_3COOH(aq)$ can be calculated using Equation 3.13. This equation can be called the rule of the independent migration of ions in the infinitely diluted solution. Note that the degree of dissociation, α, defined earlier by Equation 1.33 can be experimentally obtained for a weak electrolyte as $\alpha = \Lambda/\Lambda^0$.

It should be mentioned that while Equation 3.10 looks similar to Equation 3.13, the ionic conductivities at any given concentration are not independent, and it can be shown that both λ_+ and λ_- strongly depend on the ionic strength of the solution. The Debye–Hückel–Onsager limiting law, described in detail in Ref. [1], can predict this dependence in a region of very low concentrations.

As another example to use Equation 3.13, the limiting molar conductivity of $CaCl_2(aq)$ can be calculated using the limiting ionic conductivities of ions (Table 10.12) and then compared with Λ^0 of $CaCl_2(aq)$ given in (Table 10.11). Both values should be the same and are equal to 271.54 S cm² mol⁻¹.

3.14   TRANSPORT NUMBERS

A contribution to electric current provided by anions or cations can conveniently be defined by the transport number, t_- or t_+:

$$t_+ = \frac{I_+}{\left(I_+ + I_-\right)}, \quad t_- = \frac{I_-}{\left(I_+ + I_-\right)}, \tag{3.14}$$

where I_+ and I_- are the partial currents of cations and anions, respectively. If there are more than two ions in the solution, the transport number of the ith ion is defined as follows:

$$t_i = \frac{I_i}{\Sigma I_i}. \tag{3.15}$$

Obviously, the sum of the transport numbers of all ions in a solution is equal to 1:

$$\Sigma t_i = 1. \tag{3.16}$$

The transport numbers can also be estimated if the molar conductivities of all ions (λ_i) and their molar concentrations (c_i) are available:

$$t_i = \frac{\lambda_i c_i}{\Sigma \lambda_i c_i}. \tag{3.17}$$

The transport number is a measurable value using a variety of well-developed methods [1]. One of the most popular methods is using a Hittorf cell [1]. Johann Wilhelm Hittorf introduced his transport cell in 1853. In the Hittorf cell, electrolysis is carried out using three compartments: anode, central, and cathode. Measuring the change of concentration in the anode and cathode compartments allows for the determination of transport numbers. Equation 3.17 is the background for estimating the ionic conductivities tabulated in (Table 10.12) as soon as the transport numbers are experimentally measured.

Johann Wilhelm Hittorf (1824–1914) was a German physicist who contributed to the theory of ionic transport in electrolyte solutions due to migration of the oppositely charged ions to anode and cathode. He introduced the ion transport numbers and the first method for their measurements.

REFERENCES

A.J. Bard, G. Inzelt, and F. Scholz, *Electrochemical Dictionary*, Springer, Berlin, Germany, 2008.
http://en.wikipedia.org/wiki/Johann_Wilhelm_Hittorf

TABLE 3.2

The Transport Numbers of HCl(aq) and KCl(aq) over a Wide Concentration Range [1]

b (mol kg^{-1})	t+ (HCl, aq)	t+ (KCl, aq)
0	0.821	0.490
0.1	0.831	0.490
0.2	0.834	0.489
0.5	0.838	0.489
1.0	0.841	0.488

Transport numbers do not significantly depend on concentration, and therefore it is a reasonable approximation that

$$t_i \approx t_i^0 = \frac{v_i \lambda_i^0}{\sum v_i \lambda_i^0}, \tag{3.18}$$

where t_i^0 is the transport number of the ith ion in the infinitely diluted solution. The advantage here is that in order to calculate t_i^0, only the limiting ionic conductivities should be available, while to exactly estimate t_i, both the individual ionic conductivities and their concentrations should be known. As an example, the transport numbers of two 1–1 electrolytes, KCl(aq) and HCl(aq), over a wide concentration range are presented in Table 3.2 [1].

It can be seen that when concentration changes from 0 to 1 mol kg^{-1}, the cation transport number, t_+, in HCl(aq) is changed by 2.4% and in KCl(aq) only by 0.4%.

3.15   DIFFUSION AND HYDRATION OF IONS IN INFINITELY DILUTED SOLUTION

The limiting ionic mobility, $u_i^0 = \lambda_i^0/(F|z_i|)$, is an important transport characteristic of the ith ion when the solution is at infinite dilution, meaning that there are no interactions between ions but the ions still interact with solvent (water) molecules. This suggests that there should be a relationship between the limiting ionic mobility (conductivity) and its diffusion coefficient when the ion is moving in the solvent. Such a relationship was proposed by Albert Einstein in 1905:

$$D_i^0 = \frac{RTu_i^0}{F|z_i|} = \frac{RT\lambda_i^0}{(Fz_i)^2}, \tag{3.19}$$

where D_i^0 is the limiting ionic diffusion coefficient available from (Table 10.12). As an example, D_i^0 of Li$^+$(aq), Na$^+$(aq), and K$^+$(aq) at 25°C are, respectively, 1.029×10^{-5}, 1.334×10^{-5}, and 1.957×10^{-5} cm^2 s^{-1}. Note that the limiting ionic diffusion coefficients allow to calculate the limiting electrolyte diffusion coefficient $D^0 = (z_+ + |z_-|) D_+^0 D_-^0 / (z_+ D_+^0 + |z_-| D_-^0)$ [1]. An example of calculating D^0 of CaCl$_2$(aq) is presented in Section 3.19.

Albert Einstein (1879–1955) was a German-born American theoretical physicist. He was awarded the Nobel Prize in Physics in 1921. In electrochemical science, a number of equations bear Einstein's name, for example, Nernst–Einstein equation showing the relationship between conductivity and diffusion coefficient or the correlation between diffusion coefficient and viscosity, which is known as the Stokes–Einstein equation.

REFERENCES

A.J. Bard, G. Inzelt, and F. Scholz, *Electrochemical Dictionary*, Springer, Berlin, Germany, 2008.
http://en.wikipedia.org/wiki/Albert_Einstein

The diffusion coefficient can traditionally be used to explain the ability of transport and therefore the hydration of an ion in aqueous solution. Hydration is some orientation of water molecules around an ion due to molecular interactions. As a result, a hydration shell forms around the ion so that an ion will move in the solution surrounded by a number of water molecules rather than alone (see Figure 3.9). By analyzing the data on D_i^0 in (Table 10.12), we can see that the hydration shell of $Li^+(aq)$ should be larger than that of $Na^+(aq)$ and the hydration shell of $K^+(aq)$ is smaller than that of $Na^+(aq)$. This is due to the values of the diffusion coefficients given earlier and a well-known fact that a Li ion is smaller than a Na ion and a K ion is larger than a Na ion while the charge of these three ions is the same. Generally, a larger charge and smaller radius of an ion allows keeping more hydrated molecules around the ion. Therefore, the data on the limiting diffusion coefficients and conductivities for $Li^+(aq)$, $Na^+(aq)$, and $K^+(aq)$ can be explained if hydration of ions is taken into account; hydration of $Li^+(aq)$ is larger than that of $Na^+(aq)$, and hydration of $Na^+(aq)$ is larger than that of $K^+(aq)$. In other words, the hydrodynamic radius of $Li^+(aq)$ is the largest one, and the mobility is the smallest one from the three ions considered earlier. Still, $H^+(aq)$ and $OH^-(aq)$ ions show the unusually high limiting ionic diffusion coefficient (Table 10.12), and this fact cannot be explained by the hydrodynamic approach described earlier.

3.16   PROTON CONDUCTIVITY

Because the proton in water is strongly hydrated, the hydronium ion, $H_3O^+(aq)$, is formed in an aqueous solution, and conductivity in $H^+(aq)$-containing solutions has a special mechanism. There is not any hydrodynamic movement of the hydronium ion when a potential is applied. Actually, the charge is moving from one hydronium structure to another one. This type of conductivity is called the hopping (Grotthuss) mechanism and is schematically illustrated in Figure 3.21.

Freiherr Christian Johann Dietrich Theodor von Grotthuss (1785–1822) was a Lithuania-born German chemist who introduced a number of new theories including the first theory of electrolysis known as the Grotthuss mechanism of proton conductance via hopping process of electrolytic conductivity.

REFERENCES

A.J. Bard, G. Inzelt, and F. Scholz, *Electrochemical Dictionary*, Springer, Berlin, Germany, 2008.
http://en.wikipedia.org/wiki/Theodor_Grotthuss

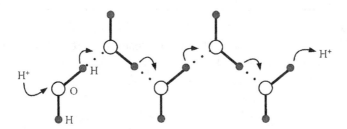

FIGURE 3.21 Hopping (Grotthuss) mechanism of conductivity in a H^+(aq)-containing solution.

Theodor von Grotthuss proposed the hopping mechanism in 1805. The Grotthuss mechanism results in a significant increase of conductivity (diffusivity) in strong bases and acids. Conductivity of an OH^-(aq) ion can be considered as H^+(aq) conductivity but in the opposite direction due to the ionization of water, $H_2O(l) = H^+(aq) + OH^-(aq)$, and negative charge of the hydroxyl ion. From Table 10.12, the limiting ionic conductivities of H^+(aq) and OH^-(aq) ions at 25°C are, respectively, 349.65 and 198.0 S cm^2 mol^{-1}, and these values are much larger than limiting ionic conductivities of Li^+(aq) [38.66 S cm^2 mol^{-1}], Na^+(aq) [50.08 S cm^2 mol^{-1}], and K^+(aq) [73.78 S cm^2 mol^{-1}].

3.17 WALDEN'S RULE

In the beginning of the nineteenth century, Paul Walden suggested that the product of the ionic mobility and solvent viscosity is a constant value and does not significantly depend on temperature and pressure:

$$u_i \eta = \frac{|z_i|e}{6\pi r_i} = \text{const},$$ (3.20)

as long as the ion radius is not changing and the mechanism of the ion movement is purely hydrodynamic. In Equation 3.20, which is directly coming from Equation 3.2, the ionic conductivity can be used instead of the ionic mobility using the following equation:

$$\lambda_i = Fu_i|z_i|.$$ (3.21)

Walden's rule can be obtained by simply combining Equations 3.20 and 3.21:

$$\lambda_i \eta = z_i^2 Fe/(6\pi r_i) = \text{const},$$ (3.22)

which is clearly related to the well-known Stokes–Einstein equation:

$$D_i^0 = \frac{kT}{6\pi r_i \eta}.$$ (3.23)

Note that Equation 3.23 is a combination of Equations 3.19 and 3.22.

TABLE 3.3

Conductivity (k) of 0.1 mol kg^{-1} KCl(aq), Viscosity of Water η, and Their Product Divided by Density of Water ρ

t (°C)	0	10	20	30	40	50
κ (µS cm^{-1})	7117	9292	11,616	14,059	16,591	19,181
η (µPa s)	1791	1306	1002	797	653	547
ρ (g cm^{-3})	0.9998	0.9997	0.9982	0.9956	0.9922	0.9880
$10^{-7} \kappa\eta\,\rho^{-1}$ (S µPa s cm^2 g^{-1})	1.275	1.214	1.166	1.125	1.092	1.060

Therefore, using the hydrodynamic model of mobility, we can see from Equation 3.22 that the product of ionic conductivity and viscosity should theoretically be constant and independent of temperature.

Combining Equations 3.10 and 3.22, a similar equation can be obtained for the equivalent conductivity of an electrolyte:

$$\Lambda\eta = \left[\left(v_+z_+^2/r_+\right)+\left(v_-z_-^2/r_-\right)\right]Fe/(6\pi) = \text{const.} \tag{3.24}$$

The data in Table 3.3 on conductivity, viscosity, and density in a temperature range from 0°C to 50°C are taken from Tables 10.9 and 10.22 and show a reasonable validity of the Walden's rule, because the value of ($k\eta/\rho$) changes by only 20% when temperature is changed from 0°C to 50°C.

Note that Walden's rule cannot be used for H$^+$(aq) and OH$^-$(aq) ions due to a special mechanism of transport of these ions and, therefore, Walden's rule is not applicable to strong acids and bases but can be used for many other electrolytes that do not have any H$^+$(aq) and OH$^-$(aq) ions in significant amount.

3.18 SUMMARY

- There are different types of electric conductivity. In this book, we are mainly interested in the ionic conductivity of electrolytes (liquid or solid).
- Conductivity of a separator between half cells is an important property of many electrochemical energy conversion systems. This property can significantly contribute to the electrochemical cell performance.
- Conductance of an electrolyte (in S) depends on the electrolyte conductivity (S m^{-1} or S cm^{-1}) and geometry (cross section and thickness) of the conductor (separator). A clear understanding of the difference between conductance and conductivity (or resistance (in Ω) and resistivity (in Ω m or Ω cm)) is important.
- The electric conductance in an electrolyte can be observed by measuring an ac response. The dc can be used to measure the electric conductance of an electron conductor.

- When ac is used, the typical electrochemical conductivity cell (two electrodes and an electrolyte between them) can theoretically be represented by an equivalent electric circuit, which consists of a number of resistors and capacitors.
- EIS (electrochemical impedance spectroscopy) should be used to extract the electrolyte conductance from the total impedance of the circuit.
- Conductivity of an electrolyte can be calculated if (1) the concentration of the conducting ions, (2) their charges, and (3) their mobilities (ionic conductivities) are known.
- Molar conductivity can easily be calculated if conductivity and molar concentration are known. While the equivalent conductivity is not recommended by IUPAC, a reader should know how it can be handled. The molar conductivity and equivalent conductivity for 1–1 electrolytes are equal.
- Concentration dependence of the molar conductivity is very different for strong and weak electrolytes. For dilute strong electrolytes, Kohlrausch's law prescribes a linear dependence, $\Lambda(c^{1/2})$. For weak electrolytes, there is not such a linear dependence. When concentration approaches zero, both strong and weak electrolytes achieve the molar conductivity at infinite dilution or limiting molar conductivity.
- The molar conductivity at infinite dilution, Λ^0, is an additive value and can be calculated using the ionic conductivities.
- Contribution to the total current provided by an ion (anion or cation) is conveniently defined by the transport number. The transport number is the current provided by an ion divided by the total current. The transport number of an ion can be experimentally measured, while none of the thermodynamic properties of a single ion can be experimentally obtained.
- When concentration approaches zero, the transport number achieves the transport number at infinite dilution or limiting transport number. Concentration dependence of the transport number is relatively small. The limiting transport numbers allow calculating limiting ionic conductivities.
- The limiting conductivities of ions can be discussed based on the hydrodynamic theory taking into account hydration of ions. Hydration is a possibility of an ion to keep the solvent (water) molecules in the immediate neighborhood. As a result, when the ion is moving, some solvent molecules can be transported with the ion.
- Conductivity of the aqueous proton is provided by a special hopping or Grotthuss mechanism. Due to the mechanism, the conductivity of $H^+(aq)$ and $OH^-(aq)$ is much higher than the conductivity of other singly charged ions.
- Based on a simple hydrodynamic theory, Walden's rule can be formulated— the product of ionic (molar) conductivity and viscosity is approximately constant. The rule can be used to predict the temperature dependence of the ion (molar) conductivity. Clearly, due to the hopping mechanism, Walden's rule is not applicable for $H^+(aq)$ and $OH^-(aq)$ ions.

3.19   EXERCISES

PROBLEM 1

For dilute solutions, the molar conductivity can be calculated using the Debye–Hückel–Onsager equation, which can be written for 1:1 electrolyte as [1,2]

$$\Lambda = \Lambda^0 - \left(A + B\Lambda^0\right)\left(c/c^0\right)^{1/2},$$

where c and c^0 are, respectively, the molar concentration and its standard value (1 mol L^{-1}), while A and B are theoretical parameters equal at 25°C and 1 bar as $A = 60.6$ cm^2 S mol^{-1} and $B = 0.23$. Calculate Λ for KCl(aq) and HCl(aq) for a concentration ranger from 0 to 0.1 mol L^{-1}, compare the calculated values with data given in Table 10.11, and make conclusions on a range of concentration where the Debye–Hückel–Onsager equation can be used for completely dissociated 1–1 electrolytes.

Solution:

The calculated and corresponding experimental data are given below.

c (mol L^{-1})	$\Lambda_{KCl(Calc)}$ (cm^2 S mol^{-1})	$\Lambda_{KCl(Exp)}$ (cm^2 S mol^{-1})	$\Lambda_{HCl(Calc)}$ (cm^2 S mol^{-1})	$\Lambda_{HCl(Exp)}$ (cm^2 S mol^{-1})
0	149.79	149.79	425.95	425.95
0.0005	147.66	147.74	422.40	422.53
0.001	146.78	146.88	420.94	421.15
0.005	143.07	143.48	414.74	415.59
0.01	140.28	141.20	410.09	411.80
0.02	136.35	138.27	403.52	407.04
0.05	128.54	133.30	390.49	398.89
0.1	119.73	128.90	375.81	391.13

If we assume that the difference between the calculated and experimental data should not be larger than 2 cm^2 S mol^{-1}, then the Debye–Hückel–Onsager equation is valid for a concentration range up to 0.01 mol L^{-1} for both electrolytes.

PROBLEM 2

0.5 L of Millipore water is mixed at 25°C with 100 mL of 0.0001 molal CH$_3$COOH(aq) solution. What is the conductivity of this mixture?

Solution:

When 100 mL of 0.0001 mol kg^{-1} CH$_3$COOH(aq) solution is added to 0.5 L of Millipore water, the total concentration of the acid becomes 1.6667×10^{-5} mol kg^{-1}. This number can be easily calculated using material of Chapter 1.

The speciation in the 1.6667×10^{-5} mol kg^{-1} CH$_3$COOH(aq) solution is defined by the following two reactions:

$$CH_3COOH(aq) = H^+(aq) + CH_3COO^-(aq),$$

$$H_2O(l) = H^+(aq) + OH^-(aq).$$

Excluding $H_2O(l)$, the molal concentrations of four species, $b_{H^+(aq)}$, $b_{CH_3COO^-(aq)}$, $b_{CH_3COOH(aq)}$, and $b_{OH^-(aq)}$ should be calculated using following four equations:

$$K_a = \frac{b_{H^+(aq)} b_{CH_3COO^-(aq)}}{b_{CH_3COOH(aq)}} = 10^{-4.756},$$

$$K_w = b_{H^+(aq)} b_{OH^-(aq)} / x_{H_2O} = 10^{-13.908},$$

$$b_{CH_3COO^-(aq)} + b_{CH_3COOH(aq)} = 1.6667 \times 10^{-5} \, \text{mol kg}^{-1},$$

$$b_{H^+(aq)} = b_{CH_3COO^-(aq)} + b_{OH^-(aq)},$$

where K_a and K_w are the dissociation constant of the acetic acid and water, respectively, available in Tables 10.6 and 10.15.

Solving the four equations given above, four concentrations can be numerically calculated as follows:

$$b_{H^+(aq)} = 1.04465 \times 10^{-5} \, \text{mol kg}^{-1}, \, b_{CH_3COO^-(aq)} = 1.04453 \times 10^{-5} \, \text{mol kg}^{-1},$$

$$b_{CH_3COOH(aq)} = 6.22142 \times 10^{-6} \, \text{mol kg}^{-1}, \, b_{OH^-(aq)} = 1.18313 \times 10^{-9} \, \text{mol kg}^{-1}.$$

Molarity of the three ions responsible for ionic conductivity can be calculated using the density of water $\rho_{H_2O(l)} = 0.99705 \, \text{g cm}^{-3}$ (Table 10.22):

$$c_{H^+(aq)} = c_{CH_3COO^-(aq)} = b_{H^+(aq)} \rho_{H_2O(l)} = 1.0414 \times 10^{-5} \, \text{mol L}^{-1},$$

$$c_{OH^-(aq)} = b_{OH^-(aq)} \rho_{H_2O(l)} = 1.1796 \times 10^{-9} \, \text{mol L}^{-1}.$$

The number density of each ion can be calculated as

$$n_{H^+(aq)} = c_{H^+(aq)} N_A = 6.2723 \times 10^{21} \, \text{m}^{-3},$$

$$n_{OH^-(aq)} = c_{OH^-(aq)} N_A = 7.1038 \times 10^{17} \, \text{m}^{-3},$$

$$n_{CH_3COO^-(aq)} = c_{CH_3COO^-(aq)} N_A$$

$$= 1.04145 \times 10^{-5} \times 6.022 \times 10^{23}$$

$$= 6.2716 \times 10^{18} \, \text{L}^{-1} = 6.2716 \times 10^{21} \, \text{m}^{-3},$$

where N_A is the Avogadro constant.

The ionic mobility of each ions, $u_i = \lambda_i/F|z_i|$, can be calculated using ionic conductivities (Table 10.12):

$$u_{H^+ (aq)} = \frac{\lambda_{H^+}}{Fz_{H^+}} = \frac{349.65 \times 10^{-4}}{96485 \times 1} = 3.6239 \times 10^{-7} \, m^2 \, SC^{-1},$$

$$u_{OH^- (aq)} = \frac{\lambda_{OH^-}}{F|z_{OH^-}|} = \frac{198 \times 10^{-4}}{96,485 \times 1} = 2.0521 \times 10^{-7} \, m^2 \, SC^{-1},$$

$$u_{CH_3COO^- (aq)} = \frac{\lambda_{CH_3COO^-}}{F|z_{CH_3COO^-}|} = \frac{40.9 \times 10^{-4}}{96,485 \times 1} = 4.2390 \times 10^{-8} \, m^2 \, SC^{-1}.$$

Now we can calculate the electric conductivity, k, of this mixture by applying Equation 3.4 with $e = 1.602 \times 10^{-19}$ C:

$$k = e\left(n_{H^+ (aq)} u_{H^+ (aq)} z_{H^+} + n_{OH^- (aq)} u_{OH^- (aq)} |z_{OH^-}| + n_{CH_3COO^- (aq)} u_{CH_3COO^- (aq)} |z_{CH_3COO^-}|\right)$$

$$= 1.602 \times 10^{-19} \times \left(6.27232 \times 10^{21} \times 3.62388 \times 10^{-7} + 7.10379 \times 10^{17} \times 2.05213 \times 10^{-7}\right.$$

$$\left. + 6.2716 \times 10^{21} \times 4.239 \times 10^{-8}\right)$$

$$= 0.00040675 \, S/m$$

PROBLEM 3

Calculate the diffusion coefficient of $CaCl_2(aq)$ at the infinitely diluted solution at ambient conditions (25°C and 1 bar).

Solution:

The diffusion coefficient of $CaCl_2(aq)$ at the infinitely diluted solution, $D^0_{CaCl_2 (aq)}$, can be calculated by using the diffusion coefficient of the corresponding ions (Table 10.12) and applying equation given in Section 3.15:

$$D^0_{CaCl_2 (aq)} = \frac{\left(z_{Ca^{2+}} + |z_{Cl^-}|\right) D^0_{Ca^{2+} (aq)} D^0_{Cl^- (aq)}}{z_{Ca^{2+}} D^0_{Ca^{2+} (aq)} + |z_{Cl^-}| D^0_{Cl^- (aq)}}$$

$$= \frac{(2+1) \times 0.792 \times 10^{-5} \times 2.032 \times 10^{-5}}{2 \times 0.792 \times 10^{-5} + 1 \times 2.032 \times 10^{-5}} = 1.3352 \times 10^{-5} \, cm^2 \, S^{-1}.$$

REFERENCES

1. R.A. Robinson and R.H. Stokes, *Electrolyte Solutions*, Dover, New York, 2002 (originally published in 1959).
2. J. Rumble (Editor-in-Chief), *Handbook of Chemistry and Physics*, 101st ed., CRC Press, Boca Raton, FL, 2020.
3. E.R. Cohen et al., *Quantities, Units, and Symbols in Physical Chemistry*, 3rd ed., RSC Publishing, Cambridge, UK, 2007.

4 Equilibrium Electrochemistry

4.1 OBJECTIVES

The main objective of this chapter is to introduce students to one of the most important subjects of the book, equilibrium electrochemistry, which is mainly based on equilibrium thermodynamics. Equilibrium electrochemistry is usually the first and required step in analyzing any electrochemical system. How to estimate the equilibrium potential of a half-reaction and the electric potential difference of an electrochemical cell are described in this chapter. One of the most fundamental equations of electrochemical science and engineering, the Nernst equation, is introduced and employed for composing the potential–pH (Pourbaix) diagrams. Temperature dependence of the electrode potential and the cell potential difference is also described.

4.2 EQUILIBRIUM THERMODYNAMICS AND ELECTROCHEMISTRY

Although we frequently use the word "equilibrium" in our life, equilibrium is not a common state for any real system. The common thermodynamic state is called steady state when a system's parameters are not changing because the output and input fluxes of energy and mass are balanced. For example, the temperature of a human body is remarkably constant due to a delicately balanced steady state, and this temperature, 36.6°C, is quite different from any common temperature in the surroundings. Therefore, we can say that a human body is in a steady state but not in equilibrium with the environment. Still, in our electrochemical studies we can set up a system to be in equilibrium. An example was given in Section 2.12 when the equilibrium potential of the Daniell cell is measured using a high resistance electrometer to prevent any significant flow of electrons from one electrode to another one.

Now let us recall some basic thermodynamics that are needed in this chapter. If thermodynamic variables are temperature and pressure, the main characteristic for understanding whether the system (process, reaction) is in equilibrium or not is Gibbs energy (ΔG). Electrochemical cells commonly work at well-defined temperature and pressure, and therefore, ΔG is the thermodynamic function showing how far the electrochemical cell is from equilibrium. When an electrochemical cell is at equilibrium, ΔG is simply zero. Two other important thermodynamic functions to understand an electrochemical cell behavior are enthalpy (ΔH) and entropy (ΔS), which are connected to ΔG via a well-known and fundamental equation:

$$\Delta G = \Delta H - T\Delta S, \tag{4.1}$$

DOI: 10.1201/b21893-4

where T is thermodynamic temperature. Note that Equation 4.1 should be used for an isobaric ($P = $ const) and isothermal ($T = $ const) system, and this is perfectly suitable for most of the electrochemical systems that are usually operated at constant and well-defined temperature and pressure.

It is usually not easy to illustrate Equation 4.1 in a thermodynamics course, so some abstract approaches with a heat engine are used. However, in electrochemical science and engineering, all values of Equation 4.1 are well defined, and two of them, ΔG and ΔS, can experimentally be estimated using only electrochemical measurements. If the potential difference of the Daniell cell in the equilibrium mode, E_{eq}, is measured, this value can immediately be used to calculate $\Delta_r G$, the Gibbs energy of the total reaction taking place in the Daniell cell. If the temperature dependence of $\Delta_r G$ can experimentally be estimated, the entropy of the electrochemical reaction, $\Delta_r S$, can be calculated as follows:

$$\Delta_r S = -\left(\frac{\partial \Delta_r G}{\partial T}\right)_P. \tag{4.2}$$

Equations 4.1 and 4.2 have been widely used in science to estimate $\Delta_r G$, $\Delta_r S$, and $\Delta_r H$ of chemical reactions employing a variety of electrochemical cells operating in the equilibrium mode. Note that only two of the quantities of Equation 4.1 are independent, and the third one depends on two others. Also, any other thermodynamic characteristic can be defined if $\Delta_r G$ is measured as a function of temperature and pressure. For example, the isobaric heat capacity, $\Delta_r C_P$, and volume (change) of a reaction, $\Delta_r V$, can be calculated as follows:

$$\Delta_r C_P = -T\left(\partial^2 \Delta_r G / \partial T^2\right)_P, \tag{4.3}$$

$$\Delta_r V = \left(\partial \Delta_r G / \partial P\right)_T, \tag{4.4}$$

or Helmholtz energy, $\Delta_r A$, can be calculated as $\Delta_r A = \Delta_r G - P\, \Delta_r V = \Delta_r G - P (\partial \Delta_r G / \partial P)_T$.

4.3 EQUILIBRIUM BETWEEN PHASES IN AN ELECTROCHEMICAL CELL

Earlier, in this book in Section 1.10, it was shown that if two phases are in equilibrium (e.g., dissolved electrolyte and solid salt of this electrolyte or electrolyte solvent and solvent in vapor phase), the Gibbs energies of formation of the solution components should be the same in both phases. Fundamentally, there is equilibrium between phases α and β if the chemical potentials (the Gibbs energies of formation) of any of the ith component in both phases are equal:

$$\mu_i^\alpha = \mu_i^\beta \quad \text{or} \quad \Delta_f G_i^\alpha = \Delta_f G_i^\beta. \tag{4.5}$$

In an electrochemical system with phases α and β, the criterion of equilibrium should be modified due to different electric potentials φ^α and φ^β, respectively, in phases α

and β. As an example, let us consider the zinc electrode of the Daniell cell, which was introduced by John Frederic Daniell in 1836.

John Frederic Daniell (1790–1845) was an English scientist who is well known for his invention of the Daniell cell, which is one of the earliest prototypes of electric batteries. The Daniell cell consists of two metal, copper and zinc, electrodes and is a classical electrochemical system commonly used in electrochemical education.

REFERENCES

A.J. Bard, G. Inzelt, and F. Scholz, *Electrochemical Dictionary*, Springer, Berlin, Germany, 2008.
http://en.wikipedia.org/wiki/John_Frederic_Daniell

Think of a piece of zinc metal being dipped into a dilute solution of $ZnSO_4(aq)$. Between the solution and metal phases, aqueous zinc ions, $Zn^{2+}(aq)$, can be transferred. If the initial solution is extremely dilute, then the rate of transfer of ions from the metal to the solution is faster than the transfer from the solution to the metal. When $Zn^{2+}(aq)$ leaves the metal surface, electrons are left behind because they cannot enter the solution. This builds up a negative electric potential in the metal phase. After some time, an equilibrium state is reached between the so-called electrochemical potential of $Zn^{2+}(aq)$ within the metal and solution phases. The electrochemical potential of the species in each phase comprises two components: (1) the chemical potential of the species μ_i and (2) the electric potential of the phase, φ, multiplied by $z_i F$. Therefore, in an electrochemical system, there is equilibrium between phases α

and β if the electrochemical potentials of any of the ith components in both phases are equal:

$$\mu_i^\alpha + z_i F \varphi^\alpha = \mu_i^\beta + z_i F \varphi^\beta, \tag{4.6}$$

where the left-hand part represents the electrochemical potential of the ith charged species (e.g., an ion or electron) in phase α and the right-hand part is the electrochemical potential of the same ith species in phase β. When phases α and β are in equilibrium, then the difference between electric potentials is defined as follows:

$$\varphi^\alpha - \varphi^\beta = \left(\mu_i^\beta - \mu_i^\alpha\right)\!\big/\!\left(z_i F\right) = \left(\Delta_f G_i^\beta - \Delta_f G_i^\alpha\right)\!\big/\!\left(z_i F\right), \tag{4.7}$$

and is referred to as the Galvani potential difference $\Delta\varphi = \varphi^\alpha - \varphi^\beta$, named after Luigi Aloisio Galvani who was one of the first electrochemists who carried out bioelectrochemical experiments in the second half of the eighteenth century.

Luigi Aloisio Galvani (1737–1798) was an Italian scientist who was the first to introduce bioelectrochemistry to study the nature of electrical effects on animal tissues.

REFERENCES

A.J. Bard, G. Inzelt, and F. Scholz, *Electrochemical Dictionary*, Springer, Berlin, Germany, 2008.
http://en.wikipedia.org/wiki/Luigi_Galvani

Because thermodynamic properties of a single ion are not known, the absolute value of $\Delta\varphi$ is not known either. The measurable value is the difference, $E_{eq} = \Delta\varphi - \Delta\varphi'$, where $\Delta\varphi'$ is the Galvani potential difference of a conventionally selected electrode. The conventionally selected electrode is the hydrogen electrode, which consists of adsorbed molecular (or atomic) hydrogen on platinum in equilibrium with a gas phase (phase α) hydrogen with fugacity, $f_{H_2(g)}$, as well as with aqueous protons in an aqueous solution (phase β) with activity $a_{H^+(aq)}$. $\Delta\varphi'$ of the hydrogen electrode is

$$\Delta\varphi' = E_{eq\,H^+/H_2} = \frac{\mu_{H^+(aq)} - 0.5\mu_{H_2(g)}}{z_i F} = \frac{\Delta_f G_{H^+(aq)} - 0.5\Delta_f G_{H_2(g)}}{z_i F}, \tag{4.8}$$

where
   phase α is gaseous molecular hydrogen.
   phase β is an acidic aqueous solution.
   $z_i = +1$, because of the unit charge of the proton.

The stoichiometric coefficient 0.5 in Equation 4.8 is needed to keep the conservation of matter in the equilibrium reaction between $H_2(g)$ and $H^+(aq)$: $H^+(aq) + e^- = 0.5H_2(g)$. The Gibbs energy of formation of electrons in any metal is conventionally decided to be zero, and therefore, μ_{e^-} or $\Delta_f G_{e^-}$ does not appear in Equation 4.8.

4.4   GALVANI POTENTIAL OF THE HYDROGEN ELECTRODE

Based on Chapter 1, the Gibbs energies of formation of gaseous hydrogen and aqueous protons are defined as follows:

$$\Delta_f G_{H^+(aq)} = \Delta_f G^0_{H^+(aq)} + RT \ln\left(a_{H^+(aq)}\right) = \Delta_f G^0_{H^+(aq)} + RT \ln\left(\gamma_{H^+(aq)} b_{H^+(aq)}/b^0\right), \tag{4.9}$$

$$\Delta_f G_{H_2(g)} = \Delta_f G^0_{H_2(g)} + RT \ln\left(a_{H_2(g)}\right) = \Delta_f G^0_{H_2(g)} + RT \ln\left(\frac{f_{H_2(g)}}{f^0}\right), \tag{4.10}$$

where both $\Delta_f G^0_{H^+(aq)}$ and $\Delta_f G^0_{H_2(g)}$ are, in principle, not known and, therefore, are conventionally accepted to be zero at any temperature and pressure:

$$\Delta_f G^0_{H^+(aq)} = \Delta_f G^0_{H_2(g)} = 0\left(\text{at any } T \text{ and } P\right). \tag{4.11}$$

Now the Galvani potential of the hydrogen electrode $\Delta\varphi'$ can be referred to as the (equilibrium) potential of the hydrogen electrode, $E_{eq,H^+/H_2}$, and is experimentally defined as follows:

$$\Delta\varphi' = E_{eq},$$

$$= E^0_{eq,H^+/H_2} + \frac{RT}{z_i F} \ln\left[\frac{a_{H^+(aq)}}{a_{H_2(g)}^{0.5}}\right], \tag{4.12}$$

where $E^0_{eq,H^+/H_2}$ is the standard potential of the hydrogen electrode, which is mathematically equal to $E_{eq,H^+/H_2}$ when both $a_{H_2(g)} = f_{H_2(g)}/f^0$ ($f^0 = 1$ bar) and $a_{H^+ (aq)}$ equal 1. Also, $E^0_{eq,H^+/H_2}$ can be defined by the standard values of the Gibbs energies of formation as follows:

$$E^0_{eq,H^+/H_2} = \frac{\Delta_f G^0_{H^+ (aq)} - 0.5\Delta_f G^0_{H_2(g)}}{z_i F},$$ (4.13)

which equals zero for the hydrogen electrode at any temperature and pressure.

Equation 4.12 represents the well-known Nernst equation for the hydrogen electrode corresponding to the following reduction half-reaction:

$$H^+(aq) + e^- = 0.5H_2(g).$$ (4.14)

In conclusion, the Galvani potential of the hydrogen electrode defined by Equations 4.12 and 4.13 can experimentally be measured in dilute aqueous solutions at a relatively low pressure of hydrogen to accurately define $a_{H+ (aq)}$ and $a_{H_2(g)}$ in Equation 4.12 and, then, to carry out an extrapolation to the infinitely dilute solution to provide the standard state conditions of $a_{H_2(g)} = a_{H+ (aq)} = 1$. An example of such extrapolation is presented in Section 5.11 and Figure 5.3.

4.5 GIBBS ENERGY OF REACTION AND EQUILIBRIUM ELECTRODE POTENTIAL

Based on Chapter 1 and fundamentals of thermodynamics, the Gibbs energy of reaction (4.14) can be presented using the Gibbs energies of formation of gaseous hydrogen $\left(\Delta_f G_{H_2(g)}\right)$, aqueous protons ($\Delta_f G_{H^+ (aq)}$), and electrons ($\Delta_f G_{e^-}$):

$$\Delta_r G = 0.5\Delta_f G_{H_2(g)} - \Delta_f G_{H^+ (aq)} - \Delta_f G_{e^-} = 0.5\Delta_f G_{H_2(g)} - \Delta_f G_{H^+ (aq)},$$ (4.15)

where $\Delta_f G_{e^-}$ is conventionally accepted to be zero. In other words, the electron activity is always 1, and its standard Gibbs energy of formation is always zero. It is also known from thermodynamics that the Gibbs energy of a reaction ($\Delta_r G$) can generally be presented by two terms: the standard Gibbs energy of the reaction ($\Delta_r G^0$) and a logarithmic term, $RT \ln\left[\Pi a_i^{v_i}\right]$ as follows:

$$\Delta_r G = \Delta_r G^0 + RT \ln\left[\Pi a_i^{v_i}\right],$$ (4.16)

which for reaction (4.14) can be written as follows:

$$\Delta_r G = \Delta_r G^0 + RT \ln\left[\frac{\left(a_{H_2(g)}\right)^{0.5}}{a_{H^+ (aq)}}\right],$$ (4.17)

where $\Delta_r G^0 = 0.5 \Delta_f G^0_{H_2(g)} - \Delta_f G^0_{H^+ (aq)}$, taking into account that the activity of electrons is conventionally 1. Now comparing Equations 4.12 and 4.17, the fundamental relationships between E_{eq} and $\Delta_r G$ as well as between E^0_{eq} and $\Delta_r G^0$ can be obtained:

$$E_{eq} = \frac{-\Delta_r G}{nF},\qquad(4.18)$$

$$E_{eq}^0 = \frac{-\Delta_r G^0}{nF},\qquad(4.19)$$

where n is the electron number in the corresponding electrochemical half-reaction.

In summary, Equations 4.18 and 4.19 allow the equilibrium electrode potential and its standard value to be calculated using thermodynamic properties of the participating chemicals or thermodynamic properties can be obtained by carrying out the corresponding electrochemical measurements.

4.6   NERNST EQUATION

By generalizing Equation 4.12, the key equation of equilibrium electrochemistry, the Nernst equation, can be written as follows:

$$E_{eq} = E_{eq}^0 + \frac{RT}{nF}\ln\left[\frac{\Pi\left(a_{i,R}^{v_i}\right)}{\Pi\left(a_{i,P}^{v_i}\right)}\right],\qquad(4.20)$$

where in the square brackets, the activities of reactants, $a_{i,R}$, are in the numerator, activities of products $a_{i,P}$ are in the denominator, and all stoichiometric coefficients are positive numbers. Walther Hermann Nernst derived this equation in 1888. It can be easily shown that Equation 4.20 can be used for both the electrochemical half-reactions and the total reaction. In the second case, the electron number of both half-reactions should be known. Note that in an electrochemical cell, the right-hand and left-hand half-reactions should always have the same electron number.

Walther Hermann Nernst (1864–1941) was a German physical chemist who is known for his theories behind the calculation of chemical affinity as embodied in the third law of thermodynamics, for which he won the 1920 Nobel Prize in Chemistry. Nernst also made fundamental contributions to the theory of electrolyte solutions. He is most known for developing the Nernst equation, one of the most fundamental equations of equilibrium electrochemistry.

REFERENCES

A.J. Bard, G. Inzelt, and F. Scholz, *Electrochemical Dictionary*, Springer, Berlin, Germany, 2008.
http://en.wikipedia.org/wiki/Walther_Nernst

Here, we provide a recipe to properly compose a Nernst equation for an electrochemical cell:

1. Write down the electrochemical half-reactions showing chemicals, stoichiometric coefficients (v_i), and phases of all chemical components (reactants and products).
2. Define n, ensuring that the amount of electrons in both half-reactions is same.
3. Define all concentrations and activity coefficients for chemical components.
4. Calculate E_{eq}^0 from the standard Gibbs energy of reaction $\Delta_r G^0$ by employing Equation 4.19. $\Delta_r G^0$ can be calculated as follows:

$$\Delta_r G^0 = \Sigma v_i \Delta_f G_i^0 \left(\text{products}\right) - \Sigma v_i \Delta_f G_i^0 \left(\text{reactants}\right). \tag{4.21}$$

Using Equation 4.21, one should remember that the standard Gibbs energy of formation of any element in its reference form is zero as well as the standard Gibbs energy of formation of $H^+(aq)$ is zero.
5. Calculate the equilibrium (open circuit) potential, E_{eq}, of the electrochemical cell using Equation 4.20.

An example of applying this recipe is given in Section 4.11. Note that for simplicity in the following text, E will be used in place of E_{eq} and E^0 in place of E_{eq}^0.

4.7   STANDARD HYDROGEN ELECTRODE (SHE)

The electrode shown in Figure 4.1 is the most important one for all of equilibrium electrochemistry. The potential of this electrode is stable and reproducible. The electrochemical half-reaction taking place at the electrode is $H^+(aq) + e^- = 0.5H_2(g)$, which is the reaction of the reduction of protons, $H^+(aq)$, to molecular hydrogen, $H_2(g)$.

In the SHE, platinum is used because it is one of the most chemically stable and noncorrosive metals. In addition, it has high catalytic activity to speed up the electrochemical half-reaction and shows excellent reproducibility of the electrode potential. The SHE surface should be platinized (i.e., covered with platinum black) to increase the real electrode surface and capability to adsorb hydrogen on the electrode

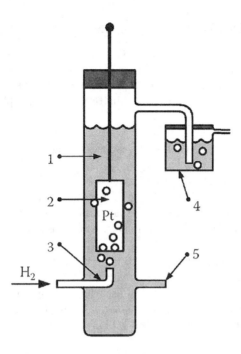

FIGURE 4.1 Standard hydrogen electrode (SHE): (1) acidic solution with the hypothetical unit activity of H^+(aq), (2) platinized platinum electrode, (3) hydrogen blow, (4) hydro-seal for the prevention of air interference, and (5) a salt bridge through which another electrode can be attached to make an electrochemical cell.

surface. Hydrogen is not an electrically conductive substance and, therefore, should be adsorbed on the electrically conductive material (Pt) to be able to participate in forming the electrode potential.

The IUPAC definition of the SHE is as follows [1]: "The SHE consists of a Pt electrode in contact with a solution of H^+ at unit activity and saturated with H_2 gas with a fugacity referred to the standard pressure of 10^5 Pa." Clearly, the solution and H_2 gas are hypothetical in this definition, so nonideality of both should be taken into account if the SHE is to be used in an experiment. For the H_2 gas, it can easily be done using the H_2 fugacity coefficients to be calculated from the van der Waals equation with constants available from (Table 10.24). For the aqueous solution, a few experiments should be carried out using relatively dilute acidic solutions (e.g., HCl(aq)), and then, an extrapolation should be carried out to the infinitely diluted solution as described in Section 5.11.

Despite a number of advantages, the SHE has some disadvantages that do not allow for routine use in common electrochemical studies. The main disadvantages are as follows: (1) the electrode design is relatively complicated, (2) SHE is quite expensive due to using Pt, (3) Pt surface can be easily contaminated by impurities in H_2(g) and/or in aqueous solution, and (4) H_2 is quite a hazardous substance. Therefore, other reference electrodes were developed such as Ag/AgCl (silver–silver chloride) and Hg/Hg_2Cl_2 (calomel) electrodes.

4.8   Ag/AgCl REFERENCE ELECTRODE

The silver–silver chloride electrode, shown in Figure 4.2, is the most practical reference electrode and is widely used in a variety of electrochemical studies. The potential of this electrode is stable and reproducible, and can thermodynamically be calculated. The reduction electrochemical half-reaction taking place at the Ag/AgCl reference electrode is

$$AgCl(s) + e^- = Ag(s) + Cl^-(aq),$$

so a chloride ion of an aqueous solution must be in contact with the electrode for proper operation. One key advantage of the Ag/AgCl reference electrode is that it can be easily and safely prepared in the lab with a reasonable potential reproducibility of around ±1 mV or even less. This precision is usually sufficient for the most applications.

To make the electrode using a thermal method, silver powder, a powder of silver chloride, and a silver wire are the only materials required. Anodic electrochemical deposition of silver chloride can also be used for making this reference electrode. The Ag/AgCl reference electrode is commercially available and if properly maintained can provide a precision of around ±0.1 mV.

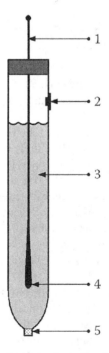

FIGURE 4.2   Reference Ag/AgCl electrode: (1) Ag(s) wire, (2) filling hole, (3) KCl(aq) solution (usually saturated), (4) Ag(s) wire coated with AgCl(s), (5) ceramic, quartz, or glass fiber junction.

4.9 HARNED CELL

If an electrochemical cell consists of an SHE and a Ag/AgCl electrode imbedded in the same solution of HCl(aq), that is, a cell without transfer, then this cell is called the Harned cell. The Harned cell can be used for measuring the standard thermodynamic properties and mean activity coefficients of HCl(aq) without any complications/uncertainties that can appear due to a salt bridge and corresponding diffusion potential in the cell with transfer. The traditional electrochemical diagram of the Harned cell is as follows:

$$Cu(s)|Pt(s)|H_2(g)|HCl(aq)|AgCl(s)|Ag(s)|Cu(s),$$

and the new diagram can be presented as

$$\begin{array}{cc} H_2(g) & AgCl(s) \\ Cu(s) \mid Pt(s) \times HCl(aq) \times Ag(s) \mid Cu(s). \end{array}$$

It can be seen from this diagram that the cell electrolyte, HCl(aq), has a cation, $H^+(aq)$, which is electrochemically active with respect to the left-hand electrode and an anion, $Cl^-(aq)$, which is electrochemically active with respect to the right-hand electrode. The reduction half-reactions in the cell were given in the previous sections and are $H^+(aq) + e^- = 0.5H_2(g)$ for the left-hand (L) electrode and $AgCl(s) + e^- = Ag(s) + Cl^-(aq)$ for the right-hand (R) electrode. The corresponding Nernst equations for these half-reactions are ($n = 1$)

$$E_L = E^0_{H^+(aq)/H_2(g)} + \frac{RT}{F} \ln \left[\frac{a_{H^+(aq)}}{\left(a_{H_2(g)}\right)^{0.5}} \right].$$

$$E_R = E^0_{AgCl(s)/Ag(s),Cl^-(aq)} + \frac{RT}{F} \ln \left(\frac{a_{AgCl(s)}}{a_{Cl^-(aq)} a_{Ag(s)}} \right).$$

The potential difference, E, between the right-hand and left-hand electrodes is as follows:

$$E = E_R - E_L = E^0_{AgCl(s)/Ag(s),Cl^-(aq)} - E^0_{H^+(aq)/H_2(g)} + \frac{RT}{F} \ln \left(\frac{a^{0.5}_{H_2(g)}}{a_{Cl^-(aq)} a_{H^+(aq)}} \right),$$

or

$$E = E^0 + \frac{RT}{F} \ln \left[\frac{a^{0.5}_{H_2(g)}}{\left(\gamma_{\pm} b_{HCl(aq)} / b^0 \right)^2} \right], \tag{4.22}$$

where $E^0 = E^0_{AgCl(s)/Ag(s),Cl^-(aq)} - E^0_{H^+(aq)/H_2(g)}$ and activities of pure solid phases are equal to 1, that is, $a_{AgCl(s)} = a_{Ag(s)} = 1$. Equation 4.22 can be used either to experimentally find E^0 by carrying out the measurements in very dilute solutions where the mean activity coefficient can be modeled by Debye–Hückel equation or for measuring the mean activity coefficient if E^0 can be calculated using the corresponding Gibbs energies of formations (Tables 10.3 through 10.5) as follows:

$$E^0 = -\Delta_r G_R^0 / F + \Delta_r G_L^0 / F$$

$$= \left(1/F\right)\left(\Delta_f G_{AgCl(s)}^0 - \Delta_f G_{Ag(s)}^0 - \Delta_f G_{Cl^-(aq)}^0 + 0.5\Delta_f G_{H2(g)}^0 - \Delta_f G_{H+(aq)}^0\right)$$

$$= \left(1/F\right)\left(\Delta_f G_{AgCl(s)}^0 - \Delta_f G_{Cl^-(aq)}^0\right) = 0.2218\,V.$$

4.10 ELECTROCHEMICAL SERIES

The electrochemical series (Table 10.13) or, in fact, a list of the standard electrode potentials for reduction half-reactions can be very useful to quickly calculate the standard value of the potential. An example of the electrochemical series just for a few electrochemical half-reactions taken from Table 10.13 is given in Table 4.1.

TABLE 4.1

Selected Standard Electrode Potentials in Alphabetical Order and the Order from the Most Positive in the Top to the Most Negative in the Bottom for Demonstrating So-Called Electrochemical Series

Reduction Half-Reactions	E^0 (V)
E^0 Values Ordered Alphabetically	
$Ag^+(aq) + e^- = Ag(s)$	+0.800
$AgCl(s) + e^- = Ag(s) + Cl^-(aq)$	+0.222
$Ag_2O(s) + H_2O(l) + 2e^- = 2Ag(s) + OH^-(aq)$	+0.342
$Al^{3+}(aq) + 3e^- = Al(s)$	−1.676
$Au^+(aq) + e^- = Au(s)$	+1.692
$Cd^{2+}(aq) + 2e^- = Cd(s)$	−0.403
$Cl_2(g) + 2e^- = 2Cl^-(aq)$	+1.358
$Cu^+(aq) + e^- = Cu(s)$	+0.521
$Cu^{2+}(aq) + e^- = Cu^+(aq)$	+0.153
$Cu^{2+}(aq) + 2e^- = Cu(s)$	+0.342
$Fe^{2+}(aq) + 2e^- = Fe(s)$	−0.447
$Fe^{3+}(aq) + 3e^- = Fe(s)$	−0.037
$Fe^{3+}(aq) + e^- = Fe^{2+}(aq)$	+0.771
$[Fe(CN)_6]^{3-} + e^- = [Fe(CN)_6]^{4-}$	+0.358
$2H^+(aq) + 2e^- = H_2(g)$	+0.000
$Hg_2Cl_2(s) + 2e^- = 2Hg(s) + 2Cl^-(aq)$	+0.268
$HgO(s) + H_2O(l) + 2e^- = Hg(l) + 2OH^-(aq)$	+0.098

(Continued)

TABLE 4.1 (*Continued*)

Selected Standard Electrode Potentials in Alphabetical Order and the Order from the Most Positive in the Top to the Most Negative in the Bottom for Demonstrating So-Called Electrochemical Series

Reduction Half-Reactions	E^0 (V)
$0.5O_2(g) + 2H^+(aq) + 2e^- = H_2O(l)$	+1.229
$Ti^{2+}(aq) + 2e^- = Ti(s)$	−1.628
$Zn^{2+}(aq) + 2e^- = Zn(s)$	−0.762

E^0 Values Ordered from the Most Positive in the Top to the Most Negative in the Bottom	
$Au^+(aq) + e^- = Au(s)$	+1.692
$Cl_2(g) + 2e^- = 2\,Cl^-(aq)$	+1.358
$0.5O_2(g) + 2H^+(aq) + 2e^- = H_2O(l)$	+1.229
$Ag^+(aq) + e^- = Ag(s)$	+0.800
$Fe^{3+}(aq) + e^- = Fe^{2+}(aq)$	+0.771
$Cu^+(aq) + e^- = Cu(s)$	+0.521
$[Fe(CN)_6]^{3-} + e^- = [Fe(CN)_6]^{4-}$	+0.358
$Ag_2O(s) + H_2O(l) + 2e^- = 2Ag(s) + OH^-(aq)$	+0.342
$Cu^{2+}(aq) + 2e^- = Cu(s)$	+0.342
$Hg_2Cl_2(s) + 2e^- = 2Hg(s) + 2Cl^-(aq)$	+0.268
$AgCl(s) + e^- = Ag(s) + Cl^-(aq)$	+0.222
$Cu^{2+}(aq) + e^- = Cu^+(aq)$	+0.153
$HgO(s) + H_2O(l) + 2e^- = Hg(l) + 2OH^-(aq)$	+0.098
$2H^+(aq) + 2e^- = H_2(g)$	+0.000
$Fe^{3+}(aq) + 3e^- = Fe(s)$	−0.037
$Cd^{2+}(aq) + 2e^- = Cd(s)$	−0.403
$Fe^{2+}(aq) + 2e^- = Fe(s)$	−0.447
$Zn^{2+}(aq) + 2e^- = Zn(s)$	−0.762
$Ti^{2+}(aq) + 2e^- = Ti(s)$	−1.628
$Al^{3+}(aq) + 3e^- = Al(s)$	−1.676

Note: The full list of reactions is given in Table 10.13.

The potential difference in an electrochemical cell at 25°C and 1.013 bar (1 atm) can be calculated by simply subtracting the left-hand standard electrode potential from the right-hand one. For example, using Table 4.1 data, the standard value of the potential difference of the Harned cell, E^0, can be calculated as follows:

$$E^0 = E^0_{AgCl(s)/Ag(s),Cl^-(aq)} - E^0_{H^+(aq)/H_2(g)} = 0.222 - 0.0 = 0.222\,V.$$

The obtained value is practically the same as previously obtained in Section 4.9 (0.2218 V). However, the thermodynamically calculated standard potentials can be slightly different from the standard potentials in Table 10.13 for two reasons: (1) marginally different pressures (1 and 1.013 bar) are used in these calculations, and (2) different sources of the reference data usually give slightly different results.

An important conclusion to be made here is related to the definition of the standard electrode potential given in the IUPAC manual [1]:

The standard potential of an electrochemical reaction, abbreviated as standard potential, is defined as the standard potential of a hypothetical cell, in which the electrode (half-cell) on the left of the cell diagram is the SHE and the electrode at the right is the electrode in question.

Note that E^0 of a half-reaction (or a total electrochemical reaction) as an intensive variable does not depend on the number of electrons used in the half-reaction.

4.11 CALCULATION OF EQUILIBRIUM POTENTIAL OF THE DANIELL CELL

Let us implement the earlier-described definitions and approach to calculate the equilibrium potential of the Daniell cell at 25°C and 1 bar when concentrations of the cell electrolytes are well defined and relatively low in order to use the Debye–Hückel theory for calculating the individual activity coefficients of the electrochemically active ions. In this example, the Daniel cell diagram is as follows:

$$Cu(s)|Zn(s)|ZnSO_4\left(aq, 0.001\,mol\,kg^{-1}\right)||CuSO_4\left(aq, 0.005\,mol\,kg^{-1}\right)|Cu(s).$$

For the Daniel cell, there is no need to use the new type of diagram because (1) only three species participate in both half-reactions, (2) two of these three, metal atoms and electrons, in each of the half-reactions are in the same phase, and (3) there are only two-phase boundaries in this electrochemical cell. The double vertical line in the diagram shows that the potential difference between $ZnSO_4$ (aq, 0.001 mol kg^{-1}) and $CuSO_4$ (aq, 0.005 mol kg^{-1}), called the diffusion potential (see Chapter 5), is somehow eliminated.

Our calculations will include a number of steps:

Step 1: Write the half-reactions in the reduction direction for both the left-hand (L) and right-hand (R) half-reactions:

$$(L)\,Zn^{2+}(aq) + 2e^- = Zn(s),$$

$$(R)\,Cu^{2+}(aq) + 2e^- = Cu(s).$$

Step 2: Make sure to balance the number of electrons, n, and chemical atoms in each reaction and have the same number of electrons in each of the half-reactions:
Here, the number of electrons $n = 2$.
Step 3: Define the species concentrations and activity coefficients:
Because of the low concentrations of electrolytes, we can calculate the activity coefficients of Zn^{2+}(aq) and Cu^{2+}(aq) using the Debye–Hückel

limiting law (Chapter 1) and then calculate the ion activities ignoring any association between ions:

$$\gamma_{Cu^{2+}(aq)} = 0.515, \gamma_{Zn^{2+}(aq)} = 0.743,$$

$$a_{Cu^{2+}(aq)} = \frac{b_{Cu^{2+}(aq)}\gamma_{Cu^{2+}(aq)}}{b^0} = \frac{0.005 \times 0.515}{1} = 2.575 \times 10^{-3},$$

$$a_{Zn^{2+}(aq)} = \frac{b_{Zn^{2+}(aq)}\gamma_{Zn^{2+}(aq)}}{b^0} = \frac{0.001 \times 0.743}{1} = 0.743 \times 10^{-3}.$$

The activities of both metals, as pure substances, are equal to 1 because both the molar fraction and activity coefficient of a pure chemical equal 1:

$$a_{Zn(s)} = x_{Zn(s)}\gamma_{Zn(s)} = 1,$$

$$a_{Cu(s)} = x_{Cu(s)}\gamma_{Cu(s)} = 1.$$

Step 4: Calculate the standard electrode potential, E^0, for each electrode from the standard Gibbs energy of reaction $\Delta_r G^0$, or find E^0 in the electrochemical series. Using (Table 10.13 or 4.1), the standard electrode potentials are obtained:

$$E^0_{Zn^{2+}(aq)/Zn(s)} = -0.7618\,V \quad \text{and} \quad E^0_{Cu^{2+}(aq)/Cu(s)} = 0.3419\,V.$$

Using Table 10.5, the standard electrode potentials can also be calculated as follows:

$$\Delta_f G^0_{Zn^{2+}(aq)} = -147.1\,kJ\,mol^{-1},$$

$$\Delta_f G^0_{Zn(s)} = 0,$$

$$\Delta_r G^0_{Zn^{2+}(aq)/Zn(s)} = 147.1\,kJ\,mol^{-1},$$

$$E^0_{Zn^{2+}(aq)/Zn(s)} = -147,100/(2 \times 96,485) = -0.7623\,V.$$

$$\Delta_f G^0_{Cu^{2+}(aq)} = 65.5\,kJ\,mol^{-1},$$

$$\Delta_f G^0_{Cu(s)} = 0,$$

$$\Delta_r G^0_{Cu^{2+}(aq)/Cu(s)} = -65.5\,kJ\,mol^{-1},$$

$$E^0_{Cu^{2+}(aq)/Cu(s)} = 65,500/(2 \times 96,485) = 0.3394\,V.$$

Note that a small difference between thermodynamically calculated E^0 values and those collected from the electrochemical series is unavoidable due to errors in the data employed for obtaining these values. The common assumption is that 1–2 mV difference is quite acceptable and corresponds to the uncertainty in electrochemical studies of equilibrium electrochemical cells, that is potentiometric measurements (see Chapter 5).

Step 5: Calculate the equilibrium (open circuit) potential, E, of the electrochemical cell as the difference between the right-hand, E_R, and left-hand, E_L, electrode potentials:

$$E_R = E^0_{Cu^{2+}(aq)/Cu(s)} + \frac{RT}{2F} \ln \left[\frac{a_{Cu^{2+}(aq)}}{a_{Cu(s)}} \right]$$

$$E_L = E^0_{Zn^{2+}(aq)/Zn(s)} + \frac{RT}{2F} \ln \left[\frac{a_{Zn^{2+}(aq)}}{a_{Zn(s)}} \right]$$

$$
\begin{aligned}
E = E_R - E_L &= \left(E^0_{Cu^{2+}(aq)/Cu(s)} - E^0_{Zn^{2+}(aq)/Zn(s)} \right) \\
&\quad + \left[RT/(2F) \right] \ln \left[a_{Cu^{2+}(aq)} a_{Zn(s)} \Big/ \left(a_{Cu(s)} a_{Zn^{2+}(aq)} \right) \right] \\
&= E^0 + \left[RT/(2F) \right] \ln \left[a_{Cu^{2+}(aq)} a_{Zn(s)} \Big/ \left(a_{Cu(s)} a_{Zn^{2+}(aq)} \right) \right] \\
&= E^0 + \left[RT/(2F) \right] \ln \left(a_{Cu^{2+}(aq)} / a_{Zn^{2+}(aq)} \right).
\end{aligned}
\tag{4.23}
$$

The substitution of all necessary values into Equation 4.23 allows calculating the potential of the Daniell cell schematically shown by the electrochemical diagram given in the beginning of this section:

$$E = (0.3394 + 0.7623) + \frac{RT}{2F} \ln \left[\frac{2.575 \times 10^{-3}}{0.743 \times 10^{-3}} \right] = 1.1017 + 0.0160 = 1.1177\,\text{V}.$$

4.12 NERNST EQUATIONS FOR TYPICAL ELECTRODES

Following is a list of the most common types of electrodes along with their Nernst equations.

1. Metal electrodes, for example, $Cu^{2+}(aq) + 2e^- = Cu(s)$, with the electrode potential as

$$E = E^0_{Cu^{2+}(aq)/Cu(s)} + \frac{RT}{2F} \ln \left(a_{Cu^{2+}(aq)} \right).
\tag{4.24}$$

2. Redox electrodes, for example, $Fe^{3+}(aq) + e^- = Fe^{2+}(aq)$, with the electrode potential as

$$E = E^0_{Fe^{3+}(aq)/Fe^{2+}(aq)} + \frac{RT}{F} \ln \left[\frac{a_{Fe^{3+}(aq)}}{a_{Fe^{2+}(aq)}} \right].
\tag{4.25}$$

3. Gas electrodes, for example, $Cl_2(g) + 2e^- = 2Cl^-(aq)$, with the electrode potential as

$$E = E^0_{Cl_2(g)/Cl^-(aq)} + \frac{RT}{2F} \ln \left[\frac{f_{Cl_2(g)}/f^0}{a_{Cl^-(aq)}} \right]. \tag{4.26}$$

4. Metal/salt electrodes, for example, $Hg_2Cl_2(s) + 2e^- = 2Hg(l) + 2Cl^-(aq)$, with the electrode potential as

$$E = E^0_{Hg_2Cl_2(s)/Hg(l),Cl^-(aq)} + \frac{RT}{2F} \ln \left[\frac{1}{\left(a_{Cl^-(aq)}\right)^2} \right]. \tag{4.27}$$

5. Metal/oxide electrodes, for example, $HgO(s) + H_2O(l) + 2e^- = Hg(l) + 2OH^-(aq)$, with the electrode potential as

$$E = E^0_{HgO(s),H_2O(l)/Hg(l),OH^-(aq)} + \frac{RT}{2F} \ln \left[\frac{a_{H_2O(l)}}{\left(a_{OH^-(aq)}\right)^2} \right]. \tag{4.28}$$

The standard electrode potentials of these electrodes (half-reactions) can be found in Table 4.1. Note that $a_{H_2O(l)}$ in Equation 4.28 is very close to 1 in dilute solution. However, in highly concentrated solutions, $a_{H_2O(l)}$ can significantly deviate from 1, and this should be taken into account in calculating the electrode potential. Also, in Equation 4.26, $f_{Cl_2(g)}/f^0$, can be close to 1 if the gas electrode is used at ambient pressure.

4.13   ANOTHER EXAMPLE OF ESTIMATING OF E^0 AND E

Let us estimate E and E^0 of the H_2/Cl_2 galvanic cell shown in Figure 2.2. The traditional electrochemical diagram

$$Cu(s)|Pt(s)|H_2(g)|HCl(aq)|Cl_2(g)|Pt(s)|Cu(s),$$

does not show the three-phase boundaries at both electrodes, but the new diagram

$$\begin{array}{ccc} & H_2(g) & Cl_2(g) \\ \text{Anode } (-) \text{ Pt(s)} & \times \quad HCl(aq) \quad \times & Pt(s) \ (+) \text{ Cathode,} \end{array}$$

definitely does and suggests the following half-reactions:

$$(L)\ 2H^+(aq) + 2e^- = H_2(g),$$

$$(R)\ Cl_2(g) + 2e^- = 2Cl^-(aq).$$

The total reaction in the galvanic cell can be obtained as the difference between the right-hand and left-hand half-reactions:

$$Cl_2(g) + H_2(g) = 2Cl^-(aq) + 2H^+(aq).$$

The standard cell potential can be found using Table 4.1 as follows:

$$E_L^0 = E_{H^+(aq)/H_2(g)}^0 = 0.0\,V,$$

$$E_R^0 = E_{Cl_2(g)/Cl^-(aq)}^0 = 1.358\,V,$$

$$E^0 = E_{Cl_2(g)/Cl^-(aq)}^0 - E_{H^+(aq)/H_2(g)}^0 = 1.358 - 0.0 = 1.358\,V.$$

The expression for the cell potential difference, $E = E_R - E_L$, of the H_2/Cl_2 galvanic cell is as follows:

$$E = E_R - E_L = E^0 + \left[RT/(2F) \right] \ln\left[a_{Cl_2(g)} a_{H_2(g)} \Big/ \left(b_{HCl(aq)} \gamma_{\pm} / b^0 \right)^4 \right],$$

where $a_{Cl_2(g)} = f_{Cl_2(g)}/f^0$ and $a_{H_2(g)} = f_{H_2(g)}/f^0$, $\gamma_{\pm}$ is the mean activity coefficient of HCl(aq). Derivation of the above equation is given in Section 4.19.

Note that the potential difference of the H_2/Cl_2 galvanic cell can be calculated at any concentration of HCl(aq) where $\gamma_{\pm}$ is available. For example, in Table 10.17, such values can be found at concentrations up to $10\,mol\,kg^{-1}$, and at this concentration, $\gamma_{\pm} = 10.4$, which suggests the HCl(aq) activity equals 104. The fugacity of $Cl_2(g)$ or $H_2(g)$ at ambient pressure is very close to the partial pressure, so the partials pressures of $Cl_2(g)$ or $H_2(g)$ can be used instead of their fugacities, that is, $f_{Cl_2(g)} = p_{Cl_2(g)} \gamma_{Cl_2(g)} \approx p_{Cl_2(g)}$ and $f_{H_2(g)} = p_{H_2(g)} \gamma_{H_2(g)} \approx p_{H_2(g)}$ without making any significant error.

In conclusion, if partial pressures of $Cl_2(g)$ or $H_2(g)$ and the molality of HCl(aq) are known, the cell potential of the H_2/Cl_2 galvanic cell can be precisely calculated using the experimentally obtained mean activity coefficient of HCl(aq), which is known in a wide range of concentrations.

4.14 CALCULATION OF E^0 AT ELEVATED TEMPERATURE

Most of the electrochemical energy conversion systems such as batteries, fuel cells, and electrolyzers should be operating at temperatures above or below 25°C. Using the well-known approach for calculating the standard Gibbs energy of reaction at an elevated (reduced) temperature, the electrode potential difference can be defined applying the following methods:

Method 1: Use the standard Gibbs energies of formation that are already calculated over a wide range of temperatures (Table 10.4), and apply

Equation 4.21 for calculating $\Delta_r G^0$ and, then, Equation 4.19 for calculating E^0. This approach is particularly simple and requires only making a linear interpolation of the data from (Table 10.4) to a temperature of interest. The disadvantage of the method is that there are a limited number of chemicals given in Ref. [2]. Also, the standard Gibbs energies of formation at elevated temperature are usually calculated for chemicals in gas and solid phases but not for aqueous species.

Method 2: If the temperature is below about 150°C, the approximation that the standard heat capacity of reaction does not change, $\Delta_r C_P^0 = \Delta_r C_{P.T_0}^0$, can be used. In this case, Equation 4.21 can be combined with Equations 4.1 through 4.3 to derive the following expression:

$$\Delta_r G^0 \approx \Delta_r G_{T_0}^0 - \Delta_r S_{T_0}^0 (T - T_0) + \Delta_r C_{P.T_0}^0 \left[(T - T_0) - T \ln\left(T/T_0\right)\right], \quad (4.29)$$

where $T_0 = 298.15$ K, and $\Delta_r G_{T_0}^0$, $\Delta_r S_{T_0}^0$, and $\Delta_r C_{P.T_0}^0$ can be calculated using Table 10.3 for gases, liquids, and solids and Table 10.5 for aqueous species. Note that Equation 4.29 is easy to obtain from Equation 4.30 assuming that the standard heat capacity of reaction is not a function of temperature.

Method 3: If the temperature is above 150°C, the most accurate thermodynamic equation should be used:

$$\Delta_r G^0 = \Delta_r G_{T_0}^0 - \Delta_r S_{T_0}^0 (T - T_0) + \int_{T_0}^{T} \Delta_r C_P^0 dT - T \int_{T_0}^{T} \left(\Delta_r C_P^0/T\right) dT, \quad (4.30)$$

where $\Delta_r C_P^0$ is a function of temperature, and therefore, the temperature dependence of $\Delta_r C_P^0(T)$ should be known for the whole temperature range from T_0 to T. Note that methods 1 and 3 are equivalent and should give the same results inside of the calculation uncertainty if an accurate $\Delta_r C_P^0(T)$ dependence is known. Derivation of Equation 4.30 is shown in Section 4.19.

Method 4: If the temperature is slightly elevated above 25°C but still below about 75°C, the two last terms in Equation 4.30 can be canceled out because of a similar magnitude and opposite sign of the last two terms:

$$\Delta_r G^0 \approx \Delta_r G_{T_0}^0 - \Delta_r S_{T_0}^0 (T - T_0) = \Delta_r H_{T_0}^0 - \Delta_r S_{T_0}^0 T. \quad (4.31)$$

Note that the right-hand part of Equation 4.31 can be very confusing. It could misguidedly appear that $\Delta_r G^0$ can be calculated at any elevated temperature using only $\Delta_r H_{T_0}^0$ and $\Delta_r S_{T_0}^0$. However, Equation 4.31 usually should not be used at temperatures above about 75°C.

4.15 TEMPERATURE DEPENDENCE OF THE STANDARD ELECTRODE POTENTIAL

There might be a misinterpretation in calculating the standard electrode potential at an elevated temperature. The issue will be illustrated using an example. Let us assume we need to calculate E^0 of the following half-reaction:

$$Cl_2(g) + 2e^- = 2Cl^-(aq),$$

at a temperature T, which is above 25°C. Based on the definition of the standard reduction potential, the reaction

$$Cl_2(g) + H_2(g) = 2Cl^-(aq) + 2H^+(aq),$$

should actually be considered rather than the chlorine reduction half-reaction. The standard electrode potential can be calculated using Equation 4.19, where the Gibbs energy of the reaction can be calculated using one of the methods described earlier using one of the approaches represented by Equations 4.21, 4.29, 4.30, and 4.31. In all these equations,

$$\Delta_r G_{T_0}^0 = 2\Delta_f G^0\left(Cl^-, aq, 298.15\,K\right) - \Delta_f G^0\left(Cl_2, g, 298.15\,K\right),$$

where $\Delta_f G^0(H_2) = \Delta_f G^0(H^+, aq) = 0$ at any temperature.

$$\Delta_r S_{T_0}^0 = 2S^0\left(Cl^-, aq, 298.15\,K\right) - S^0\left(Cl_2, g, 298.15\,K\right) - S^0\left(H_2, g, 298.15\,K\right),$$

where $S^0(H^+, aq, 298.15\,K) = 0$.

$$\Delta_r C_{P.T_0}^0 = 2C_P^0\left(Cl^-, aq, 298.15\,K\right) - C_P^0\left(Cl_2, g, 298.15\,K\right) - C_P^0\left(H_2, g, 298.15\,K\right),$$

where C_P^0 (H$^+$, aq, 298.15 K) = 0.

A possible mistake is when calculating $\Delta_r S_{T_0}^0$ and $\Delta_r C_{P.T_0}^0$, the values of S^0(H$_2$, g, 298.15 K) and C_P^0 (H$_2$, g, 298.15 K) are not taken into account assuming that the half-reaction, $Cl_2(g) + 2e^- = 2Cl^-(aq)$, is used instead of the whole reaction of the cell with SHE, $Cl_2(g) + H_2(g) = 2Cl^-(aq) + 2H^+(aq)$.

Note that in the preceding equations, the absolute values of entropy and isobaric heat capacity are used. As it is known from thermodynamics, the absolute values of Gibbs energy and enthalpy are not known, but the absolute values of their temperature and pressure derivatives are known. Also, we should keep in mind that the Gibbs energies and enthalpies of formation of all chemical elements in their reference state conventionally equal zero at any temperature. In addition, the absolute values of entropy, heat capacity, etc. (all other derivatives) of chemical elements in their reference state are known and depend on temperature. Finally, let us be reminded that all thermodynamic functions of aqueous proton, H$^+$(aq), and electron, e$^-$, are conventionally equal to zero at any temperature and pressure.

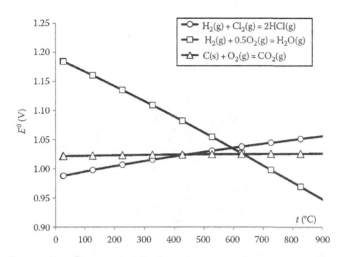

FIGURE 4.3 Temperature dependence of the standard OCP for H_2/O_2, C/O_2, and H_2/Cl_2 fuel (galvanic) cells. The calculations were carried out using the data from Table 10.4.

4.16 TEMPERATURE DEPENDENCE OF OPEN CIRCUIT AND DECOMPOSITION POTENTIALS

Equilibrium electrochemistry allows us to calculate the standard values of the open circuit potential (OCP = equilibrium potential difference) of a fuel cell and the decomposition potential (DP = −OCP) of an electrolytic cell if the thermodynamic properties required for such calculations are available. The equilibrium electrochemical calculations should be done first before any other calculations or even experimental measurements to see any thermodynamic constraints of the electrochemical system. As an example, Figure 4.3 shows results of such calculations for three fuel cell reactions over a wide temperature range from ambient up to 900°C.

Note that the temperature dependence of the standard OCP and DP is very important for estimating the efficiency of the electrochemical systems vs. efficiency of a heat engine. For example, the theoretical (thermodynamic) efficiency of a H_2/O_2 fuel cell decreases when temperature increases, and the theoretical efficiency of a C/O_2 fuel cell does not significantly change over a wide temperature range. In contrast, the thermodynamic efficiency of a heat engine always increases when temperature increases. The efficiency of fuel and electrolytic cells will be considered later in Chapter 8.

4.17 POTENTIAL–PH (POURBAIX) DIAGRAM

Knowledge of the standard electrode potentials along with applying the Nernst equation allows the construction of a special type of stability diagrams, which are called Potential–pH or Pourbaix diagrams due to Marcel Pourbaix who developed them in the 1950s and the 1960s [3].

Marcel Pourbaix (1904–1998) was a Belgian physical chemist who is known for inventing the (electric) potential–pH, better known as Pourbaix, diagrams in the 1930s. In 1963, Pourbaix produced the "Atlas of Electrochemical Equilibria," which contained potential–pH diagrams for all elements known at that time.

REFERENCES

A.J. Bard, G. Inzelt, and F. Scholz, *Electrochemical Dictionary*, Springer, Berlin, Germany, 2008.
http://en.wikipedia.org/wiki/Marcel_Pourbaix, https://www.geni.com/people/Marcel-Jean-Nestor-Pourbaix/6000000032913298725

Figure 4.4 shows a Pourbaix diagram for copper. Let us find out how this diagram can be constructed.

To construct this diagram, the following steps should be taken:

1. The upper dash line (1) corresponds to the Nernst equation, $E(pH)$ for the $O_2(g)$, $H^+(aq)/H_2O(l)$ half-reaction, $0.5O_2(g) + 2H^+(aq) + 2e^- = H_2O(l)$, with the following Nernst equation:

$$Eh = E^0_{O_2(g),\, H^+(aq)/H_2O(l)} - \left(\frac{2.303RT}{F}\right) pH,$$

where Eh is a symbol introduced to be used in constructing the electric potential–pH diagrams.

2. The lower dash line (2) corresponds to the Nernst equation, $E(pH)$ for the $H^+(aq)/H_2(g)$ half-reaction, $2H^+(aq) + 2e^- = H_2(g)$, with the following Nernst equation:

$$Eh = -\left(2.303RT/F\right)pH.$$

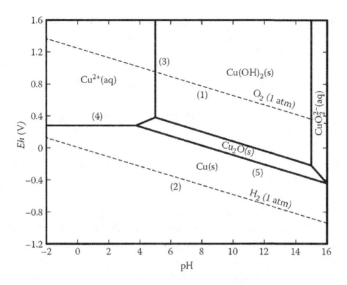

FIGURE 4.4 Simplified potential–pH (*Eh*-pH) Pourbaix diagram for Cu.

3. Any vertical line corresponds to a reaction without electrons involved. For example, line (3) corresponds to a hydrolysis reaction:

$$Cu^{2+}(aq) + 2H_2O(l) = Cu(OH)_2(s) + 2H^+(aq),$$

with the thermodynamic constant $K = a_{Cu(OH)_2(S)} \left[a_{H^+(aq)} \right]^2 / \left\{ a_{Cu^{2+}(aq)} \left[a_{H_2O(l)} \right]^2 \right\}$, which can be calculated using the standard Gibbs energies of formation [2]. Assuming $a_{Cu(OH)_2(s)} = a_{H_2O(l)} = 1$, and $a_{Cu^{2+}(aq)} = 1$ (conventionally), pH $= -\log_{10}(a_{H+(aq)})$ can be calculated and in Figure 4.4 equals about 5. Note that the activity of aqueous species in a Pourbaix diagram can also be equal to 10^{-6}, 10^{-4}, 10^{-2} [3], or any other desirable value. However, most commonly, 1 is taken as an activity of all aqueous species for composing a Pourbaix diagram.

4. A horizontal line corresponds to a reaction without any H⁺(aq) or OH⁻(aq) being involved. For example, line (4) corresponds to an electrochemical half-reaction:

$$Cu^{2+}(aq) + 2e^- = Cu(s),$$

with the following Nernst equation:

$$Eh = E^0_{Cu^{2+}(aq)/Cu(s)} + \frac{RT}{2F} \ln\left(a_{Cu^{2+}(aq)} \right), \qquad (4.24b)$$

where the standard electrode potential $E^0_{Cu^{2+}(aq)/Cu(s)}$ at 25°C is given in Table 4.1 and Table 10.13, and the activity of $a_{Cu^{2+}(aq)}$ can be taken to be equal to 1 as described earlier in Step 3.

5. An inclined line corresponds to the reaction when both electrons and protons (or OH⁻(aq) ions) are involved. For example, line (5) corresponds to an
electrochemical half-reaction:

$$Cu_2O(s) + 2H^+(aq) + 2e^- = 2Cu(s) + H_2O(l),$$

with a Nernst equation showing the same slope as in Step 2 because
$a_{Cu_2O(s)} = a_{Cu(s)} = a_{H_2O(l)} = 1$.

Note that $a_{H_2O(l)}$ should not be exactly 1 but very close to 1 if concentrations of all
dissolved species are low.

Each of the lines described in Steps 1 through 5 corresponds to the equilibrium
between the neighboring species involved in the corresponding to the line reaction,
and in each area on the diagram, some Cu species, solid and/or aqueous, are thermodynamically stable. For example, water is stable in the region between two dashed
lines (1) and (2) and is not stable in the regions above or below these lines. Above
dashed line (1), $O_2(g)$ evolution takes place, and below dash line (2), $H_2(g)$ evolution
takes place. In the area below lines (4) and (5), Cu(s) is thermodynamically stable and
will not be transformed to any other species until Eh and pH correspond to this area.
A demonstration of constructing a Pourbaix diagram is presented in Section 4.19.

Note that all statements made thus far are correct when (1) there are not any
kinetic limitations and (2) there not any thermodynamically unstable species. These
two conditions are unfortunately quite common and should be seriously taken into
account when analyzing Pourbaix diagrams. For example, in the region above line (1)
at pH = 7 and Eh = 1.2 V, the oxygen evolution will not immediately take place due to
a relatively low rate of the oxidation reaction of water, $H_2O(l) = 0.5O_2(g) + 2H^+(aq) +
2e^-$. To speed up this reaction, an additional potential, called overpotential, should be
applied, and this will be considered in Chapters 6 and 8. Also, $Cu(OH)_2(s)$ could be
thermodynamically less stable than $CuO(s)$, so the diagram given in Figure 4.4 may
represent a metastable equilibrium.

4.18 SUMMARY

- The Nernst equation is one of the most fundamental equations in electrochemical science and engineering. The reader should learn how this equation can be constructed and used for a variety of reactions.
- The Nernst equation consists of two terms. The first term, the standard electrode potential, does not depend on concentration, but the second one does.
Both of the terms depend on temperature, pressure, and concentration scale.
- The standard electrode potential can be calculated using either thermodynamic data or found in a table of standard electrode potentials, so-called
electrochemical series. The tabulated standard electrode potentials are
available only at a temperature of 25°C and a pressure of 1.013 bar (not
1 bar).
- The standard reduction potential of the hydrogen electrode is conventionally set equal to zero.

- The absolute value of the electric potential of a phase (Galvani potential) is not known. The absolute value of the difference in the electric potentials between two phases (Galvani potential difference) is not known either. The Galvani potential difference for the hydrogen electrode in the standard state is conventionally set equal to zero.
- The conditions for the standard hydrogen electrode (SHE) are as follows: The activity of $H^+(aq)$ is 1, and the fugacity of hydrogen is 1 bar (not 1.013 bar) at any temperature and pressure. The state is entirely hypothetical and cannot be experimentally realized. An extrapolation using real experimental data should be carried out to achieve this standard state.
- The reader should be familiar with how to construct the Nernst equation for a number of electrodes discussed in this book as well as how to calculate the standard electrode potentials for these electrodes.
- The reader should learn and clearly understand how to calculate the potential difference of an electrochemical cell using IUPAC recommendations [1]. The Daniell cell is used in this book as an example. It is suggested practicing other electrochemical cells (e.g., H_2/Cl_2) for calculating the potential difference at given concentrations of participating chemicals (see Section 4.19).
- How to calculate the equilibrium potential of a half-cell and the equilibrium potential difference of a cell at an elevated temperature is briefly described in this book.
- Temperature dependence of the OCP/DP is very important for estimating the efficiency of the electrochemical energy conversion systems. OCP and electrode potential difference are synonyms in this book.
- The electrochemical series (table of the standard electrode potentials) can be very useful for a quick calculation at 25°C. However, if the electrochemical calculations are supposed to be carried out at elevated temperatures, thermodynamic properties of chemicals of the electrochemical reaction should be used.
- Knowledge of the standard electrode potentials and the use of the Nernst equation allow constructing a special type of stability diagram, which is called the potential–pH or Pourbaix diagrams. These diagrams are correct when there are not any kinetic limitations and there not any thermodynamically unstable species.

4.19 EXERCISES

PROBLEM 1

Using the following equations:

$$\Delta_r G_T^0 = \Delta_r H_T^0 - T\Delta_r S_T^0, \tag{1}$$

$$\Delta_r G_{T_0}^0 = \Delta_r H_{T_0}^0 - T_0\Delta_r S_{T_0}^0, \tag{2}$$

$$\frac{\partial^2 \Delta_r G_T^0}{\partial T^2} = -\Delta_r C_p^0 / T, \tag{3}$$

$$\frac{\partial \Delta_r G_T^0}{\partial T} = -\Delta_r S_p^0, \qquad (4)$$

derive Equation 4.30.

Solution:

Subtracting Equation 2 from Equation 1, the difference between the standard Gibbs energy of reaction at temperature T and T_0, $\Delta_r G_T^0 - \Delta_r G_{T_0}^0$, is

$$\Delta_r G_T^0 - \Delta_r G_{T_0}^0 = \Delta_r H_T^0 - \Delta_r H_{T_0}^0 - T\Delta_r S_T^0 + T_0 \Delta_r S_{T_0}^0. \qquad (5)$$

Now, we add $(+T\Delta_r S_{T_0}^0)$ and $(-T\Delta_r S_{T_0}^0)$ to Equation 5 on the right:

$$\Delta_r G_T^0 - \Delta_r G_{T_0}^0 = \Delta_r H_T^0 - \Delta_r H_{T_0}^0 - T\Delta_r S_T^0 + T_0 \Delta_r S_{T_0}^0 + T\Delta_r S_{T_0}^0 - T\Delta_r S_{T_0}^0, \qquad (6)$$

which after simple rearrangement is

$$\Delta_r G_T^0 - \Delta_r G_{T_0}^0 = \Delta_r H_T^0 - \Delta_r H_{T_0}^0 - T\left(\Delta_r S_T^0 - \Delta_r S_{T_0}^0\right) - \Delta_r S_{T_0}^0 \left(T - T_0\right). \qquad (7)$$

Taking into account Equations 3 and 4, the difference between the standard enthalpy of reaction at temperature T and T_0, $\Delta_r H_T^0 - \Delta_r H_{T_0}^0$, is related to the standard heat capacity of the reaction as

$$\Delta_r H_T^0 - \Delta_r H_{T_0}^0 = \int_{T_0}^{T} \Delta_r C_p^0 dT, \qquad (8)$$

and the difference between the standard entropy of reaction at temperature T and T_0, $\Delta_r S_T^0 - \Delta_r S_{T_0}^0$, is also related to the standard heat capacity of the reaction as

$$\Delta_r S_T^0 - \Delta_r S_{T_0}^0 = \int_{T_0}^{T} \left(\Delta_r C_p^0 / T\right) dT. \qquad (9)$$

When Equations 8 and 9 are substituted to Equation 7, Equation 4.30 is obtained.

PROBLEM 2

Show how to derive the equation for the potential difference, $E = E_R - E_L$, of the cell given in Section 4.13. Calculate E if concentration of HCl(aq) is 5 mol kg^{-1} and partial pressure of each gas, Cl_2(g) and H_2(g), is 2 bar.

Solution:

The reduction half-reactions for both the left hand (L) and right hand (R) can be written as follows:

$$(L)\ 2H^+(aq) + 2e^- = H_2(g),$$

$$(R)\, Cl_2(g) + 2e^- = 2Cl^-(aq).$$

The corresponding Nernst equations for these half-reactions are

$$E_L = E^0_{H^+(aq)/H_2(g)} + \frac{RT}{2F}\ln\left[\frac{\left(a_{H^+(aq)}\right)^2}{a_{H_2(g)}}\right],$$

$$E_R = E^0_{Cl_2(g)/Cl^-(aq)} + \frac{RT}{2F}\ln\left[\frac{a_{Cl_2(g)}}{\left(a_{Cl^-(aq)}\right)^2}\right].$$

The expression for the cell potential difference, $E = E_R - E_L$, of the H_2/Cl_2 galvanic cell can be obtained as follows:

$$E = E^0 + \frac{RT}{2F}\ln\left[\frac{a_{Cl_2(g)}a_{H_2(g)}}{\left(b_{HCl(aq)}\gamma_\pm/b^0\right)^4}\right],$$

where $E^0 = E^0_{Cl_2(g)/Cl^-(aq)} - E^0_{H^+(aq)/H_2(g)}$. Now if $p_{Cl_2(g)} = p_{H_2(g)} = 2$ bar, $b_{HCl(aq)} = 5$ mol kg^{-1}, $E^0_{H^+(aq)/H_2(g)} = 0\,V$, $E^0_{Cl_2(g)/Cl^-(aq)} = 1.358$ (Table 10.13) and $\gamma_\pm = 2.38$ (Table 10.17), then

$$E = (1.358 - 0.0) + \frac{8.314 \times 298.15}{2 \times 96,485} \times \ln\left[\frac{2 \times 2}{(5 \times 2.38)^4}\right] = 1.2488\,V.$$

PROBLEM 3

Applying the appropriate Nernst equations, construct a simplified Pourbaix diagram of Zn(s)-H_2O(l) system using the following three half-reactions given in Table 10.13 as

$$Zn^{2+}(aq) + 2e^- = Zn(s), \tag{1}$$

$$Zn(OH)_2(s) + 2e^- = Zn(s) + 2OH^-(aq), \tag{2}$$

$$ZnO_2^{2-}(aq) + 2H_2O(l) + 2e^- = Zn(s) + 4OH^-(aq), \tag{3}$$

assuming that all activities, except H^+(aq), equal 1. Describe all horizontal, vertical, and inclined lines in a similar manner as presented in Section 4.17.

Solution:

The Nernst equation corresponding the half-reaction (1) is

$$Eh(1) = E^0_{Zn^{2+}(aq)/Zn(s)} + \frac{RT}{2F}\ln\left(a_{Zn^{2+}(aq)}\right),$$

which gives a horizontal line with (Table 10.13)

$$Eh(1) = -0.762\,\text{V}. \tag{4}$$

The appropriate Nernst equation corresponding the half-reaction (2) can be constructed if $2H^+(aq)$ is added to both sides of this half-reaction:

$$Zn(OH)_2\,(s) + 2H^+(aq) + 2e^- = Zn(s) + 2H_2O(l),$$

where it was taken into account that $2OH^-(aq) + 2H^+(aq) = 2H_2O(l)$.

Now, the Nernst equation corresponding half-reaction (2) can be written as

$$Eh(2) = E^0_{Zn(OH)_2(s),H^+(aq)/Zn(s),H_2O(l)} + \frac{RT}{2F}\ln\left(a_{H^+}\right)^2$$

$$= E^0_{Zn(OH)_2(s),H^+(aq)/Zn(s),H_2O(l)} - 0.0592\,\text{pH}.$$

To calculate the standard electrode potential, $E^0_{Zn(OH)_2(s),H^+(aq)/Zn(s),H_2O(l)}$, the Gibbs energy formation of $Zn(OH)_2(s)$ and $H_2O(l)$ should be known. From Table 10.3 $\Delta_f G^0_{H_2O(l)} = -237.1$ kJ mol^{-1}, $\Delta_f G^0_{Zn(s)} = 0$ kJ mol^{-1} by convention and $\Delta_f G^0_{Zn(OH)_2\,(s)} = -553.5$ kJ mol^{-1} can be found in Ref. [2]. The calculated $E^0_{Zn(OH)_2(s),H^+(aq)/Zn(s),H_2O(l)} = -0.411\,\text{V}$ and, therefore, the inclined line corresponding to half-reaction (3) is

$$Eh(2) = -0.411 - 0.0592\,\text{pH}. \tag{5}$$

The appropriate Nernst equation corresponding the half-reaction (3) can be constructed in a similar way as for half-reaction (2):

$$ZnO_2^{2-}(aq) + 4H^+(aq) + 2e^- = Zn(s) + 2H_2O(l).$$

From Ref. [3] $\Delta_f G^0_{ZnO_2^{2-}(aq)} = -389.2$ kJ mol^{-1} and the calculated $E^0_{ZnO_2^{2-}(aq),H^+(aq)/Zn(s),H_2O(l)} = +0.440\,\text{V}$, so

$$Eh(3) = 0.440 - 0.118\,\text{pH}. \tag{6}$$

To estimate the positions of the two vertical lines, Equations 4–6 should be used. The vertical line which separates $Zn^{2+}(aq)$ and $Zn(OH)_2(s)$ areas corresponds to pH = 5.93, which can be received by combining Equations 4 and 5. The vertical line which separates $Zn(OH)_2(s)$ and $ZnO_2^{2-}(aq)$ corresponds to pH = 14.47, which can be obtained by combining Equations 5 and 6.

The one horizontal, two inclined, and two vertical lines can be placed on the given below Eh-pH diagram to obtain a simplified Pourbaix diagram of the $Zn(s)$-$H_2O(l)$ system. The dashed lines corresponding to the oxygen reduction reaction and the proton reduction reaction (see Section 4.17) are shown in this figure as well.

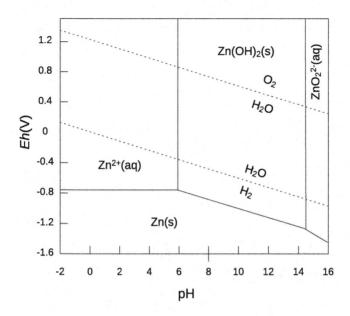

REFERENCES

1. E.R. Cohen et al., *Quantities, Units, and Symbols in Physical Chemistry*, 3rd ed., RSC Publishing, Cambridge, UK, 2007.
2. J. Rumble (Editor-in-Chief), *Handbook of Chemistry and Physics*, 101st ed., CRC Press, Boca Raton, FL, 2020.
3. M. Pourbaix, *Atlas of Electrochemical Equilibria in Aqueous Solutions*, NACE, Houston, TX, 1974.

5 Electrochemical Techniques I

5.1 OBJECTIVES

The main objective of this chapter is to describe how a number of electrochemical techniques can be used to study equilibrium electrochemical cells. A few electrochemical sensors will be discussed in this chapter. Both the physical design and electrochemistry of some sensors are covered. The problem of diffusion potential will also be presented. Some examples of potentiometry, which is based on the OCP measurement, will be provided.

5.2 POTENTIOMETRIC MEASUREMENTS

A high resistance electrometer should be used in all potentiometric measurements such as the measurement of the open circuit potential (OCP) of the Daniell cell (Figure 2.9). Failure to use an appropriate electrometer will lead to incorrect measurements. A simple electrometer usually has insufficiently high resistance, and therefore, some current will go through the circuit and electrodes. This is not allowed in any potentiometric measurements.

The liquid junction (diffusion) potential should always be a concern until a cell without transfer is tested. Note that the terms "liquid junction potential" and "diffusion potential" are used interchangeably in literature. Measurements, calculations, and minimization of the liquid junction potential formed at the interface of two solutions are some of the topics to be addressed in this chapter.

A variety of commercial electrometers can be used for potentiometric measurements. However, the internal resistance of the electrometer should be above 10^{13} Ω. At such a resistance value, the current passing through the electrodes of a cell potential of about 1 V will be 10^{-13} A, which is a negligible value.

5.3 COMMERCIAL Ag/AgCl REFERENCE ELECTRODE

The Ag/AgCl reference electrode, shown in Figure 4.2, is one of the most popular and reliable electrodes with the following electrochemical half-reaction:

$$AgCl(s) + e^- = Ag(s) + Cl^-(aq),$$

and the corresponding Nernst equation:

$$E = E^0_{AgCl(s)/Ag(s),Cl^-(aq)} - \frac{RT}{F} \ln \left[a_{Cl^-(aq)} \right], \tag{5.1}$$

DOI: 10.1201/b21893-5

where we assume that activities of both AgCl(s) and Ag(s) equal 1. Using the electrochemical series, Table 4.1, the standard electrode potential of the half-reaction is 0.222 V at 25°C and 1.013 bar. The electrode potential, E, of the Ag/AgCl electrode depends on the activity of the chloride ions, $a_{Cl^- (aq)}$, which is a product of molality, $b_{Cl^- (aq)}$, and its activity coefficient, $\gamma_{Cl^- (aq)}$, which is unknown for any concentration above 0.05 mol kg^{-1} (see Chapter 1) but can theoretically be calculated using Equation 1.26 this concentration. However, it can be reasonably assumed that the activity coefficients of the cation and the anion in KCl(aq) are very similar because of their similar size, magnitude of the ions charge, and number of electrons, 18 in both ions. In this case, we can use Equation 1.17 and reliably define $\gamma_{Cl^- (aq)}$ as the mean activity coefficient of KCl(aq), $\gamma_\pm$, which is tabulated in Table 10.17 over a wide concentration range. In KCl(aq) solution saturated with KCl(s), where concentration of KCl(aq) is 4.766 mol kg^{-1} (Table 10.19), the mean activity coefficient can be found from Table 10.17 by a simple interpolation and equals 0.591. Using Equation 5.1, the KCl(aq) solubility, and the mean activity coefficient provided earlier, the Ag/AgCl electrode potential can be calculated and equals 0.195 V. Similarly, the electrode potential calculated using 1 mol L^{-1} KCl(aq) is 0.237 V and using 0.1 mol L^{-1} KCl(aq) is 0.289 V. The details of these calculations are left for the reader as an exercise in using the Nernst equation.

A widely available commercial Ag/AgCl electrode can work at temperatures up to about 80°C, and its electrode potentials at elevated temperatures can be calculated quite accurately. It is important to know how the potential of the reference electrode changes with temperature. When analyzing the Nernst equation for the Ag/AgCl electrode, Equation 5.1, the variables that depend on temperature are the standard electrode potential, $E^0_{AgCl(s)/Ag(s),Cl^- (aq)}$, and the Cl$^-$(aq) activity, $a_{Cl^- (aq)}$. The standard electrode potential depends on temperature because the thermodynamic components used to calculate the standard electrode potential are a function of temperature. The activity depends on temperature because the activity coefficient is a function of temperature. However, there are not any standard electrode potentials tabulated at temperatures other than 25°C [1], and therefore, the Ag/AgCl standard electrode potential at elevated temperatures should be calculated using thermodynamic data from Tables 10.3 and 10.5 with an approximation that the heat capacity of the reaction is constant. The KCl(aq) activity coefficients at elevated temperatures cannot be collected from Ref. [1] but can be found in some published papers. Calculation of the standard electrode potential of Ag/AgCl electrode and the mean activity coefficient of KCl(aq) at an elevated temperature are presented in Section 5.15.

Regarding the electrode design, which can be seen in Figure 4.2, there are two holes in the glass electrode. The upper hole at the top of the electrode is for replacing the internal solution. When the electrode is used for some time, the internal reference solution of the electrode changes in concentration and composition. This change comes from the diffusion of the species between the internal reference solution and the test solution through the porous plug shown in Figure 4.2. The porous plug is used to delay the exchange of ions between the test and reference solutions. However, because the porous plug only delays the exchange, after some time, the solutions' diffusion will eventually change the potential of the reference electrode. At this point, the reference solution should be replaced with a fresh KCl(aq) solution of a known concentration. It is assumed that precision of the commercial and properly maintained Ag/AgCl reference electrode should be around ±0.1 mV.

5.4 COMMERCIAL CALOMEL REFERENCE ELECTRODE

The calomel reference electrode is a metal/salt-type electrode, so electrochemistry of the Hg/Hg_2Cl_2 and Ag/AgCl electrodes is similar. The electrochemical half-reaction of the Hg/Hg_2Cl_2 reference electrode is

$$Hg_2Cl_2(s) + 2e^- = 2Hg(l) + 2Cl^-(aq),$$

and Equation 4.27 shows the Nernst equation of this electrode.

Similar to the Ag/ACl reference electrode, the electrode potential of the calomel electrode at a concentration of KCl(aq) equal (1) 0.1 mol L^{-1}, (2) 1.0 mol L^{-1}, (3) saturated with respect to KCl(s) can be calculated and compared with values given in (Table 10.13). The calomel and silver/silver chloride electrodes have similar precision and stability. A disadvantage of the calomel reference electrode is the use of Hg, which is a particularly hazardous chemical. The design of the Hg/Hg_2Cl_2 reference electrode is shown in Figure 5.1.

There are other reference electrodes that are less favored but can be useful in some circumstances. For example, if a testing solution is significantly basic, e.g., KOH(aq), it might be useful to consider using mercury/mercury oxide reference electrode, which is discussed in Section 5.15.

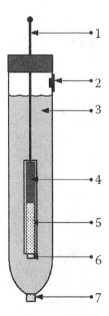

FIGURE 5.1 Schematic of calomel reference electrode: (1) platinum wire, (2) filling hole, (3) KCl(aq) solution (usually saturated), (4) liquid mercury, and (5) mercury–calomel (Hg/ Hg_2Cl_2) paste, (6) glass frit, and (7) ceramic, or quartz, or glass porous plug.

5.5 pH GLASS ELECTRODE

The pH glass electrode is one of the most popular electrodes for pH measurements in aqueous solutions. The electrochemistry of the glass electrode is relatively complicated and does not allow simple derivation of a Nernst equation similar to the Ag/AgCl or Hg/Hg$_2$Cl$_2$ electrode. Still, the electrode potential of a glass electrode similarly depends on the activity of H$^+$(aq) as follows:

$$E = E^0_{glass} + \theta \log_{10}\left[a_{H^+(aq)} \right], \qquad (5.2)$$

where the Nernstian slope $\theta \approx (\ln 10)\, RT/F = 2.303\, RT/F$ and E^0_{glass} cannot be accurately calculated but could be changed when the material of the glass membrane (see Figure 5.2) is changed. A possible design of the glass electrode is illustrated in Figure 5.2.

Any glass electrode should be calibrated using a few buffer solutions to develop a calibration curve (or better a straight line). The glass electrode cannot provide reliable results when the solution is (1) highly acidic, (2) highly basic, and (3) a highly concentrated brine. There are some glass-type electrodes (ion-selective) that are capable of measuring the activity of ions other than H$^+$(aq).

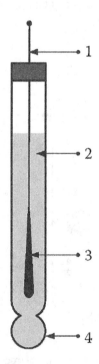

FIGURE 5.2 Schematic of pH glass electrode: (1) Pt wire, (2) a buffered Cl(aq)-containing solution, (3) Ag/AgCl electrode, and (4) H$^+$(aq) exchange glass membrane.

5.6   LIQUID JUNCTION (DIFFUSION) POTENTIAL

Let us consider an electrochemical cell that consists of two Ag/AgCl electrodes immersed into HCl(aq) solutions with different concentrations. The solutions are connected by a separator that (1) delays the mixing of the solutions for sufficient time to allow measurements and (2) allows electric contact via ion transport between the solutions. The electrochemical diagram of such a cell is

$$Ag(s)|AgCl(s)|HCl(aq, I)|HCl(aq, II)|AgCl(s)|Ag(s),$$

and the Nernst equation is as follows:

$$E = -\left[\frac{RT}{F}\right]\ln\left[\frac{(a_{Cl^-(aq,II)})}{(a_{Cl^-(aq,I)})}\right] + E_{diff}, \qquad (5.3)$$

where the first term is the Nernst equation for the cell potential, $E = E_R - E_L$, defined in a similar way as it was demonstrated in Chapter 4, and the second term represents the liquid junction or diffusion potential, E_{diff}, which is formed due to different concentrations of ions in the two neighboring solutions of HCl(aq). Generally, the diffusion potential is formed due to (1) different concentrations, (2) different compositions, and (3) both (1) and (2). The diffusion potential can properly be explained using nonequilibrium thermodynamics, a discipline that is out of the scope of this book. This theory provides an equation to calculate the diffusion potential as follows [2]:

$$E_{diff} = -\left(\frac{RT}{F}\right)\sum_i\left[\int_{a_i(I)}^{a_i(II)}\left(\frac{t_i}{z_i}\right)da_i\right], \qquad (5.4)$$

where a summation should be done over all ions of both solutions and integration should be made over a range of activities specified by Solution I (left-hand side of the liquid junction) and Solution II (right-hand side of the liquid junction).

5.7   HENDERSON EQUATION

Using (1) the Debye–Hückel–Onsager theory (Chapter 3) for calculating ionic conductivities and transport numbers, (2) using the Debye–Hückel theory (Chapter 1) for calculating individual activity coefficients, and (3) knowing the solution speciation (Section 1.15), it is possible to calculate E_{diff} using Equation 5.4 when the electrolyte concentration is quite low. Otherwise, corresponding experimental data are needed. Examples of such data are given in Table 5.1 [3] for a case when solutions are different but their concentrations are the same, 0.1 mol L^{-1}.

When a number of assumptions are made, an analytical expression, which is well known as the Henderson equation [2], can be obtained:

TABLE 5.1

Experimental [3] and Calculated (Using Equation 5.6) Values of the Diffusion Potential

Solution I (0.1 mol L^{-1})	Solution II (0.1 mol L^{-1})	E_{diff} (mV) (Observed)	E_{diff} (mV) (Calculated)
HCl	KCl	26.8	28.5
HCl	NaCl	33.2	33.4
HCl	LiCl	35.0	36.1
HCl	NH$_4$Cl	28.4	28.6
KCl	NaCl	6.4	4.9
KCl	LiCl	8.8	7.6
KCl	NH$_4$Cl	2.2	0.0
NaCl	LiCl	2.6	2.8
NaCl	NH$_4$Cl	−4.2	−4.8
LiCl	NH$_4$Cl	−6.9	−7.6

$$E_{diff} = -\frac{RT}{F}\left[\frac{\sum \frac{\left[a_{i(\mathrm{II})} - a_{i(\mathrm{I})}\right]u_i|z_i|}{z_i}}{\sum \left[a_{i(\mathrm{II})} - a_{i(\mathrm{I})}\right]u_i|z_i|}\right]\ln\left\{\frac{\sum a_{i(\mathrm{II})}u_i|z_i|}{\sum a_{i(\mathrm{I})}u_i|z_i|}\right\}. \tag{5.5}$$

The assumptions in deriving Equation 5.5 are as follows: (1) linear dependence of concentrations with distance between electrolytes (I) and (II), (2) nonideality is ignored in estimating the transport numbers, and (3) ionic mobility is independent of concentration.

5.8 CALCULATION OF THE DIFFUSION POTENTIAL

While Equation 5.4 can be used to calculate the diffusion potential, it is not common practice due to a lack of information for such calculations. Equation 5.5 can be used instead, which requires a simple computer code to avoid mistakes when using an equation with a number of summations.

In some cases, Equation 5.5 can be simplified when one of the ions in solutions (I) and (II) are the same and molar concentrations of both electrolytes are the same:

$$E_{diff} = \pm\left(\frac{RT}{F}\right)\ln\left(\frac{\Lambda_{\mathrm{II}}}{\Lambda_{\mathrm{I}}}\right), \tag{5.6}$$

where the positive sign corresponds to a junction with a common cation in both neighboring solutions, and the negative sign applies to a case with a common anion. Λ_{II} and Λ_{I} are molar conductivities of solutions (II) and (I), respectively. Using Equation 5.6, E_{diff} can be calculated for junctions shown in Table 5.1, taking molar conductivities from Table 10.11. As can be seen from the table, the calculated and observed [3] values are in good agreement, within about 1 mV, in most cases.

Another simplification of Equation 5.5 can be achieved if electrolytes of the liquid junction are the same but concentrations of solutions (I) and (II) are different:

$$E_{diff} = \left[(t_+ - t_-)\right]\left[\frac{RT}{F}\right]\ln\left(\frac{a_{\pm,I}}{a_{\pm,II}}\right), \tag{5.7}$$

where

t_+ and t_- are transport numbers of the cation and the anion, respectively.

$a_{\pm,I}$ and $a_{\pm,II}$ are mean activities of electrolyte in solutions (I) and (II), respectively.

Equation 5.7 can now be used for calculating the diffusion potentials in Equation 5.3. If we assume that concentrations of HCl(aq, I) and HCl(aq, II) are, respectively, 0.1 and 0.5 mol kg^{-1}, using transport numbers for H$^+$(aq) and Cl$^-$(aq) from Table 3.2, and mean activity coefficients from Table 10.17, the E_{diff} value can be calculated as -26.9 mV. This is a pronounced value comparing with the first term of Equation 5.3, which is -40.1 mV.

Equation 5.7 also shows that using a salt bridge with an electrolyte that has similar cation and anion conductivities is beneficial for minimizing the diffusion potential. This is why KCl(aq) is suggested to be used as the salt bridge electrolyte (see Section 2.12), in which limiting ionic conductivities of K$^+$(aq) and Cl$^-$(aq) are very similar and, respectively, 73.48 and 76.31 cm^2 S mol^{-1} (Table 10.12).

5.9 MINIMIZATION OF THE DIFFUSION POTENTIAL IN A CELL

Whenever possible, the diffusion potential of a cell should be minimized. There are two common approaches to minimize E_{diff}: (1) use a salt bridge with a high concentration of KCl (solubility of KCl at 25°C is about 4.8 mol kg^{-1}), and (2) use the same background electrolyte in both half-cells along with low concentrations of the electrochemically active species. The complication with using a background electrolyte is that the background ions must be electrochemically inactive. Oxygen-based ions like SO_4^{2-} (aq) or ClO_4^- (aq) are often preferable. Figure 2.9 illustrates the first approach. In the Daniell cell, the concentration of KCl(aq) in the salt bridge might be as much as 4 mol L^{-1} and concentrations of ZnSO$_4$(aq) and CuSO$_4$(aq) can be as low as 0.1 mol L^{-1}. The calculation of the diffusion potential using the Henderson equation (5.5) shows that in such a cell, the diffusion potential is less than 1 mV. This can be confirmed by measurements comparing them with corresponding calculations. Clearly, the two diffusion potentials in the Daniell cell shown in Figure 2.9 are approximately canceled out because they have similar magnitudes and opposite signs. The double vertical line, ||, in electrochemical diagrams shows that the diffusion potential in this cell is practically eliminated and should not be taken into account in any calculations. However, if one uses a dilute solution of KCl(aq) in the salt bridge, the diagram should not have the double vertical line. In this case, two diffusion potentials must be calculated and taken into account.

Let us consider a cell with two identical Ag/AgCl electrodes. KCl(aq) with different concentrations is used for making the salt bridge, which is in contact with HCl(aq, 0.1 mol L^{-1}) on the left and with NaCl(aq, 0.1 mol L^{-1}) on the right:

$$Ag(s)|AgCl(s)|HCl(aq)|KCl(aq)|NaCl(aq)|AgCl(s)|Ag(s).$$

Results of calculations using the Henderson equation for HCl(aq, 0.1 mol L^{-1}) | KCl(aq) and KCl(aq) | NaCl(aq, 0.1 mol L^{-1}) junctions are given in Table 5.2. Also, in the table is the diffusion potential of HCl(aq, 0.1 mol L^{-1}) | KCl(aq) | NaCl(aq, 0.1 mol L^{-1}) calculated as the sum of diffusion potential of HCl(aq, 0.1 mol L^{-1}) | KCl(aq) and KCl(aq) | NaCl(aq, 0.1 mol L^{-1}).

It can be seen from Table 5.2 that when the concentration of KCl(aq) in the salt bridge is 4.2 mol L^{-1}, the preceding cell electrochemical diagram given can be simplified as

$$Ag(s)|AgCl(s)|HCl\left(aq, 0.1\,mol\,L^{-1}\right)\|NaCl\left(aq, 0.1\,mol\,L^{-1}\right)|AgCl(s)|Ag(s),$$

where the double vertical line shows that the diffusion potential in this cell is minimized down to an acceptable value of 3.1 mV. Whether the value of 3 mV is acceptable or not should be specified when using a cell with the salt bridge.

When the same diagram is shown with a single line instead of the double line

$$Ag(s)|AgCl(s)|HCl\left(aq, 0.1\,mol\,L^{-1}\right)|NaCl\left(aq, 0.1\,mol\,L^{-1}\right)|AgCl(s)|Ag(s),$$

we should expect a diffusion potential of about 33.4 mV (see Table 5.1), which cannot be ignored in any potentiometric measurements.

TABLE 5.2
Diffusion Potentials of HCl(aq, 0.1 mol L^{-1}) | KCl(aq) and KCl(aq) | NaCl(aq, 0.1 mol L^{-1}) Junctions and Potential of HCl(aq, 0.1 mol L^{-1}) | KCl(aq) and KCl(aq) | NaCl(aq, 0.1 mol L^{-1}) Junctions

	E_{diff} (mV)		
KCl(aq) Concentration (mol L^{-1})	HCl(aq, 0.1 mol L^{-1}) | KCl(aq)	KCl(aq) | NaCl(aq, 0.1 mol L^{-1})	HCl(aq, 0.1 mol L^{-1}) | KCl(aq) | NaCl(aq, 0.1 mol L^{-1})
0.1	26.8	4.4	31.2
1	9.1	−0.1	9.0
2	6.4	−0.9	5.6
3	5.3	−1.2	4.1
4.2	4.6	−1.5	3.1

5.10   KINDS OF POTENTIOMETRY

Potentiometry is a powerful electrochemical method when the OCP is measured using a high-resistance electrometer. This method is a common tool for studying equilibrium systems in electrochemical science and engineering. In nonequilibrium studies, an OCP measurement is usually the first step to find out if a cell of interest works properly. Note that during any potentiometric measurements, due to the absence of current in the circuit, any undesirable potential drops inside the phases are absent. In nonequilibrium electrochemical system, the potential drop between working and counter electrodes (called IR drop) is a serious issue to be taken into account experimentally and/or theoretically.

The main applications of potentiometry are related to measuring the following values:

1. Standard electrode potential
2. Mean activity coefficient
3. Solubility product (equilibrium constant) of slightly soluble salts in water
4. Dissociation constant of weak electrolyte, including the ionization constant of water
5. Standard thermodynamic properties of reaction, including Gibbs energy and entropy
6. pH
7. Concentration of ions by titration
8. Corrosion potential.

In this chapter, we will briefly consider only two of these applications, (1) and (6). Reference [4] can be used for learning some other applications of potentiometry.

5.11   MEASUREMENT OF THE STANDARD ELECTRODE POTENTIAL

Let us consider the Harned cell, which is a cell without transfer ($E_{diff} = 0$), to show how a standard electrode potential can be obtained by measuring the concentration dependence of the cell potential and, then, making an extrapolation to an infinitely dilute solution. The electrochemical diagram and the corresponding Nernst equation (4.22) for the Harned cell were considered in Chapter 4. Equation 4.22, which can, first, be simplified if the activity of $H_2(g)$ equals 1 ($p_{H2} = 1$ bar and fugacity coefficient is very close to 1) and, second, is rearranged as follows:

$$E' = E + \left(\frac{2RT}{F}\right)\ln\left(\frac{b_{HCl(aq)}}{b^0}\right) = E^0_{AgCl(s)/Ag(s),Cl^- (aq)} - \left(\frac{2RT}{F}\right)\ln\left(\gamma_\pm\right). \qquad (5.8)$$

The standard electrode potential of Ag/AgCl, $E^0_{AgCl(s)/Ag(s),Cl^- (aq)}$, can be obtained by a linear extrapolation shown in Figure 5.3 from a region of dilute solutions at $b_{HCl(aq)} < 0.05\,\text{mol kg}^{-1}$, where the activity coefficients can be calculated using the Debye–Hückel theory. Alternatively, as shown in Figure 5.3, a wider range of concentrations can be used if independently obtained experimental data on HCl(aq) mean activity coefficients are available (Table 10.17).

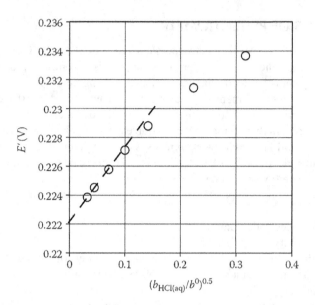

FIGURE 5.3 Determination of the standard electrode potential of Ag/AgCl using extrapolation to infinite dilute solution of $E' = E + (2RT/F) \ln (b_{HCl(aq)}/b_0)$ to $b_{HCl(aq)} = 0$.

5.12 MEASUREMENT OF pH

pH is an important characteristic of any aqueous solution and shows its acidity. In industry, measurement of pH is an important task to make sure that a process of interest is under control. Traditionally, it is believed that pH measurement is a simple task. If conditions are ambient and the solution is almost pure water with a small amount of additives, pH measurements are truly quite simple. However, if temperature and pressure deviate from ambient conditions and/or the solution is highly acidic, basic, or with high concentration of salts, pH measurements are already not a simple task. These complications originate from the cross sensitivity of the glass membrane to other ionic species present in common solutions. For example, when alkali ions such as lithium and sodium are in concentrations greater than 0.1 mol kg^{-1}, a component of the glass probe potential difference from cross sensitivity to alkali ions becomes significant. This additional potential from the alkali ions can decrease the estimated pH calculation in a range of 0.01–0.5 pH units, depending on the concentration and ionic species. In this book, we will consider only simple cases. If one has (1) a glass electrode, (2) a reference electrode, and (3) a high-resistance electrometer, the OCP of the following cell, E_S, can be measured using a standard (usually buffer) solution:

$$Cu(s)|Ag(s)|AgCl(s)|\text{Standard solution}|\text{Glass electrode}|Cu(s).$$

Then, the standard solution can be replaced with a test solution, and OCP of the cell, E_X, should be measured. Assuming the Nernst equation for the glass electrode is given by Equation 5.2, the slope is the Nernstian ($\theta = 2.303\ RT/F$), and pH$_S$ of the standard solution is known, pH$_X$ of the test solution can be defined as follows:

$$pH_X = pH_S + \left[\frac{F}{(2.303RT)}\right](E_S - E_X). \tag{5.9}$$

When a number of buffer solutions are available, it is safer to assume that the slope θ can slightly deviate from Nernstian and, therefore, should be found using two or more standard solutions. If two standard solutions S1 and S2 are used, then the Nernstian slope can be found from the following equation:

$$pH_{S1} = pH_{S2} + \left(\frac{1}{\theta}\right)(E_{S2} - E_{S1}). \tag{5.10}$$

Then, pH_X can be defined by an equation similar to Equation 5.10 replacing [$F/(2.303RT)$] with ($1/\theta$):

$$pH_X = pH_S + \left(\frac{1}{\theta}\right)(E_S - E_X). \tag{5.11}$$

More details on pH measurements can be found in Refs. [1] and [5].

5.13 COMBINED pH/REFERENCE ELECTRODE SENSOR AND PH METER

These days, the use of a pH glass indicator electrode and an Ag/AgCl reference electrode along with a high-resistance electrometer is not common practice. Instead, a combined pH/silver–silver chloride sensor shown in Figure 5.4 is commonly used

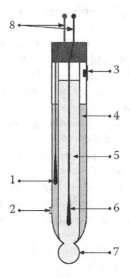

FIGURE 5.4   Schematic of a combined pH/silver–silver chloride sensor. (1) Ag/AgCl reference electrode, (2) salt bridge, (3) filling hole, (4) reference electrode solution, (5) internal buffer solution, (6) Ag/AgCl internal electrode, (7) pH-sensitive glass membrane, and (8) Pt wires.

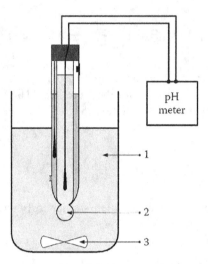

FIGURE 5.5 Schematic of a system for measuring pH: (1) test solution, (2) combined pH glass/silver–silver chloride sensor (see Figure 5.4), and (3) magnetic stirrer.

along with a compact pH meter as shown in Figure 5.5. The pH meter, which is actually a high-resistance electrometer, has a capability for temperature correction, which is important because the Nernstian slope [$2.303RT/F$] depends on temperature.

5.14 SUMMARY

- The hydrogen electrode is the most important one in studying the fundamentals of electrochemical science and engineering. The reader of the book should be familiar with the main features of this electrode.
- The standard hydrogen electrode (SHE) is a hypothetical sensor. The potential of the SHE can be obtained by extrapolation to the infinitely dilute solution, and the fugacity of hydrogen should be well defined.
- While the hydrogen electrode can be constructed, most practical reference electrodes are silver/silver chloride and calomel reference electrodes.
- The reader of the book should clearly understand the design and electrochemistry of the silver/silver chloride reference electrode.
- It is important to be able to calculate the equilibrium potential of the reference electrode as a function of temperature.
- The calomel reference electrode is still a common reference electrode in electrochemical labs. The reader of the book should be familiar with the design of the calomel reference electrode and be able to calculate its equilibrium potential.
- Most of electrochemical cells have liquid junction (diffusion) potentials. Electrochemical cells without transfer are free of it. An example of an electrochemical cell without transfer is the famous Harned cell.
- The reader should know which properties should be available to calculate the diffusion potential. There are two cases when calculation of the

diffusion potential is quite simple. Calculation of the diffusion potential using the Henderson equation requires it to be very accurate in estimating a number of sums.

- An electrochemical method, called potentiometry, is used to measure an equilibrium potential difference between a reference and an indicator electrode. To correctly measure an equilibrium potential, a high-resistance electrometer should be used.

- There are a number of applications of potentiometry. In this book, the reader will learn only two of them: (1) estimation of the standard electrode potential and (2) pH measurements.

5.15   EXERCISES

PROBLEM 1

Equations 1.26–1.29 can be used to calculate the activity coeffect of low-concentration aqueous solutions at elevated temperature because A_{DH} and B_{DH} can be theoretically calculated using the following equations [6]:

$$A_{DH} = \frac{2.303 \times 1.8246 \times 10^6 \times \rho_w^{0.5}}{(\varepsilon_w T)^{1.5}},$$

$$B_{DH} = \frac{50.29 \times \rho_w^{0.5}}{(\varepsilon_w T)^{0.5}},$$

where ρ_w is water density in g cm^{-3}, ε_w is water relative permittivity (dimensionless), and units of A_{DH} and B_{DH} are, respectively, (kg mol^{-1})$^{0.5}$ and (kg mol^{-1})$^{0.5}$ (Å)$^{-1}$. Calculate the mean activity coefficient of KCl(aq) at 80°C and 1 bar at molality equals 0.005 mol kg^{-1}.

Solution:

Using Equation 1.27, the mean activity coefficient can be obtained as

$$\ln \gamma_\pm = -\frac{A_{DH}|z_+ z_-| I_b^{0.5}}{\left(1 + B_{DH} \mathring{a} I_b^{0.5}\right)}.$$

In Table 10.22, the density and relative permittivity (dielectric constant) of water at 80°C can be found as 971.79 kg m^{-3} = 0.97179 g cm^{-3} and 60.898, respectively. Therefore, the parameter A_{DH} and B_{DH} can be calculated as

$$A_{DH} = \frac{2.303 \times 1.8246 \times 10^6 \times \rho_w^{0.5}}{(\varepsilon_w T)^{1.5}} = \frac{2.303 \times 1.8246 \times 10^6 \times 0.97179^{0.5}}{(60.898 \times 353.15)^{1.5}}$$

$$= 1.3134 (\text{kg mol}^{-1})^{0.5},$$

$$B_{DH} = \frac{50.29 \times 0.97179^{0.5}}{(60.898 \times 353.15)^{0.5}} = 0.3381 (\text{kg mol}^{-1})^{0.5} \, (\text{Å})^{-1}.$$

The ionic strength of the $0.005 \, \text{mol kg}^{-1}$ KCl(aq) on the molal concentration scale can be obtained with Equation 1.15:

$$I_b = \frac{1}{2}\left(b_+ z_+^2 + b_- z_-^2\right) = \frac{1}{2}(0.005 \times 1 + 0.005 \times 1) = 0.005 \, \text{mol kg}^{-1}.$$

By substituting the calculated values of A_{DH}, B_{DH}, and I_b to Equation 1.27, $\ln\gamma_{\pm}$ can be calculated as

$$\ln\gamma_{\pm} = -\frac{1.3134 \times |1 \times (-1)| \times 0.005^{0.5}}{\left(1 + 0.3381 \times 4.5 \times 0.005^{0.5}\right)} = -0.083853,$$

in which $\mathring{a}$ is taken as $4.5 \, \text{Å}$ (see Section 1.12). Then, the mean activity coefficient of $0.005 \, \text{mol kg}^{-1}$ KCl(aq) at $80°C$ and 1 bar can be obtained as

$$\gamma_{\pm} = e^{-0.0838531} = 0.9196.$$

PROBLEM 2

Calculate the standard electrode potential of the Ag/AgCl electrode at a temperature of $80°C$ at 1 bar. What is the electrode potential if the electrode is immersed in $0.005 \, \text{mol kg}^{-1}$ KCl(aq) solution?

Solution:

As described in Section 4.10, the standard electrode potential of the Ag/AgCl electrode is defined as the standard potential difference of the following hypothetical cell:

$$\text{Cu(s)}|\text{Pt(s)}|\text{H}_2(\text{g, ideal})|\text{HCl (aq, standard)}|\text{AgCl(s)}|\text{Ag(s)}|\text{Cu(s)}.$$

The electrochemical half-reactions on the right hand (RH) and left hand (LH) of the cell are as follows:

$$\text{RH}: \text{AgCl(s)} + e^- = \text{Ag(s)} + \text{Cl}^-(\text{aq}),$$

$$\text{LH}: \text{H}^+(\text{aq}) + e^- = 0.5\text{H}_2(\text{g}),$$

giving the total cell reaction as difference between the RH half-reaction and the LH half-reaction:

$$0.5\text{H}_2(\text{g}) + \text{AgCl(s)} = \text{Ag(s)} + \text{Cl}^-(\text{aq}) + \text{H}^+(\text{aq}).$$

The standard potential difference of the cell, E_T^0, can be obtained by calculating the standard Gibbs energy of the total cell reaction, $\Delta_r G_T^0$, at temperature $T = 353.15$ K:

$$E_T^0 = -\frac{\Delta_r G_T^0}{nF},$$

where $\Delta_r G_T^0$ can be calculated using Equation 4.29:

$$\Delta_r G_T^0 = \Delta_r G_{T_0}^0 - \Delta_r S_{T_0}^0 \left(T - T_0\right) + \Delta_r C_{P T_0}^0 \left[\left(T - T_0\right) - T \ln\left(\frac{T}{T_0}\right)\right],$$

with $T_0 = 298.15$ K. Using data from Tables 10.3 and 10.5,

Chemicals	$\Delta_f G^0$ (kJ mol^{-1})	S^0 (J mol^{-1} K^{-1})	C_p^0 (J mol^{-1} K^{-1})
H_2(g)	0	130.7	28.8
AgCl(s)	−109.8	96.3	50.8
Ag(s)	0	42.6	25.4
Cl$^-$(aq)	−131.2	56.5	−136.4
H$^+$(aq)	0	0	0

The values of $\Delta_r G_{T_0}^0$, $\Delta_r S_{T_0}^0$, and $\Delta_r C_{P \, T_0}^0$ of the total cell reaction can be calculated as follows:

$$\Delta_r G_{T_0}^0 = \Delta_f G^0\left(\text{Ag,s},T_0\right) + \Delta_f G^0\left(\text{Cl}^-,\text{aq},T_0\right) + \Delta_f G^0\left(\text{H}^+,\text{aq},T_0\right)$$

$$- \Delta_f G^0\left(\text{AgCl,s},T_0\right) - 0.5 \times \Delta_f G^0\left(\text{H}_2,\text{g},T_0\right)$$

$$= -131.2 - (-109.8) = -21.4\,\text{kJ mol}^{-1} = -21{,}400\,\text{J mol}^{-1},$$

$$\Delta_r S_{T_0}^0 = s^0\left(\text{Ag,s},T_0\right) + S^0\left(\text{Cl}^-,\text{aq},T_0\right) + S^0\left(\text{H}^+,\text{aq},T_0\right)$$

$$- S^0\left(\text{AgCl,s},T_0\right) - 0.5 \times S^0\left(\text{H}_2,\text{g},T_0\right)$$

$$= 42.6 + 56.5 - 96.3 - 0.5 \times 130.7 = -62.55\,\text{J mol}^{-1}\,\text{K}^{-1},$$

$$\Delta_r C_{P T_0}^0 = C_p^0\left(\text{Ag,s},T_0\right) + C_p^0\left(\text{Cl}^-,\text{aq},T_0\right) + C_p^0\left(\text{H}^+,\text{aq},T_0\right)$$

$$- C_p^0\left(\text{AgCl,s},T_0\right) - 0.5 \times C_p^0\left(\text{H}_2,\text{g},T_0\right)$$

$$= 25.4 - 136.4 - 50.8 - 0.5 \times 28.8 = -176.2\ \text{J mol}^{-1}\,\text{K}^{-1}.$$

By substituting the calculated values of $\Delta_r G_{T_0}^0$, $\Delta_r S_{T_0}^0$, and $\Delta_r C_{P \, T_0}^0$ to Equation 4.29, $\Delta_r G_T^0$ can be calculated as

$$\Delta_r G_T^0 = -21,400 - (-62.55) \times (353.15 - 298.15) + (-176.2) \times [(353.15 - 298.15)$$

$$- 353.15 \times \ln\left(\frac{353.15}{298.15}\right)] = -17,116\,\mathrm{J\,mol^{-1}}.$$

The standard potential difference of the hypothetical cell, E_T^0, can be calculated as

$$E_T^0 = -\frac{-17,116}{1 \times 96,485} = 0.1774\ \mathrm{V},$$

and, therefore, the standard electrode potential of Ag/AgCl electrode at temperature of 80°C at 1 bar $E^0_{\mathrm{AgCl(s)/Ag(s),Cl^-(aq),T}} = 0.1774\,\mathrm{V}$ as well.
Using Nernst equation, the electrode potential of the cell $E = E_{RH} - E_{LH}$ is

$$E = E^0_{\mathrm{AgCl(s)/Ag(s),Cl^-(aq),T}} - \frac{RT}{F}\ln\left[a_{\mathrm{Cl^-(aq)}}\right]$$

$$= E^0_{\mathrm{AgCl(s)/Ag(s),Cl^-(aq),T}} - \frac{RT}{F}\ln\left[b_{\mathrm{Cl^-(aq)}}\gamma_{\mathrm{Cl^-(aq)}}\right].$$

Due to the low concentration of KCl(aq), $0.005\,\mathrm{mol\,kg^{-1}}$, the activity coefficient of Cl⁻(aq), $\gamma_{\mathrm{Cl^-(aq)}}$, at temperature of 80°C at 1 bar can be calculated as it was done in Problem 1:

$$\gamma_{\mathrm{Cl^-(aq)}} = \gamma_{\pm,\mathrm{KCl(aq)}} = 0.9196.$$

The electrode potential of the cell as well as the electrode potential of Ag/AgCl electrode at temperature of 80°C at 1 bar with KCl(aq) solution of $0.005\,\mathrm{mol\,kg^{-1}}$ is

$$E = 0.1774 - \frac{8.314 \times 298.15}{96,485}\ln(0.005 \times 0.9196) = 0.3157\,\mathrm{V}.$$

PROBLEM 3

Calculate the electrode potential of HgO(s), H_2O(l)/Hg(l), OH⁻(aq) electrochemical couple (electrode) at 25°C and 1 bar if the electrode is immersed in $5\,\mathrm{mol\,kg^{-1}}$ KOH(aq) solution.

Solution:

The electrode half reaction of the HgO(s), H_2O(l)/Hg(l), OH⁻(aq) electrochemical couple (electrode) is as follows (Section 4.12):

$$\mathrm{HgO(s) + H_2O(l) + 2e^- = Hg(s) + 2OH^-(aq)}.$$

The Nernst equation corresponding to the half-reaction above is (Equation 4.28)

$$E = E^0_{HgO(s),H_2O(l)/Hg(s),OH^-(aq)} - \frac{RT}{2F} \ln\left[\frac{a^2_{OH^-(aq)}}{a_{H_2O(l)}}\right]$$

$$= E^0_{HgO(s),H_2O(l)/Hg(s),OH^-(aq)} - \frac{RT}{2F} \ln\left[\frac{b^2_{OH^-(aq)}\gamma^2_{OH^-(aq)}}{x_{H_2O(l)}\gamma_{H_2O(l)}}\right],$$

where $E^0_{HgO(s),\,H_2O(l)/Hg(s),OH^-(aq)} = 0.0977\,V$ (Table 10.13), $\gamma_{OH^-(aq)} \approx \gamma_{\pm,KOH(aq)} = 1.697$ (Table 10.17), $x_{H_2O(l)}$ is the mole fraction of $H_2O(l)$ and equals 0.9174 as can be calculated using an equation given in Table 1.1., and $\gamma_{H_2O(l)}$ can be assumed to be 1 (Section 1.7). Now the electrode potential can be calculated as

$$E = 0.0977 - \frac{8.314 \times 298.15}{2 \times 96,485} \ln\left[\frac{(5 \times 1.697)^2}{0.9174}\right] = 0.04166\,V.$$

PROBLEM 4

Apply the Henderson equation (5.5) to show how the diffusion potential changes at 25°C and 1 bar for the junction:

Solution I [HCl(aq), b] | Solution II [KCl(aq), 0.1 mol kg^{-1}]

if molal concentration of HCl(aq), b, decreases from 0.1 to 0.001 mol kg^{-1}. Carry out your calculations ignoring nonideality in both solutions.

Solution:

Both HCl(aq) and KCl(aq) at ambient conditions are strong electrolytes, so for Solution II

$$b_{Cl^-(II)} = b_{K^+(II)} = b_{KCl(aq)} = 0.1\,mol\,kg^{-1},$$

and for solution I

$$b_{Cl^-(I)} = b_{H^+(I)} = b_{HCl(aq)} = b.$$

While in both solutions, water ionization provides H$^+$(aq) and OH$^-$(aq) ions, their concentration is very small due to the small value of the ionization constant of water (Table 10.6).

If nonideality can be ignored, then Equation 5.5 can be presented as

$$E_{diff} = -\frac{RT}{F}\left[\frac{\sum\frac{\left[b_{i(II)} - b_{i(I)}\right]u_i|z_i|}{z_i}}{\sum\left[b_{i(II)} - b_{i(I)}\right]u_i|z_i|}\right]\ln\left(\frac{\sum b_{i(II)}u_i|z_i|}{\sum b_{i(I)}u_i|z_i|}\right),$$

where the ionic mobilities, u_i, can be calculated using the limiting conductivities λ_i^0 available in Table 10.12 as follows (see Section 3.6):

$$u_{H^+} \approx u_{H^+}^o = \frac{\lambda_{H^+}^o}{F z_{H^+}} = \frac{349.65 \times 10^{-4}}{96{,}485} = 3.624 \times 10^{-7} \, \text{m}^2 \, \text{SC}^{-1},$$

$$u_{K^+} \approx u_{K^+}^o = \frac{\lambda_{K^+}^o}{F z_{K^+}} = \frac{73.48 \times 10^{-4}}{96{,}485} = 7.909 \times 10^{-8} \, \text{m}^2 \, \text{SC}^{-1},$$

$$u_{Cl^-} \approx u_{Cl^-}^o = \frac{\lambda_{Cl^-}^o}{F |z_{Cl^-}|} = \frac{76.31 \times 10^{-4}}{96{,}485} = 7.616 \times 10^{-8} \, \text{m}^2 \, \text{SC}^{-1}.$$

A numerical example of calculating E_{diff} for $b_{Cl^- \, (II)} = b_{H^+ \, (II)} = 0.1 \, \text{mol kg}^{-1}$ is given as follows:

$$E_{diff} = -\frac{8.314 \times 298.15}{96{,}485}$$

$$\left[\frac{(0-0.1) \times 3.624 \times 10^{-7} \times |1|}{1} + \frac{(0.1-0) \times 7.909 \; 10^{-8} \times |1|}{1} + \frac{(0.1-0.1) \times 7.616 \; 10^{-8} \times |-1|}{-1} }{(0-0.1) \times 3.624 \times 10^{-7} \times |1| + (0.1-0) \times 7.909 \; 10^{-8} \times |1| + (0.1-0.1) \times 7.616 \; 10^{-8} \times |-1|} \right]$$

$$\times \ln \left(\frac{0 \times 3.624 \times 10^{-7} \times |1| + 0.1 \times 7.909 \; 10^{-8} \times |1| + 0.1 \times 7.616 \; 10^{-8} \times |-1|}{0.1 \times 3.624 \times 10^{-7} \times |1| + 0 \times 7.909 \; 10^{-8} \times |1| + 0.1 \times 7.616 \; 10^{-8} \times |-1|} \right)$$

$$= 0.02685 \, \text{V} = 26.85 \, \text{mV}.$$

In a similar manner, E_{diff} was calculated for all other concentrations of HCl(aq) and is given in the following table:

$b_{HCl(aq)}$ (mol kg^{-1})	E_{diff} (mV)
0.1	26.85
0.05	19.96
0.02	12.91
0.01	9.09
0.005	6.43
0.002	4.33
0.001	3.50

The calculated results show that in this junction the diffusion potential decreases with the decrease of $b_{HCl(aq)}$.

REFERENCES

1. J. Rumble (Editor-in-Chief), *Handbook of Chemistry and Physics*, 101st ed., CRC Press, Boca Raton, FL, 2020.
2. A.J. Bard and L.R. Faulkner, *Electrochemical Methods: Fundamentals and Applications*, Wiley, New York, 2001.
3. D.A. MacInnes and Y.L. Yeh, The potentials at the junctions of monovalent chloride solutions. *J. Am. Chem. Soc.*, **43**, 1921, 2563–2573.
4. R. Holze, *Experimental Electrochemistry. A Laboratory Textbook*, Wiley-VCH, Weinheim, Germany, 2009.
5. E.R. Cohen et al., *Quantities, Units, and Symbols in Physical Chemistry*, 3rd ed., RSC Publishing, Cambridge, UK, 2007.
6. R.A. Robinson and R.H. Stokes, *Electrolyte Solutions: Second Revised Edition*, Dover, New York, 2002.

6 Electrochemical Kinetics

6.1 OBJECTIVES

The main objective of this chapter is to introduce students to the processes and their explanations when current passes through an electrochemical cell. This chapter will cover both electrochemical processes and transport phenomena, which occur in the electrolyte and at the electrolyte/electrode interface due to the current in the electrochemical cell. Homogeneous electrochemical reactions are not considered in this book. Overpotentials of the electrochemical cell and half-reactions are introduced. Mechanisms of the charge and mass transfer processes are considered, and corresponding equations describing them are presented and analyzed. Some simplifications and generalizations of the fundamental equations are also given.

6.2 CONCEPT OF ELECTROCHEMICAL CELL OVERPOTENTIAL

In Chapter 2, we briefly discussed the concept of electrochemical cells with and without current in the electrochemical cell circuit. Figures 2.9 through 2.11 summarize the previous discussions on this topic. In this section, we will focus on connecting the knowledge from Chapter 2 to a more in-depth discussion on electrochemical kinetics. Note that Equations 2.8 and 2.12, respectively, corresponding to Figures 2.10 and 2.11, were written with an assumption that the electrochemical reaction will not create any resistance in addition to the electrolyte resistance, which is due to the electrolyte resistivity. Based on the diagram in Figure 2.12, E_{EQ}, E_{GC}, I_{GC} are always >0 and E_D, E_{EC}, I_{EC} are always <0. However, R_i should include a resistance related to the rate of the electrodic electrochemical reactions, transport of electrochemically active species, and other processes occurring at the solution/electrode interface. If such process produces some additional resistance, an extra potential drop takes place and is called overpotential (η).

The overpotential for galvanic and electrolytic cells can be defined as deviation of the cell potential from its equilibrium value, excluding the potential drop due to the cell solution resistance R_s:

$$\eta_{GC} = E_{GC} - E_{EQ} + I_{GC}R_s, \tag{6.1}$$

$$\eta_{EC} = E_{EC} - E_D - I_{EC}R_s. \tag{6.2}$$

It can be seen that with this interpretation, overpotential of a cell is always a negative value. Furthermore, the cell overpotential consists of contributions from both electrodes, and individual electrode contributions cannot be separated without using a third electrode. We will discuss, in the following section, how the overpotential of a single electrode can be defined. An illustration of using Equation 6.1 will be given in Section 8.9 where a fuel cell electrochemical thermodynamics and kinetics are discussed.

DOI: 10.1201/b21893-6

It is interesting to note that when we move from equilibrium electrochemistry to electrochemical kinetics, we are in a better position to discuss the properties of a single electrode. In equilibrium thermodynamics, identifying a thermodynamic property for a single species is an impossible task. For example, to measure the potential absolute value of a single electrode is impossible. However, when we switch to studying irreversible, transport, and kinetic properties, the measurement for single species and single electrodes is possible. When the interest is to study the properties of a single electrode, a three-electrode cell must be used.

6.3 OVERPOTENTIAL OF A SINGLE ELECTRODE

As shown in Figure 6.1, a three-electrode electrochemical cell should be used to measure the overpotential of a single working electrode (WE).

The cell consists of three electrodes: a WE, a counter electrode (CE), and a reference electrode (RE). Each of the electrodes should have some special features. The WE surface should carefully be prepared, usually polished. CE should have a large surface and should not degrade, so CE is usually made by a Pt mesh or coil. The RE should have Nernstian behavior and a stable potential.

Figure 6.2 shows how a three-electrode cell can be designed. The potential difference between WE and RE should be measured using a high-resistance electrometer

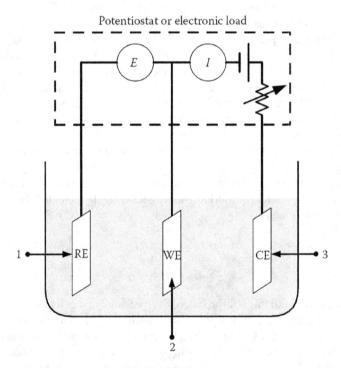

FIGURE 6.1 Schematic of a three-electrode electrochemical cell: (1) reference electrode (RE), (2) working electrode (WE), and (3) counter electrode (CE).

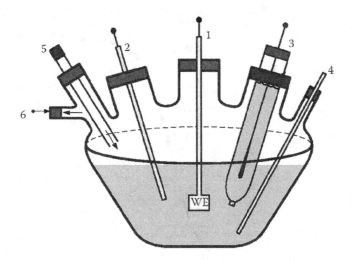

FIGURE 6.2 A three-electrode cell for overpotential measurement: (1) WE, (2) CE, (3) Ag/AgCl RE, (4) thermometer, (5) and (6) argon gas flow ports for purging, blanketing, and venting.

(voltmeter), and the current flowing between WE and CE should be measured using a very low resistance electrometer (ammeter).

If the resistance of the voltmeter is sufficiently high, effectively no current will pass between RE and WE. If the electrometer's resistance is too small, the current passing between these two electrodes will lead to errors in the potential difference measurements. In Figure 6.1, a battery and a variable resistor are used to supply a particular potential difference between WE and CE. However, in a modern laboratory, an electronic load is used instead of battery, variable resistor, and ammeter. A potentiostat can also be used, which is a convenient electronic system to carry out a variety of electrochemical measurements. The only disadvantage of using a potentiostat is that it is usually quite expensive. Note that most of the potentiostats are limited by a relatively low maximum current and voltage while an electronic load could have extended ranges of these parameters.

6.4   POLARIZATION CURVE OF A SINGLE ELECTRODE

A single WE can have an equilibrium potential or can be under an applied potential, negative or positive. For example, if a negative terminal of a battery is connected to the electrode, the electrode potential is negative, or we say that the electrode is negatively polarized. Similarly, a positive applied potential will provide a positive polarization. In electrochemical science and engineering, polarization means an applied potential, which is different from the equilibrium potential. While applying different potentials, the current flowing through the electrode can be measured using the three-electrode system shown in Figure 6.1.

When a positive overpotential is applied, the current is considered to be positive, and when a negative overpotential is applied, the current is negative. When the

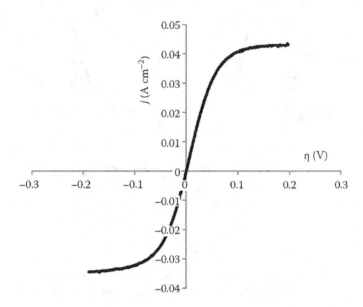

FIGURE 6.3   Polarization curve of a single $Cu^{2+}(aq)/Cu^+(aq)$ redox electrode taken at ambient temperature and pressure, initial concentration of $CuCl_2(aq)$ and $CuCl(aq)$ of 0.1 mol kg^{-1}, and concentration of $HCl(aq)$ of 8 mol kg^{-1} [1]: $Cu^+(aq) \rightarrow Cu^{2+}(aq) + e^-$ reaction takes place when the current and overpotential are positive and $Cu^{2+}(aq) + e^- \rightarrow Cu^+(aq)$ reaction occurs when the current and overpotential are negative.

dependence of the current vs. overpotential is plotted, it is called the polarization curve. A polarization curve of the $Cu^{2+}(aq)/Cu^+(aq)$ single redox electrode in highly concentrated $HCl(aq)$ is shown in Figure 6.3 [1].

As a reminder, the anodic polarization is always related to an oxidation reaction (e.g., $Cu^+(aq) \rightarrow Cu^{2+}(aq) + e^-$) and cathodic polarization related to the reduction reaction (e.g., $Cu^{2+}(aq) + e^- \rightarrow Cu^+(aq)$). As can be seen from the polarization curve of the $Cu^{2+}(aq)/Cu^+(aq)$ electrode, the current–potential relationship can be highly nonlinear. Our further goal will be to find out where the shape of the polarization curve is coming from.

When we consider a single electrode, the current density, j (usually in A cm^{-2}), can be more useful than just current. However, it is a challenge to estimate a real electrode surface vs. a geometrical area. Still, the electrode geometrical area is commonly used to calculate j.

The difference between the applied potential, E_{app}, and the equilibrium potential, E, is called the overpotential of a single electrode, $\eta = E_{app} - E$, which can be either positive or negative.

We assume here that the resistance between WE and CE is negligible. Otherwise, the solution resistance, R_s, should be experimentally measured and taken into account in estimating the single electrode overpotential, so the product of current and R_s, so-called IR drop, will not be included to the overpotential of the single electrode. One of the best methods to measure R_s is electrochemical impedance spectroscopy which will be considered in Section 7.10.

6.5 MECHANISM OF ELECTROCHEMICAL REACTION

The main goal of the electrochemical kinetics is to find a relationship between the electrode overpotential and current density due to an applied potential. There can be a number of processes contributing to the overpotential, and some examples are given in Figure 6.4.

Two main contributions to the overpotential will be discussed in this book in some detail. The first one is the charge (electron) transfer overpotential, which is due to a particular rate of the electrochemical reaction and takes place just at the electrode–solution interface. The second one is the mass transfer overpotential, which is due to delivering reactants to the electrochemical reaction interface or due to transporting products to the bulk solution. Other physicochemical processes taking place in the Nernst diffusion layer (e.g., chemical reactions and adsorption/desorption) can also contribute to the electrode overpotential, but they will not be discussed in this book. Note that chemical reactions occurring in the bulk solution should be taken into account to correctly estimate the concentration of the reduced, R_{bulk}, and oxidized, O_{bulk}, species.

Electrode overpotential is generally an additive function of all processes mentioned earlier, and each of them is described by its own overpotential, η_i. The total overpotential of a single electrode is a sum of all contributions:

$$\eta = \Sigma \eta_i, \tag{6.3}$$

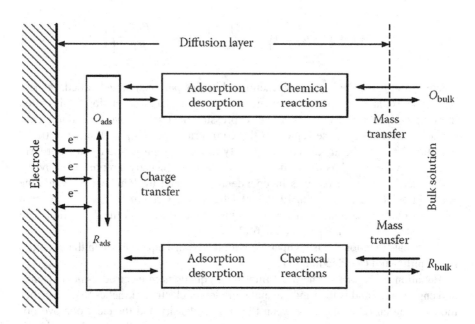

FIGURE 6.4 Possible processes in considering the mechanism of an electrochemical reaction. (Adapted from Figure 1.3.6 in A.J. Bard and L.R. Faulkner, *Electrochemical Methods: Fundamentals and Applications*, Wiley, New York, 2001.)

which in the case of two processes, charge transfer and mass transfer, can be simplified as follows:

$$\eta = \eta_{ct} + \eta_{mt}. \tag{6.4}$$

Let us consider the theoretical background of η_{ct} and η_{mt}.

6.6 CHARGE TRANSFER OVERPOTENTIAL: BUTLER–VOLMER EQUATION

Let us consider an electrochemical half-reaction as

$$Ox + ne^- = Red. \tag{6.5}$$

Using a IUPAC convention that the cathodic current and overpotential are negative while the anodic current and overpotential are positive [2], taking into account that the cathodic direction in Reaction (6.5) is from the left to the right and anodic is from the right to the left, employing the transition state theory for kinetics of the electrochemical half-reaction [3], and assuming that the electric potential affects the energy barrier of the electrochemical reaction [3], an equation describing the current density, j, as a function of overpotential, η, for an elementary (one-step) electrochemical reaction (6.5) can be derived as follows [3,4]:

$$j = j_a + j_c = j_o \left[\exp\left\{ \frac{(1-\beta)nF\eta}{RT} \right\} - \exp\left\{ \frac{-\beta nF\eta}{RT} \right\} \right], \tag{6.6}$$

where j_o and β are the key electrochemical kinetics parameters and called, respectively, the exchange current density and the symmetry factor. Equation 6.6 is called the Butler–Volmer equation and allows expressing the current–overpotential dependence in a relatively close region to the equilibrium point ($\eta = 0$ and $j = 0$). In Equation 6.6, j_a is the anodic current density (positive) represented by the first term in Equation 6.6, and j_c is the cathodic current density (negative) represented by the second term. Equation 6.6 was independently obtained by John Alfred Valentine Butler in 1924, Max Volmer in 1930, and Tibor Erdey-Grúz in 1930. An illustration of how Equation 6.6 can be derived and understood based on the Transition Sate Theory (TST) is presented in Section 6.17.

Note that Equation 6.6 is compatible with the Nernst equation when the potential is at equilibrium (see Section 6.17).

Equation 6.6 is one of the most important equations in electrochemical science and engineering and is the fundamental equation of electrode kinetics that describes the exponential relationship between the current density and the electrode overpotential. One of the key assumptions in deriving Equation 6.6 is that the electrochemically active solution at the electrode is extensively stirred and the current is kept so low that the concentrations at the site of the electrochemical reactions (electrode

John Alfred Valentine Butler (1899–1977) was an English physical chemist, who greatly contributed to theoretical electrochemistry. Particularly, he contributed to developing a relationship between electrochemical kinetics and thermodynamics. He is best known for his contribution to the development of the famous Butler–Volmer equation.

REFERENCES

A.J. Bard, G. Inzelt, and F. Scholz, *Electrochemical Dictionary*, Springer, Berlin, Germany, 2008.
http://en.wikipedia.org/wiki/John_Alfred_Valentine_Butler, https://www.jstor.org/stable/769843?seq=1#metadata_info_tab_contents

Max Volmer (1885–1965) was a German physical chemist, who made important contributions to electrochemistry, in particular on electrode kinetics. He codeveloped the Butler–Volmer equation.

REFERENCES

A.J. Bard, G. Inzelt, and F. Scholz, *Electrochemical Dictionary*, Springer, Berlin, Germany, 2008.

http://en.wikipedia.org/wiki/Max_Volmer

Tibor Erdey-Grúz (1902–1976) was a Hungarian physical chemist, who is most known for his work on transport processes in electrolyte solutions. He worked with Max Volmer in the development of the relationship between current and electrode potential known as the Erdey-Grúz–Volmer equation, which was, in fact, a prototype of the Butler–Volmer equation.

REFERENCES

A.J. Bard, G. Inzelt, and F. Scholz, *Electrochemical Dictionary*, Springer, Berlin, Germany, 2008.

http://en.wikipedia.org/wiki/Tibor_Erdey-Gr%C3%BAz, https://alchetron.com/Tibor-Erdey-Gr%C3%BAz

surface) do not substantially differ from the bulk values. Precisely, the equation can be used in a region of the charge transfer overpotential so that in Equation 6.6, η_{ct} should be used instead of the total overpotential η. Let us discuss two key parameters, j_o and β, introduced in Equation 6.6.

6.7 SYMMETRY FACTOR, β

Figure 6.5 [3] roughly demonstrates the background of the Transition State Theory (TST) of chemical kinetics for a reduction reaction, including an electron: Ox + e⁻ → Red. For the reduction reaction to take place, some Gibbs energy, G, should be provided to overcome the energy barrier and reach an activated complex located on the top of the G-Reaction coordinate curve. The same approach can be used to explain the reverse oxidation reaction Red $\rightarrow$ Ox + e$^-$. In Figure 6.5, the energy barrier for the anodic reaction is larger than that for the cathodic reaction, and the Gibbs energy barrier is symmetrical when $\beta = 0.5$. β is an indicator of the symmetry of the Gibbs energy barrier and ranges from 0 to 1. The symmetry factor can be experimentally obtained, but in some studies, the energy barrier is immediately assumed to be symmetrical, $\beta = 0.5$, due to the scarcity of the electrochemical kinetics data available to reliably estimate β.

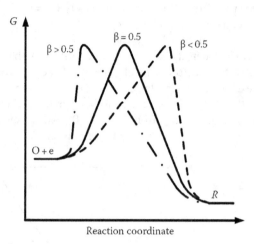

FIGURE 6.5   The influence of symmetry factor (β) on the shape of the energy barrier of an electrochemical reduction (Ox + e$^-$ $\rightarrow$ Red) and oxidation (Red $\rightarrow$ Ox + e$^-$) reactions.

6.8   EXCHANGE CURRENT DENSITY, J_O

Let us consider the anodic and cathodic terms of Equation 6.6:

$$j_a = j_o \exp\left\{ \frac{(1-\beta)nF\eta}{RT} \right\}, \tag{6.7}$$

$$j_c = -j_o \exp\left\{ \frac{-\beta nF\eta}{RT} \right\}. \tag{6.8}$$

Equations 6.6 through 6.8 are graphically represented in Figure 6.6. Equation 6.6 is plotted at $j_o = 0.05$ A cm^{-2}, $\beta = 0.2, 0.5$, and 0.8. When $\beta = 0.5$ (solid curve #3), the polarization curve is symmetrical, so at a given magnitude of the overpotential, the magnitudes of the anodic and cathodic current densities are the same. When $\beta = 0.8$ (dotted curve #5), the polarization curve is not symmetrical, and at a given magnitude of the overpotential, the magnitude of the cathodic current density is larger than the magnitude of the anodic current density. When $\beta = 0.2$ (dash-dot curve #2), the polarization curve is not symmetrical either, and at a given magnitude of the overpotential, the magnitude of the cathodic current density is smaller than the magnitude of the anodic current density. Solid curve #1 represents Equation 6.7, which is the dependence of the anodic current (only positive) from the overpotential (either positive or negative), and solid curve #4 represents Equation 6.8, which is the dependence of the cathodic current (only negative) from the overpotential (either positive or negative).

As can be seen, when an electrode reaction is in a dynamic equilibrium (the net current density, j, is zero), the magnitude of the cathodic current density ($j_c = -0.05$ A cm^{-2}) is equal to that of the anodic current density ($j_a = 0.05$ A cm^{-2}), and both the magnitudes are equal to the exchange current density $j_o = |j_c| = j_a$. Clearly, the exchange current density can only be a positive value.

The exchange current density is a key characteristic of the electrode reaction and shows the rate of the electrochemical reaction. If the electrochemical reaction is faster, then the exchange current density is larger, and when the electrochemical reaction is slower, the exchange current density is smaller. The exchange current density, j_o, can approximately range from 10^{-20} to 10^{-1} A cm^{-2} but cannot be a negative value. One of the fastest electrochemical reactions, $Ag^+(aq) + e^- = Ag(s)$, has an

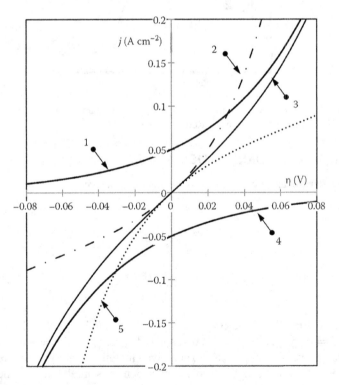

FIGURE 6.6 Current density vs. overpotential using Equations 6.6 through 6.8 to clarify their behavior plotted at $j_o = 0.05$ A cm^{-2} and temperature of 298.15 K: (1) Equation 6.7 at $\beta = 0.5$, (2) Equation 6.6 at $\beta = 0.2$, (3) Equation 6.6 at $\beta = 0.5$, (4) Equation 6.8 at $\beta = 0.5$, (5) Equation 6.6 at $\beta = 0.8$. (Adapted from Figure 4.6 in C.H. Hamann, et al., *Electrochemistry*, 2nd edn., Wiley-VCH, Weinheim, Germany, 2007.)

exchange current density of 0.15 A cm^{-2} in 1 mol L^{-1} HClO$_4$(aq). Due to such a large exchange current density, this reaction is used in the silver coulometer for determining the total amount of electricity passed by precipitating Ag(s) and weighing it after the test.

6.9 EXPERIMENTAL DATA ON THE EXCHANGE CURRENT DENSITY AND SYMMETRY COEFFICIENT

The amount of data on the exchange current density and symmetry factor is very limited due to experimental difficulties and insufficient studies in this area of electrochemical science and engineering. Some of these data are presented in Table 6.1 [4]. The fastest electrochemical reaction reported in Table 6.1 is Ag$^+$(aq) + e$^-$ = Ag(s) in 1 mol L^{-1} HClO$_4$(aq) solution with $j_o = 0.15$ A cm^{-2}. Other relatively fast reactions are Cd^{2+}(aq) + 2e$^-$ = Cd(Hg) (Cd(Hg) is Cd-Hg amalgam, which is a metal alloy) with $j_o = 0.025$ A cm^{-2}, Zn^{2+}(aq) + 2e$^-$ = Zn(Hg) (Zn(Hg) is Zn-Hg amalgam) with $j_o = 5.5$ 10^{-3} A cm^{-2}, K$_3$Fe(CN)$_6$(aq) + e$^-$ = K$_4$Fe(CN)$_6$(aq) on Pt with $j_o = 5$ 10^{-3}A cm^{-2}, Cd^{2+}(aq) + 2e$^-$ = Cd(s) with $j_o = 1.5$ 10^{-3} A cm^{-2}, and H$_2$O(l) + e$^-$ = 0.5H$_2$(g) + OH$^-$ (aq) and H$^+$(aq) + e$^-$ = 0.5H$_2$(g) both on Pt with $j_o = 10^{-3}$ A cm^{-2}. The slowest electrochemical reaction reported in Table 6.1 is H$^+$(aq) + e$^-$ = 0.5H$_2$(g) on Hg with $j_o = 10^{-12}$ A cm^{-2}. A comparison of the exchange current density of the H$^+$/H$_2$ electrode on Pt and Hg at the same H$_2$SO$_4$(aq) concentration and temperature allows concluding that the electrochemical kinetics of a reaction can significantly depend on the material of the electrode. An important reaction for the fuel cell and water electrolyzer, reduction reaction of oxygen, 2H$^+$(aq) + 0.5O$_2$(aq) + 2e$^-$ = H$_2$O(l) on Pt has $j_o = 10^{-6}$ A cm^{-2}, which is relatively slow and, therefore, a challenging issue for these electrochemical systems. It can be seen from Table 6.1 that the kinetic parameters depend on the electrode material and electrolyte composition. In addition, j_o and β depend on temperature and pressure, and therefore, when reporting these values, the electrode, solution composition, temperature, and pressure should be clearly specified.

In Table 6.1, the standard value of the exchange current density, j_o^0, can be defined using the following equation:

$$j_o = j_o^0 \left[\frac{c_{Ox}^{(1-\beta)}}{c^0} \frac{c_{Red}^{\beta}}{c^0} \right] = Fk^0 \left[\frac{c_{Ox}^{(1-\beta)}}{c^0} \frac{c_{Red}^{\beta}}{c^0} \right], \qquad (6.9)$$

where k^0 is the standard rate constant (in mol (cm^{2}s)$^{-1}$) of Reaction 6.5 and $j_o^0 = Fk^0$ is the standard exchange current density, which is defined as the exchange current density when molar concentrations of both Ox (c_{Ox}) and Red (c_{Red}) equal 1 mol L^{-1}. To experimentally obtain j_o^0, j_o should be measured as a function of c_{Ox} and c_{Red}, and then, an extrapolation should be made to $c_{Ox} = c_{Red} = 1$ mol L^{-1}. Note that the effect of the activity coefficients is completely ignored in Equation 6.9, which could be an issue if highly concentrated solutions are considered. The derivation of Equation 6.9 can be found in the literature [3,4].

TABLE 6.1

Experimental Values of Exchange Current Density, Standard Exchange Current Density, and Symmetry Coefficient

Electrode and Concentration of Aqueous Species	Aqueous Electrolyte	t (°C)	Electrode	j_o (A cm^{-2})	j_o^o (A cm^{-2})	β
Fe^{3+}/Fe^{2+} (0.005 mol L^{-1})	1 mol L^{-1} H$_2$SO$_4$	25	Pt	2×10^{-3}	4×10^{-1}	0.42
K$_3$Fe(CN)/K$_4$Fe(CN)$_6$ (0.02 mol L^{-1})	0.5 mol L^{-1} K$_2$SO$_4$	25	Pt	5×10^{-3}	5	0.51
Ag$^+$/Ag (0.001 mol L^{-1})	1 mol L^{-1} HClO$_4$	25	Ag	1.5×10^{-1}	13.4	0.35
Cd^{2+}/Cd (0.01 mol L^{-1})	0.4 mol L^{-1} K$_2$SO$_4$	25	Cd	1.5×10^{-3}	1.9×10^{-2}	0.45
Cd^{2+}/Cd(Hg) (0.0014 mol L^{-1})	0.5 mol L^{-1} Na$_2$SO$_4$	25	Cd(Hg)	2.5×10^{-2}	4.8	0.2
Zn^{2+}/Zn(Hg) (0.02 mol L^{-1})	1 mol L^{-1} HClO$_4$	25	Zn(Hg)	5.5×10^{-3}	0.10	0.25
Ti^{4+}/Ti^{3+} (0.001 mol L^{-1})	1 mol L^{-1} C$_2$H$_4$O$_2$	0	Pt	9×10^{-4}	0.9	0.45
H$_2$O/H$_2$, OH$^-$	1 mol L^{-1} KOH	25	Pt	10^{-3}	10^{-3}	0.5
H$^+$/H$_2$	1 mol L^{-1} H$_2$SO$_4$	25	Hg	10^{-12}	10^{-12}	0.5
H$^+$/H$_2$	1 mol L^{-1} H$_2$SO$_4$	25	Pt	10^{-3}	10^{-3}	0.5
H$_2$O/O$_2$, OH$^-$	1 mol L^{-1} KOH	25	Pt	10^{-6}	10^{-6}	0.7
H$^+$, O$_2$/H$_2$O	1 mol L^{-1} H$_2$SO$_4$	25	Pt	10^{-6}	10^{-6}	0.75

Source: Adapted from C.H. Hamann, et al., *Electrochemistry*, 2nd edn., Wiley-VCH, Weinheim, Germany, 2007.

6.10   SIMPLIFICATIONS OF THE BUTLER–VOLMER EQUATION

We will consider here three important cases to simplify Equation 6.6: (1) η is very close to 0, (2) $\eta \ll 0$, and (3) $\eta \gg 0$:

1. At a very small overpotential ($|\eta| < 10\,\mathrm{mV}$), the exponents in Equation 6.6 are small enough to expand both of them taking into account that $e^x \approx 1 + x$:

$$j = \frac{j_o n F \eta}{RT}. \tag{6.10}$$

   Equation 6.10 shows that in this case, the current density is proportional to overpotential. It is simply Ohm's law. Therefore, measuring j as a function of η at $\eta < 10\,\mathrm{mV}$, the exchange current density can be estimated. A disadvantage of this approach is that the precision of the measurements should be very high. Otherwise, the accuracy of j_o will be low. The value of $\eta/j = r = RT/(Fnj_o)$ (see Equation 6.10) is called the area-specific charge transfer resistance with unit $\Omega\,\mathrm{cm}^2$. This case is represented by curves #2, #3, and #5 in Figure 6.6 in a region close to equilibrium ($\eta = 0$) where j is almost a linear function of η.
2. When the electrode is highly cathodically polarized, $\eta \ll 0$, the first term of Equation 6.6 is very small and the equation is simplified to

$$j = -j_o \exp\left[\frac{-\beta n F \eta}{RT}\right]. \tag{6.11}$$

   This case is represented by curve #4 in Figure 6.6 in a region of highly negative overpotential. Equation 6.11 shows that in this case, the current density is an exponential function of overpotential. This equation represents the well-known Tafel equation (see Section 6.11) for the cathodic region. Equation 6.11 is usually valid at $\eta < -30\,\mathrm{mV}$. Both j_o and β can be found using this equation. A disadvantage here is that the measurements should be carried out far from the equilibrium region, where some irreversible process could take place and create problems for defining j_o and β correctly.
3. At $\eta \gg 0$, the second term of Equation 6.6 is very small and the equation is simplified to

$$j = j_o \exp\left[\frac{(1-\beta)n F \eta}{RT}\right]. \tag{6.12}$$

   This case is represented by curve #1 in Figure 6.6 in a region of highly positive overpotential. Equation 6.12 shows that the current density is an exponential function of overpotential again. It is also the well-known Tafel equation, but for the anodic region. The equation is usually valid at $\eta > +30\,\mathrm{mV}$ with the same disadvantage as in case (2). It is shown in the following section how Equations 6.11 and 6.12 can be used to define both j_o and β.

6.11   TAFEL EQUATION

Equations 6.11 and 6.12 have been extensively used to experimentally define both j_o and β. Let us take the natural logarithm of Equations 6.11 and 6.12:

When $\eta \ll 0$ (cathodic polarization),

$$\eta = \left[\frac{RT}{\beta nF} \right] \ln j_o - \left[\frac{RT}{\beta nF} \right] \ln |j|. \tag{6.13}$$

When $\eta \gg 0$ (anodic polarization),

$$\eta = \left\{ \frac{RT}{(1-\beta)nF} \right\} \ln j - \left\{ \frac{RT}{(1-\beta)nF} \right\} \ln j_o. \tag{6.14}$$

Both Equations 6.13 and 6.14 can be presented in a general form

$$\eta = A_T + B_T \ln |j| = A_T + B'_T \log_{10} |j|. \tag{6.15}$$

which is exactly the Tafel equation. Equation 6.15 allows using the intercept A_T and the Tafel slope B_T or B'_T to estimate both j_o and β. B_T and B'_T differ by a factor of 2.303, which originates from the difference in the outputs between ln and $\log_{10}$. Equation 6.15 was experimentally obtained by Julius Tafel in 1905.

Julius Tafel (1862–1918) was a Swiss physical chemist, who made a number of contributions, including an experimentally discovered relationship between the rate of an electrochemical reaction and the applied potential. This relation is known as the Tafel equation.

REFERENCES

A.J. Bard, G. Inzelt, and F. Scholz, *Electrochemical Dictionary*, Springer, Berlin, Germany, 2008.
http://en.wikipedia.org/wiki/Julius_Tafel

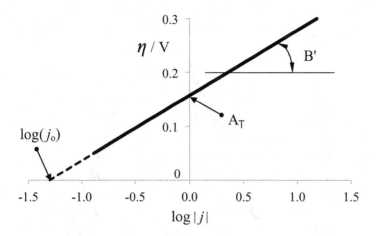

FIGURE 6.7 An example of the Tafel plot. The current density and the exchange current density are A cm^{-2}.

Note that the linear dependence between η and $\log_{10}|j|$ is applicable at overpotentials above about 30 mV. Therefore, Equation 6.15 cannot be used at current density close to 0, where the overpotential will approach infinity. This is an artifact related to the approximations that have been made deriving this equation. Use of the Tafel equation and its behavior is illustrated in Figure 6.7. Both j_o and β can be calculated using this plot.

6.12 VOLCANO PLOT

There is a well-known connection between the exchange current density of the hydrogen evolution electrochemical reaction $[H^+(aq) + e^- = (1/2)H_2(g)]$ and the binding metal–hydrogen (M–H) energy, where M is a metal. The M–H energy is defined as the bond strength obtained from the adsorption heats of hydrogen on the metals [5]. The relationship between the exchange current density and the M–H energy is called the volcano plot and originated from the well-known Sabatier principle, which declares that for obtaining high catalytic activity, the interaction between reactant (H) and catalyst (M) should be neither too strong nor too weak. As can be seen in Figure 6.8, if the M–H bond strength is either larger or smaller than 250 kJ mol^{-1}, the exchange current density is below 10^{-3} A cm^{-2}. This maximum of the volcano plot corresponds to some noble metals such as Pt, Re, Rh, and Ir. Note that the binding M–H energy is usually correlated to the catalytic activity of the metal (M). Therefore, an electrochemical approach can be used to better understand the catalytic activity of materials. The volcano plot can be explained by a competition between the surface coverage and energy of the hydrogen adsorption. On the left, due to insufficient M–H energy, the strength of adsorption to the surface is too low, and therefore, the adsorption rate is low. On the right, due to high M–H energy, the adsorption strength is too high, so the atoms fail to dissociate and the surface is fully covered with catalyst sites blocked. Apparently, Pt has the best possible combination

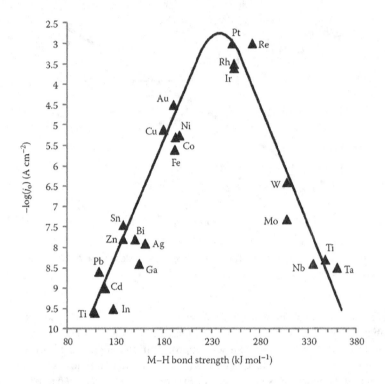

FIGURE 6.8   Volcano plot for the hydrogen evolution electrochemical reaction showing the exchange current density vs. M–H bond energy. (Adapted from S. Trasatti, *J. Electroanal. Chem.*, **39**, 1972, 163.)

between the surface coverage and surface energy and, therefore, demonstrates the highest value of the exchange current density. This is why Pt is still the best available material as electrochemical catalyst in a variety of fuel and electrolytic cells but expansive.

6.13   CONCENTRATION OVERPOTENTIAL

Thus far, it was generally assumed that the concentration of electrochemically active species at the electrode surface and that in the solution bulk are the same. However, once an electrochemical reaction is initiated, the concentration of those species will commonly be higher or lower at the electrode/electrolyte interface than that in the bulk of the solution. For example, considering copper deposition from an aqueous $CuSO_4$(aq) solution on a Cu electrode (similar to the reaction in the Daniell cell shown in Figure 2.10), a gradient of concentration of $CuSO_4$(aq) is created due to a relatively slow diffusion of Cu^{2+}(aq) to the electrode surface and a relatively fast reduction reaction, Cu^{2+}(aq) $+ 2e^- \rightarrow$ Cu(s), at the copper electrode. As a result, the molar concentration of Cu^{2+}(aq) at the surface, $c^s_{Cu^{2+}(aq)}$, will be smaller than the concentration in the bulk of the solution, $c^o_{Cu^{2+}(aq)}$.

The Nernst equation can be used to roughly describe the overpotential due to the gradient of the concentration of the electrochemically active ions between the surface and the bulk. Let us define the concentration of the oxidized species in Reaction (6.5), respectively, at the surface and in bulk as c_{Ox}^s and c_{Ox}^o. Then, the overpotential due to the concentration difference is

$$\eta = \frac{RT}{nF} \ln \frac{c_{Ox}^s}{c_{Ox}^o}. \tag{6.16}$$

where η is positive in the anodic reaction $(c_{Ox}^s > c_{Ox}^o)$ and negative in the cathodic reaction $(c_{Ox}^s < c_{Ox}^o)$.

The general problem in using these equations is to estimate the surface concentration of the electrochemically active species c_{Red}^s and c_{Ox}^s. However, there is a way to avoid the problem, and the approach is presented in the following section.

6.14 MASS TRANSFER OVERPOTENTIAL

One of the cases of mass transfer overpotential is the diffusion overpotential. This overpotential is purely due to the Fickian diffusion. The current density, j, due to Fick's first law of diffusion (one-dimensional case) is given as

$$j = nFD \left(\frac{\partial c}{\partial x} \right)_{x=0}, \tag{6.17}$$

where D is the diffusion coefficient of the aqueous electrochemically active species (e.g., Cu^{2+}(aq), $(\partial c/\partial x)_{x=0}$ is the gradient (one-dimensional) of the species concentration at the electrode surface $(x = 0)$.

Making a common approximation for $(\partial c/\partial x)_{x=0} \approx \Delta c/\delta_N$, the current density can be connected to the difference between the bulk and surface concentrations, Δc, and the Nernst diffusion layer, δ_N:

$$j \approx nFD \left[\frac{(c^s - c^o)}{\delta_N} \right], \tag{6.18}$$

which is constructed in a way that correctly provides the sign of the current density: positive for the anodic reaction and negative for the cathodic one. Note that Equations 6.17 and 6.18 are applicable only in a case when migration can be neglected (see Chapter 7 for details). An illustration of the Nernst diffusion layer thickness, δ_N, is shown in Figure 6.9 for a case of nonstationary conditions when current density is constant. In a steady state for an unstirred solution, δ_N can be on the order of 0.1 mm, and in a well-stirred solution, it can be as small as 10^{-5} mm.

As can be seen from Figure 6.9 and Equation 6.18, when $c^s = 0$, the current density cannot be changed anymore and is approaching its limiting value, j_{lim}, which is defined as follows:

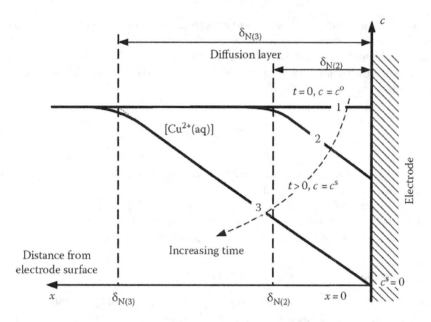

FIGURE 6.9 Schematic of the concentration gradient of $Cu^{2+}(aq)$ in the cathodic deposition of $Cu(s)$ in unstirred solutions: c^o is the bulk concentration, and c^s is the surface concentration. The numbers show the concentration gradients over time. Number 1 shows the initial time without any concentration gradient. Number 3 shows the limiting case when the surface concentration of $Cu^{2+}(aq)$ is zero. Number 2 is an intermediate case between Number 1 and Number 3.

$$j_{lim} \approx -nFD\left[\frac{c^o}{\delta_{N(3)}}\right]. \tag{6.19}$$

Combining Equations 6.18 and 6.19, we see that the surface concentration c^s can approximately be defined by the experimentally measured values of the current density and its limiting value as follows:

$$c^s = c^o\left[1-\left(\frac{j}{j_{lim}}\right)\right], \tag{6.20}$$

and, therefore, the mass transfer overpotential due to the Fickian diffusion can approximately be defined as follows:

$$\eta = \pm\frac{RT}{nF}\ln\left[1-\left(\frac{j}{j_{lim}}\right)\right], \tag{6.21}$$

which is positive for the anodic reaction and negative for the cathodic reaction. Equation 6.21 allows the estimation of the diffusion overpotential if the limiting current density is experimentally measured. While this approach is a great simplification, it is often a popular one to deal with the mass transfer overpotential.

6.15   GENERALIZED BUTLER–VOLMER EQUATION

If the mass transfer is presented by the diffusion overpotential and the electron transfer is described using the Butler–Volmer theory, these two approaches can be combined as follows:

$$j = j_o\left[\left(\frac{c_a^s}{c_a^o}\right)\exp\left\{\frac{(1-\beta)nF\eta}{RT}\right\} - \left(\frac{c_c^s}{c_c^o}\right)\exp\left\{\frac{-\beta nF\eta}{RT}\right\}\right], \tag{6.22}$$

where the factors of $\left(c_a^s/c_a^o\right)$ and $\left(c_c^s/c_c^o\right)$ were introduced in front of the corresponding exponents related to the anodic and cathodic reactions, respectively. This modification of the Butler–Volmer equation can be easily justified. When the current density goes to zero, $\left(c_a^s/c_a^o\right)$ or $\left(c_c^s/c_c^o\right)$ is approaching 1 and Equation 6.22 is converted to the original Butler–Volmer equation. Now combining Equations 6.20 and 6.22, the generalized Butler–Volmer equation can be obtained as follows [3]:

$$j = j_o\left\{\left[1-\left(\frac{j}{j_{lim,a}}\right)\right]\exp\left[\frac{(1-\beta)nF\eta}{RT}\right] - \left[1-\left(\frac{j}{j_{lim,c}}\right)\right]\exp\left[\frac{-\beta nF\eta}{RT}\right]\right\}, \tag{6.23}$$

which can easily be converted to a more convenient equation where there is not any current density on the right-hand side of the equation:

$$j = \frac{\left\{\exp\left[(1-\beta)nF\eta/(RT)\right] - \exp\left[-\beta nF\eta/(RT)\right]\right\}}{\left(1/j_o\right) + \left(1/j_{lim,a}\right)\exp\left[(1-\beta)nF\eta/(RT)\right] - \left(1/j_{lim,c}\right)\exp - \left[\beta nF\eta/(RT)\right]}, \tag{6.24}$$

and, therefore, the current density can immediately be calculated as soon as the overpotential, η, is given, the parameters of the electrochemical reaction, β, n and j_o, $j_{lim,a}$, and $j_{lim,c}$, are provided, and the temperature, T, is specified.

A graphical representation of the generalized Butler–Volmer Equation 6.24 is given in Figure 6.10. It can be seen that when overpotential is highly negative or positive, the current density approaches its limiting value to fulfill the mass transport limitation described in the previous section. The generalized Butler–Volmer equation can be used to experimentally define j_o and β when the polarization curve looks similar to the plot shown in Figure 6.10 for a wide range of current density and overpotential and also when the limiting current can experimentally be measured.

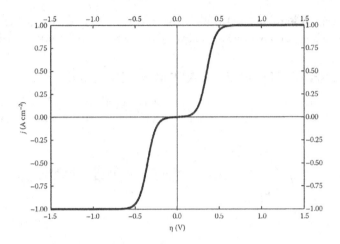

FIGURE 6.10 Current density vs. overpotential plotted using the generalized Butler–Volmer equation, Equation 6.24 with following parameters: $n = 1$, $T = 298.15$ K, $j_o = 10^{-3}$ A cm^{-2}, $j_{lim, a} = -j_{lim, c} = 1$ A cm^{-2}, $\beta = 0.5$.

6.16   SUMMARY

- In the electrochemical kinetics, the concept of the overpotential of a cell and a single electrode should be well understood.
- The cell overpotential consists of contributions from both electrodes and cannot be separated without using a third electrode.
- In order to measure the overpotential of a single electrode, a three-electrode electrochemical cell should be used. Each of the electrodes has a particular purpose and design.
- Dependence of current from potential is called the current–potential characteristic or polarization curve.
- An electrochemical reaction goes through a number of steps (in series). Each of the steps can be characterized by an overpotential. Two main contributions to the overpotential are the charge (electron) transfer and the mass transfer. The overpotential of an electrode is an additive function.
- The charge transfer overpotential is described by the Butler–Volmer equation. Two parameters of the Butler–Volmer equation are the exchange current density (j_o) and the symmetry factor β. The physical meaning of these parameters should be clearly understood.
- If an electrochemical reaction is faster (more reversible), then the exchange current density is larger. j_o can range from 10^{-20} A cm^{-2} (very slow reaction) to 10^{-1} A cm^{-2} (very fast reaction).
- β ranges from 0 to 1. If the energy barrier is symmetrical, $\beta = 0.5$, and this is a common assumption due to the scarcity of the electrochemical kinetics data.
- Simplification of the Butler–Volmer equation can lead to Ohm's law or the Tafel equation.
- High precision experimental j–η data are needed to estimate the exchange current density j_o from the j–η slope when overpotential magnitude is <10 mV.

- The area-specific charge transfer resistance is directly connected to the exchange current density.
- The Tafel equation can be used to experimentally estimate both j_o and β.
- The Tafel equation cannot be used at low overpotentials and is applicable at overpotentials above 30–50 mV.
- Only a few exchange current densities and symmetry factors are available in the literature. The values depend on the electrode material and electrolyte solution kind and concentration.
- The standard value of the exchange current density $\left(j_o^0 \right)$ is the exchange current density when both c_{Ox} and c_{Red} equal $1 \, mol \, L^{-1}$.
- The relationship between the exchange current density of the hydrogen evolution electrochemical reaction and the binding M–H energy is called the volcano plot.
- The concentration overpotential can be defined but cannot be calculated because the surface concentration of the electrochemically active species is not usually known.
- The mass transfer overpotential can be calculated using experimental data in the case of the Fickian diffusion of the electrochemically active species.
- If the concentrations of the electrochemically active species at the electrode surface and in the solution bulk are different, a concentration overpotential is developed.
- The concentration overpotential can be calculated if the concentration of the electrochemically active species at the electrode surface is known, but the surface concentration is usually unknown.
- If the concentration overpotential is due to the Fickian diffusion, the overpotential is called the diffusion overpotential and can roughly be calculated using the limiting current density. The limiting current density is defined when the surface concentration is zero.
- If mass transfer is presented by the diffusion overpotential and electron transfer is described using the Butler–Volmer theory, these two approaches can be combined and the generalized Butler–Volmer equation can be obtained.
- The reader is encouraged to practice employing the Tafel, Butler–Volmer, and generalized Butler–Volmer equations using a computer code such as Excel or Mathematica to better understand the current density–overpotential dependence.

6.17 EXERCISES

PROBLEM 1

Using Ref. [4] or any other suitable literature source, show how the Butler–Volmer Equation 6.6 can be obtained based on the TST.

Solution:

In the figure below, there is a representation of the TST applied for an electrochemical reaction (6.5):

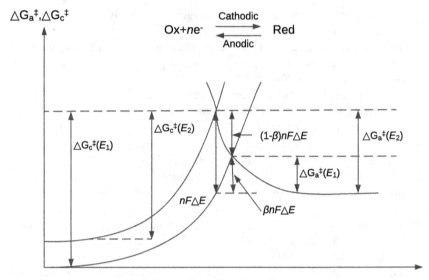

Reaction Coordinate

where the reaction extending from the left to the right is the cathodic reaction with the reaction rate k_c and Gibbs energy activation, $\Delta G_c^{\ddagger}(E)$, at a potential E. Similarly, the rate constant and Gibbs energy activation of the anodic reaction, extending from the right to the left, are, respectively, denoted at k_a and $\Delta G_a^{\ddagger}(E)$. The lower curve on the left and the curve on the right correspond to a potential E_1, while the upper curve on the left corresponds to a potential E_2. We assume that the applied potential is negative, that is cathodic, so $E_2 - E_1 = \Delta E < 0$.

The TST provides an Arrhenius type relationship between k_c and $\Delta G_c^{\ddagger}(E)$ or between k_a and $\Delta G_a^{\ddagger}(E)$ at a potential E_1 as follows:

$$k_c = k_c^0 \exp\left[-\frac{\Delta G_c^{\ddagger}(E_1)}{RT} \right],$$

$$k_a = k_a^0 \exp\left[-\frac{\Delta G_a^{\ddagger}(E_1)}{RT} \right],$$

where k_c^0 and k_a^0 are the standard rate constants of the cathodic and anodic reactions, respectively.

The relationship between the Gibbs energy activation at two potentials E_1 and E_2 can be defined using the above figure and introducing a parameter, β, which is also called the charge transfer coefficient or symmetry factor:

$$\Delta G_c^{\ddagger}(E_2) = \Delta G_c^{\ddagger}(E_1) + \beta n F \Delta E,$$

$$\Delta G_a^{\ddagger}(E_2) = \Delta G_a^{\ddagger}(E_1) - (1 - \beta) n F \Delta E.$$

Note that in the above equations, the magnitude of $\Delta G_c^{\ddagger}(E_2)$ decreases due to the negative value of ΔE and the magnitude of $\Delta G_a^{\ddagger}(E_2)$ increases.

The current density corresponding to the cathodic, $j_c(E_1)$, and the anodic $j_a(E_1)$, reactions at a potential E_1 can be defined using background of the chemical kinetics which suggest that the current density is proportional to the concentration, number of electrons, Faraday constant, and the reaction rate:

$$j_c(E_1) = -nFc_{Ox}k_c = -nFc_{Ox}k_c^0 \exp\left[-\frac{\Delta G_c^{\ddagger}(E_1)}{RT}\right],$$

$$j_a(E_1) = nFc_{Red}k_a = nFc_{Red}k_a^0 \exp\left[-\frac{\Delta G_a^{\ddagger}(E_1)}{RT}\right].$$

In the above equations the cathodic current density is negative and the anodic one is positive.

Similarly, the current density corresponding to the cathodic, $j_c(E_2)$, and anodic $j_a(E_2)$, reactions at a potential E_2 can be given as

$$j_c(E_2) = -nFc_{Ox}k_c^0 \exp\left[-\frac{\Delta G_c^{\ddagger}(E_2)}{RT}\right] = -nFc_{Ox}k_c^0 \exp\left[\frac{-\Delta G_c^{\ddagger}(E_1) - \beta nF\Delta E}{RT}\right],$$

$$j_a(E_2) = nFc_{Red}k_a^0 \exp\left[-\frac{\Delta G_a^{\ddagger}(E_2)}{RT}\right] = nFc_{Red}k_a^0 \exp\left[\frac{-\Delta G_a^{\ddagger}(E_1) + (1-\beta)nF\Delta E}{RT}\right].$$

For convenience, let us make the following simplification which will not affect the derivation result:

If we put $E_1 = 0$, then $\Delta E = E_2 \equiv E$.

$$k_c^0 \exp\left[\frac{-\Delta G_c^{\ddagger}(E_1)}{RT}\right] = k_c^{0\prime},$$

$$k_a^0 \exp\left[\frac{-\Delta G_a^{\ddagger}(E_1)}{RT}\right] = k_a^{0\prime}.$$

Now the expressions for the cathodic and anodic current densities can be presented as follows:

$$j_c(E) = -nFc_{Ox}k_c^{0\prime} \exp\left[-\frac{\beta nFE}{RT}\right],$$

$$j_a(E) = nFc_{Red}k_a^{0\prime} \exp\left[\frac{(1-\beta)nFE}{RT}\right].$$

At the equilibrium potential,E_{eq}, the two equations above will be the same in magnitude and will be equal to the exchange current density, j_o, which is always a positive value:

$$j_c(E_{eq}) = -nFc_{Ox}k_c^{0'}\exp\left[-\frac{\beta nFE_{eq}}{RT}\right] = -j_o,$$

$$j_a(E_{eq}) = nFc_{Red}k_a^{0'}\exp\left[\frac{(1-\beta)nFE_{eq}}{RT}\right] = j_o.$$

Introducing the overpotential, $\eta = E - E_{eq}$, the cathodic and anodic current densities can be presented as

$$j_c(E) = -nFc_{Ox}k_c^{0'}\exp\left[-\frac{\beta nFE_{eq}}{RT}\right]\exp\left[-\frac{\beta nF\eta}{RT}\right] = -j_o\exp\left[-\frac{\beta nF\eta}{RT}\right],$$

$$j_a(E) = nFc_{Red}k_a^{0'}\exp\left[\frac{(1-\beta)nFE_{eq}}{RT}\right]\exp\left[\frac{(1-\beta)nF\eta}{RT}\right] = j_o\exp\left[\frac{(1-\beta)nF\eta}{RT}\right].$$

The net current, $j(E)$, is the sum of the cathodic and anodic currents and, therefore, is as follows:

$$j(E) = j_o\left\{\exp\left[\frac{(1-\beta)nF\eta}{RT}\right] - \exp\left[-\frac{\beta nF\eta}{RT}\right]\right\},$$

which is the Butler–Volmer Equation 6.6.

Note that in this problem, we ignored nonideality and also assumed that the current density is sufficiently small to have the concentration of the electrochemically active species at the electrode surface the same as in the bulk of the solution.

PROBLEM 2

Show that the Nernst equation can be obtained using the TST background.

Solution

Using the cathodic and anodic current densities obtained in Problem 1 for the equilibrium potential E_{eq},

$$j_c(E_{eq}) = -nFc_{Ox}k_c^{0'}\exp\left[-\frac{\beta nFE_{eq}}{RT}\right] = -j_o,$$

$$j_a(E_{eq}) = nFc_{Red}k_a^{0'}\exp\left[\frac{(1-\beta)nFE_{eq}}{RT}\right] = j_o,$$

it is easy to see that

$$nFc_{Ox}k_c^{0'} \exp\left[-\frac{\beta nFE_{eq}}{RT}\right] = nFc_{Red}k_a^{0'} \exp\left[\frac{(1-\beta)nFE_{eq}}{RT}\right],$$

which can be converted to a logarithmic form and simplified as follows:

$$E_{eq} = \frac{RT}{nF}\ln\left(\frac{k_c^{0'}}{k_a^{0'}}\right) + \frac{RT}{nF}\ln\left(\frac{c_{Ox}}{c_{Red}}\right) = E_{eq}^0,$$

which is the Nernst equation where $E_{eq}^0 = \dfrac{RT}{nF}\ln\left(\dfrac{k_c^{0'}}{k_a^{0'}}\right)$ is the standard equilibrium potential directly related to the cathodic and anodic rate constants and the Gibbs energy activations of both reactions. Note that in this problem, we ignored nonideality, so the molar concentrations were used instead of activities.

PROBLEM 3

A table with experimental data used to construct the polarization curve in Figure 6.3 is given below.

η (V)	j (A cm^{-2})
2.79×10^{-2}	1.98×10^{-2}
2.30×10^{-2}	1.65×10^{-2}
1.89×10^{-2}	1.37×10^{-2}
1.41×10^{-2}	1.01×10^{-2}
9.06×10^{-3}	6.52×10^{-3}
4.14×10^{-3}	2.79×10^{-3}
-2.96×10^{-4}	-6.32×10^{-4}
-4.56×10^{-3}	-3.65×10^{-3}
-9.30×10^{-3}	-7.04×10^{-3}
-1.44×10^{-2}	-1.03×10^{-2}
-1.89×10^{-2}	-1.29×10^{-2}
-2.38×10^{-2}	-1.57×10^{-2}
-2.83×10^{-2}	-1.78×10^{-2}

Using data in the table and a suitable software (e.g., Mathematica or MATLAB), estimate the parameters j_o and β of the Butler–Volmer Equation 6.6.

Solution:

Equation 6.6 is given as

$$j = j_a + j_c = j_0\left[\exp\left\{\frac{(1-\beta)nF\eta}{RT}\right\} - \exp\left\{\frac{-\beta nF\eta}{RT}\right\}\right].$$

When the electrode is highly cathodically polarized, $\eta \ll 0$, the first term on the right-hand side of Equation 6.6 is neglectable. Similarly, when $\eta \gg 0$, the second term can be ignored. Therefore, η is taken in a range from -0.03 to 0.03 V to take into account both the anodic and cathodic parts of the polarization curve. According to Figure 6.3 $n = 1$ in the Butler–Volmer equation. By importing the appropriate data sets into MATLAB curve fitting tool, the fitting procedure gives $j_0 = 0.0171$ A cm^{-2} and $\beta = 0.453$. The coefficient of determination R^2 for this fit is 0.9978, and it means that the experimental current density satisfactorily follows the Butler–Volmer equation. The plot below shows both the experimental data and the fitting curve.

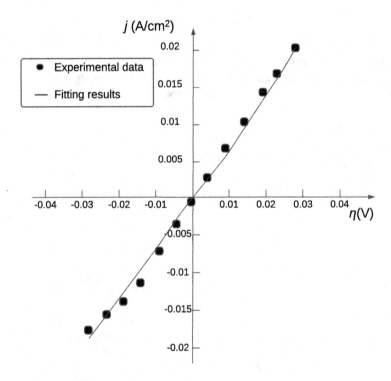

PROBLEM 4

A table with some experimental data used to construct the polarization curve shown in Figure 6.3 is given below.

Use data from this table to construct Tafel plots for both the cathodic and anodic parts of the polarization curve. Estimate j_o and β using both the anodic and cathodic polarization and compare them with the values obtained in Problem 3.

Solution:

When there is the anodic polarization, the Tafel Equation 6.14 can be written as

$$\eta = \frac{RT}{(1-\beta)nF}\ln j - \frac{RT}{(1-\beta)nF}\ln j_o = A_T + B_T{}'\log_{10}|j|.$$

Anodic	Polarization	Cathodic	Polarization
η (V) $\times 10^2$	j (A cm^{-2}) $\times 10^2$	$-\eta$ (V) $\times 10^2$	$-j$ (A cm^{-2}) $\times 10^2$
4.50	2.89	3.08	1.90
4.41	2.85	3.17	1.94
4.33	2.81	3.25	1.99
4.25	2.77	3.34	2.02
4.16	2.73	3.43	2.04
4.08	2.69	3.51	2.08
4.00	2.65	3.61	2.12
3.90	2.59	3.71	2.16
3.81	2.55	3.79	2.20
3.73	2.51	3.88	2.23
3.65	2.46	3.97	2.25
3.56	2.42	4.05	2.28
3.47	2.38	4.15	2.32
3.39	2.33	4.24	2.34
3.31	2.26	4.33	2.38
3.22	2.22	4.42	2.40
3.14	2.18	4.50	2.44
3.04	2.13		

Fitting the data given in the table, one can estimate $A_T = 0.2122$ and $B_T' = 0.1091$ with $R^2 = 0.9938$. Now β can be calculated as

$$\beta = 1 - 2.303 \times \left(\frac{RT}{nFB_T'} \right) = 1 - 2.303 \times \left(\frac{298.15 \times 8.314}{1 \times 96{,}485 \times 0.1091} \right) = 0.458$$

Then, j_o can be calculated as

$$j_o = \exp\left[\frac{(1-\beta)nFA_T}{-RT} \right] = \exp\left[\frac{(1-0.4577) \times 1 \times 96{,}485 \times 0.2122}{-(298.15 \times 8.314)} \right] = 0.0113\,\mathrm{A\,cm^{-2}}$$

The relevant Tafel plot for the anodic polarization is shown below.

$\beta = 0.4577$ obtained from this anodic polarization fit is similar to the Butler–Volmer fit $\beta = 0.453$ from Problem 3, whereas the exchange current density obtained this way $j_o = 0.01134\,\mathrm{A\,cm^{-2}}$ is slightly less than that obtained from Problem 2 ($j_0 = 0.0171$ A cm^{-2}), still having the same order of magnitude.

When there is the cathodic polarization, the Tafel Equation 6.13 can be written as

$$\eta = \frac{RT}{\beta nF} \ln j_o - \frac{RT}{\beta nF} \ln|j| = A_T + B_T' \log_{10}|j|.$$

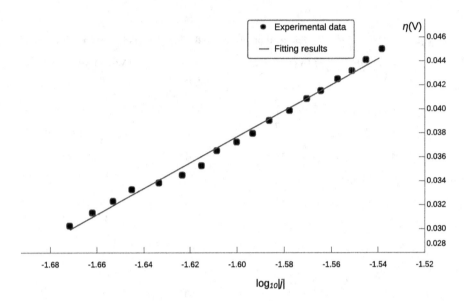

Fitting the data given in the table, one can estimate $A_T = -0.2607$ and $B_T' = -0.1340$ with $R^2 = 0.9927$. Now β can be calculated as

$$\beta = \frac{-2.303RT}{nFB_T'} = \frac{-2.303 \times 298.15 \times 8.314}{[1 \times 96,485 \times (-0.134)]} = 0.441.$$

Then, j_o can be calculated as

$$j_o = \exp\left[\frac{\beta nFA_T}{RT}\right] = \exp\left[\frac{0.4415 \times 1 \times 96,485 \times (-0.2607)}{(298.15 \times 8.314)}\right] = 0.0113\,\text{A cm}^{-2}.$$

The relevant Tafel plot for the cathodic polarization is shown below.

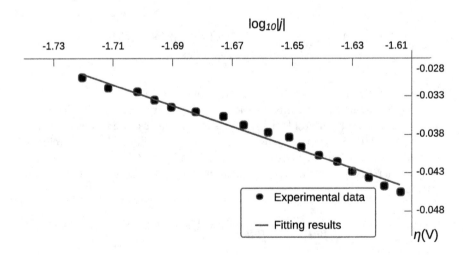

Both $\beta = 0.4415$ and the exchange current density $j_o = 0.01133\,A\,cm^{-2}$ obtained this way are slightly smaller than the Butler–Volmer fit ($\beta = 0.453$ and $j_0 = 0.0171$ A cm^{-2}) from Problem 3.

The calculated values of β and j_0 obtained in Problems 3 and 4 are summarized in the table below, showing that while there are some differences between these values using the Butler–Volmer and Tafel equations, generally, the agreement is within 30% for the exchange current density and within 4% for the symmetry factor.

	Butler–Volmer	Tafel Cathodic	Tafel Anodic
j_0	0.0171	0.0113	0.0113
β	0.453	0.441	0.458

REFERENCES

1. D.M. Hall, E.G. LaRow, R.S. Schatz, J.R. Beck, and S.N. Lvov, Electrochemical kinetics of CuCl(aq)/HCl(aq) electrolyzer for hydrogen production via a Cu-Cl thermochemical cycle. *J. Electrochem. Soc.*, **162**, 2015, F108–F114.
2. E.R. Cohen et al., *Quantities, Units, and Symbols in Physical Chemistry*, 3rd ed., RSC Publishing, Cambridge, UK, 2007.
3. A.J. Bard and L.R. Faulkner, *Electrochemical Methods: Fundamentals and Applications*, Wiley, New York, 2001.
4. C.H. Hamann, A. Hamnett, and W. Vielstich, *Electrochemistry*, 2nd ed., Wiley-VCH, Weinheim, Germany, 2007.
5. S. Trasatti, Work function, electronegativity, and electrochemical behavior of metals. III. Electrolytic hydrogen evolution in acid solutions. *J. Electroanal. Chem.*, **39**, 1972, 163–184.

7 Electrochemical Techniques II

7.1 OBJECTIVES

The main objective of this chapter is to show that transport processes can significantly contribute to the electrochemical kinetics measurements and therefore should be taken into account. Transport processes are based on hydrodynamics (fluid mechanics), which is described with mathematics of the vector analysis. Students should know what the terms "vector," "gradient," and "divergence" represent. Solutions of Fick's second law of diffusion are given as examples of hydrodynamics coupled with electrochemical kinetics. The theory and use of the rotating disk electrode (RDE) are explained. Introduction to cyclic voltammetry (CV) techniques is also given. A section on the electrochemical impedance spectroscopy (EIS) introduces the technique's background and use.

7.2 TRANSPORT PROCESSES IN ELECTROCHEMICAL SYSTEMS

Transport processes in electrochemical systems should be analyzed with vector analysis, a part of calculus. The main definitions of the vector analysis are terms like scalar, vector, gradient, divergence, and curl. The reader is encouraged to refresh their memory on definitions of these values. There are two key equations that are fundamental to transport processes in electrochemical systems. The first describes the flux vector, J_i, of the ith species [1]:

$$J_i = -D_i \text{grad}(c_i) - \frac{z_i F}{RT} D_i c_i \text{ grad}(\varphi) + c_i v, \tag{7.1}$$

due to (1) gradient of concentration of the ith species, c_i, (2) gradient of electric potential, φ, and (3) movement of the whole solution described by the velocity vector, v. The ith species is characterized by the diffusion coefficient, D_i, and its charge number, z_i. In Equation 7.1, the first term represents the Fickian diffusion, the second one describes the migration, and the last one defines the convection. Note that Equation 7.1 is valid for only dilute solutions where all effects of nonideality are ignored, and therefore, Equation 3.19 is valid. Also, for the sake of simplification, the vector values will not be shown in bold in this book.

In Equation 7.1, the unit of J_i is mol $(cm^2 s)^{-1}$, and the three vector values can be given in rectangular (Cartesian) coordinates as follows:

$$J_i = iJ_{i,x} + jJ_{i,y} + kJ_{i,z}, \tag{7.2}$$

DOI: 10.1201/b21893-7

$$v = iv_{i,x} + jv_{i,y} + kv_{i,z}, \tag{7.3}$$

$$\text{grad}(c_i) = i\left(\frac{\partial c_i}{\partial x}\right) + j\left(\frac{\partial c_i}{\partial y}\right) + k\left(\frac{\partial c_i}{\partial z}\right). \tag{7.4}$$

The second key equation is the fundamental equation for partial derivative of molar concentration, c_i, with respect to time:

$$\frac{\partial c_i}{\partial t} = -\text{div}\, J_i, \tag{7.5}$$

which is actually the differential form of the conservation law without chemical reactions taking place [1].

In some electrochemical problems (e.g., in the presence of a supporting electrolyte but not in the case of just one electrolyte), we can ignore using the second term of Equation 7.1, which represents migration due to a gradient of electric potential. Otherwise, an equation describing electroneutrality (e.g., Poisson's equation) should also be considered. Therefore, if excess of a supporting (i.e., not participating in electrochemical reactions) electrolyte is used, Equation 7.1 is simplified as follows:

$$J_i = -D_i \text{grad}(c_i) + c_i v, \tag{7.6}$$

which should be used together with the electroneutrality condition:

$$\Sigma z_i c_i = 0 (i = 1, 2, \ldots, N). \tag{7.7}$$

Note that Equations 7.1, 7.5, and 7.6 and the electroneutrality condition (7.7) can be solved together taking into account that the number of equations equals the number of ions (N), so that the total number of equations to be simultaneously solved is ($N+1$) relative to ($N+1$) unknown variables (N concentrations and a potential). Also, it is useful to recognize that if the convection does not take place ($v = 0$), then Equation 7.6 is converted to Fick's first law.

By combining Equations 7.5 and 7.6, and considering the electrolyte solution not compressible (div $v = 0$), the commonly used convective–diffusion equation can be obtained [1]:

$$\frac{\partial c_i}{\partial t} = D_i \nabla^2 c_i - v\, \text{grad}(c_i), \tag{7.8}$$

where ∇^2 is the Laplace operator or Laplacian, which in the one-dimensional case is simply the second partial derivative with respect to coordinate x, ($\partial^2/\partial x^2$). Considering the one-dimensional diffusion and convection, Equation 7.8 can be simplified as follows:

$$\frac{\partial c_i}{\partial t} = D_i\left(\frac{\partial^2 c_i}{\partial x^2}\right) - v_x\left(\frac{\partial c_i}{\partial x}\right). \tag{7.9}$$

In Equations 7.8 and 7.9, the diffusion coefficient is a constant value.

7.3 CURRENT–TIME DEPENDENCE AT CONSTANT POTENTIAL (POTENTIOSTATIC REGIME)

Using a three-electrode system and a potentiostat, described in Chapter 6, a constant potential can be set up and the current density vs. time can be measured. This organization of measurements is called the potentiostatic regime.

If there is no convection, Equation 7.8 can be used with only the first term on the right-hand side. Under these conditions and when the potential is held constant, the current measurements can be used to define (1) the Nernst diffusion layer thickness, δ_N, as a function of time, (2) the diffusion coefficient of the electrochemically active species, and concentration of the electrochemically active species at the surface, c^s. In the potentiostatic regime, the concentration of the electrochemically active species is a function of both distance from electrode, x, and time, t. Let us see how the equation for Fick's second law of diffusion

$$\frac{\partial c}{\partial t} = D\left(\frac{\partial^2 c}{\partial x^2}\right),$$
(7.10)

can be used. In Equation 7.10, c and D can be molar concentration and diffusion coefficient either for an ion or for an electrolyte. The electrolyte diffusion coefficient is defined as $D = (z_+ + |z_-|) \, D_+ D_-/(z_+ D_+ + |z_-| D_-)$ [1] where D_+ and D_- are the diffusion coefficients of the cation and the anion (Table 10.12), respectively. Note that this equation is valid in the absence of migration of the ith species due to migration in the supporting electrolyte or in a case when a single electrolyte is considered. Also, any nonideality is not taken into account.

Equation 7.10 can be solved with the following initial and boundary conditions:

$$(1)\, t = 0, x \geq 0, c = c^o; (2)\, t > 0, x \to \infty, c = c^o; (3)\, t > 0, x = 0, c = c^s.$$

The solution of Equation 7.10 is [2]

$$c(x,t) = c^s - \left(c^s - c^o\right)\mathrm{erf}\left[\frac{x}{\left(2\sqrt{Dt}\right)}\right],$$
(7.11)

where $\mathrm{erf}\left[x/\left(2\sqrt{Dt}\right)\right]$ is the error function (Table 10.28), D is the effective diffusion coefficient of a binary electrolyte as defined earlier, and c^0 is the electrolyte concentration in the bulk.

Note that when a supporting electrolyte is used, Equation 7.11 can be used for an ion. For example, solutions of Equation 7.11 for a case of Cu^{2+}(aq) in $CuCl_2$(aq) solution ($c^o = 0.05$ mol L^{-1}, $c^s = 0$ mol L^{-1}, $D = 1.258 \times 10^{-9}$ m^2 s^{-1}) are given in Figure 7.1. Note that the diffusion coefficient here is related to the whole electrolyte and is actually an average of the diffusion coefficient of the cation, D_+, and the anion, D_-, with their charges taken into account, as defined earlier.

In Figure 7.1, the distance from the electrode surface until a point where concentration is not significantly deviating from the bulk concentration is shown at three

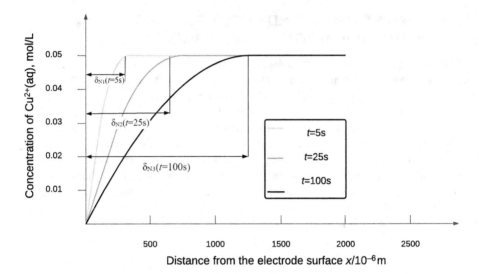

FIGURE 7.1 Concentration of Cu^{2+}(aq) in $CuCl_2$(aq) solution at an electrode surface as a function of distance (x) and time (t): $c^o = 0.05$ mol L^{-1}, $D = 1.258\ 10^{-9}$ m^2 s^{-1}.

times, 5, 25, and 100 seconds. This distance can be considered as the Nernst diffusion layer, δ_N, which can only be approximately defined. For example, if δ_N is defined for a concentration which is smaller than the bulk one by 1%, then δ_N can be approximately calculated as $3.8(Dt)^{1/2}$.

As follows from Equation 7.11, the concentration gradient at $x = 0$ is defined as follows:

$$\left(\frac{\partial c}{\partial x}\right)_{x=0} = \frac{c^o - c^s}{(\pi Dt)^{1/2}},\tag{7.12}$$

and, accordingly, the current density can be expressed as follows [1]:

$$j = -\left[nF\left(\frac{D}{t_-}\right)\right]\left(\frac{\partial c}{\partial x}\right)_{x=0} = -\left(\frac{nF}{t_-}\right)\left(\frac{D}{\pi}\right)^{1/2}\frac{c^o - c^s}{t^{1/2}}.\tag{7.13}$$

The time dependences of the current density calculated by Equation 7.13 are shown in Figures 7.2 and 7.3. The minus sign of Equation 7.13 is due to our choice of the direction of x axis starting at the surface and going to the bulk of the solution.

Here, we can conclude that (1) the Nernst diffusion layer thickness, δ_N, is a function of time as shown in Figure 7.1, (2) the dependence of the current density from time is given by Equation 7.13 and shown in Figures 7.2 and 7.3, (3) based on Equation 7.13, the diffusion coefficient of the electrolyte, D, can be found if the transport number, t_-, is known when the surface concentration is zero and the limiting current density is measured, and (4) from Equation 7.13, the surface concentration, c^s, can be obtained assuming that t_- and D are available and the current density is measured.

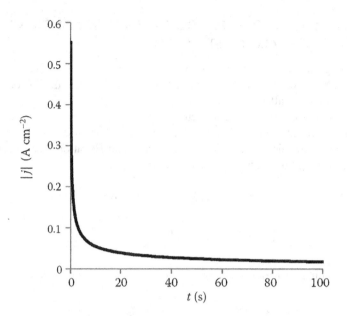

FIGURE 7.2 Magnitude of the current density vs. time at a constant surface concentration $c^s = 0.025$ mol L^{-1} for a CuCl$_2$(aq) solution with $c = 0.05$ mol L^{-1} at 298.15 K: $D_+ = 0.714 \ 10^{-9}$ m^2 s^{-1}, $t_- = 0.416$ (Table 10.12).

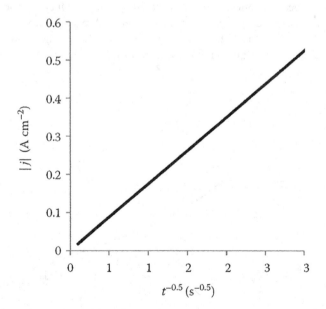

FIGURE 7.3 Magnitude of the current density vs. the inverse square root of time (from 0.1 to 100 s) at a constant surface concentration $c^s = 0.025$ mol L^{-1} for a CuCl$_2$(aq) solution with $c = 0.05$ mol L^{-1} at 298.15 K: $D_+ = 0.714 \ 10^{-9}$ m^2 s^{-1}, $t_- = 0.416$ (Table 10.12).

7.4 CONCENTRATION–TIME DEPENDENCE AT CONSTANT CURRENT (GALVANOSTATIC REGIME)

Using a three-electrode system and a potentiostat, described in Chapter 6, a constant current can be set up and the potential vs. time can be measured. This organization of measurements is called the galvanostatic regime.

Under a galvanostatic regime (the current density is constant) with a highly concentrated background electrolyte and a facile charge transfer reaction the concentration gradient at the electrode surface is constant, and Equation 7.10 can be solved with the following boundary and initial conditions [2]:

$$(1)t = 0, x \geq 0, c = c^o; (2)t > 0, x \to \infty, c = c^o; (3)t > 0,$$

$$j = -\left[nF\left(D/t_-\right)\right]\left(\partial c/\partial x\right)_{x=0}.$$

Solution of Equation 7.10 when $\left(\partial c/\partial x\right)$ is defined at $x = 0$ is as follows [1]:

$$c^s = c^o + \left[\frac{2j}{\left(nF/t_-\right)}\right]\left[\frac{t}{\pi D}\right]^{\frac{1}{2}}, \tag{7.14}$$

and the time dependence of c^s is shown in Figure 7.4.

The time to achieve zero concentration at the electrode surface, when the potential is sharply changed, is called the transition time, τ, after which the potential rapidly changes corresponding to another electrochemical reaction. An example shown in Figure 7.4 provides the transition time of 120.6 s. When the transition time is achieved, that is, $c^s = 0$, Equation 7.14 is transformed as follows:

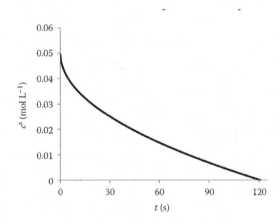

FIGURE 7.4 Surface concentration (c^s) vs. time (from 0.01 to 120.6 s) at a constant current density: $j = 0.05$ A cm^{-2} for a CuCl$_2$(aq) solution with $c^0 = 0.05$ mol L^{-1} at 298.15 K: $D_+ = 0.714\ 10^{-9}$ m^2 s^{-1}, $t_- = 0.416$ (Table 10.12).

$$j\tau^{1/2} = -\frac{nF}{2t_-}(\pi D)^{1/2}c^{o}, \qquad (7.15)$$

which shows the relationship between the diffusion coefficient, D, the transport number, t_-, and the bulk concentration, c^{o}, in the galvanostatic regime at the transition time τ. It is easy to see that Equation 7.15 can be used (1) to define the diffusion coefficient if the bulk concentration and the transport number, t_-, are known, and the current density and transition time are measured or (2) as an analytical tool to estimate the concentration of the electrochemically active species, measuring the current density and transition time if the diffusion coefficient and the transport number, t_-, are known.

7.5   EFFECT OF HYDRODYNAMICS (FLUID MECHANICS) ON ELECTROCHEMICAL REACTION

Previously, only mass transport due to diffusion was taken into account. Another influence on electrochemical process can be due to convection. Convection is the collective movement of ensembles of molecules within a fluid or a gas. To take into account convection, Equation 7.8 should be used for multidimensional processes or Equation 7.9 for a one-dimensional case.

In Equations 7.8 and 7.9, the first term is responsible for the Fickian diffusion and the second one for the convection. It is logical to assume that when convection in an electrochemical cell is changed, the Nernst diffusion layer is also changed, and therefore, we should find out how the cell current depends on hydrodynamics (fluid mechanics). A simple and common rule is that if stirring is more intensive, the Nernst diffusion layer is smaller.

If the solution flow is parallel to the electrode surface and perpendicular to the diffusion direction, the Prandtl boundary layer thickness, δ_{Pr}, and the Nernst diffusion layer thickness, δ_N, in a steady state are related as follows [2]:

$$\delta_N \approx \delta_{Pr}\left[\frac{\eta}{D\rho}\right]^{-1/3}, \qquad (7.16)$$

where η is viscosity in kg m^{-1} s^{-1}, ρ is density in kg m^{-3}, and D is diffusion coefficient in m^2 s^{-1}.

Figure 7.5 illustrates a physical meaning of both the Prandtl boundary layer and the Nernst diffusing layer.

The Prandtl layer represents the hydrodynamic layer where the flow speed of the solution is changing, and the Nernst diffusion layer represents the layer where the concentration of electrochemically active species is changing. At the distance below δ_{Pr}, the velocity of solution is gradually reducing down to zero, and at the distance below δ_N, the concentration of an electrochemically active species is gradually reducing down to c^s.

The reader may want to better understand Figure 7.5 and Equation 7.16 by calculating the ratio of the Nernst diffusion layer to the Prandtl layer for an aqueous solution species at a given temperature and pressure using the viscosity and density

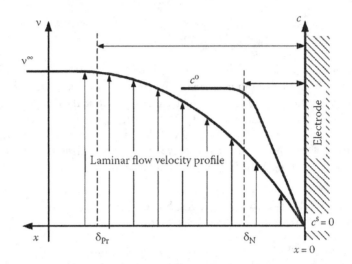

FIGURE 7.5 Illustration of the Nernst diffusion layer, δ_N, and the Prandtl boundary layer, δ_{Pr}; v^∞ is the solutions velocity at (theoretically) infinite distance from the electrode, and c^0 is the bulk solution concentration. δ_N is shown for the limiting current when the surface concentration $c^0 = 0$.

of water from Table 10.22 and an electrolyte diffusion coefficient from Table 10.12. The result of such calculations should show that at ambient conditions, δ_N is about ten times smaller than δ_{Pr} for such an electrolyte as $CuCl_2(aq)$.

7.6 ROTATING DISK ELECTRODE AND LIMITING CURRENT

It is not a simple task to estimate the Nernst diffusion layer thickness, δ_N. In the previous example presented by Figure 7.5, when solution flow is parallel to the electrode surface and perpendicular to the diffusion direction, the diffusion layer thickness is changing with the length of the electrode, so the electrode is not equally accessible in the direction of the solution flow. Commonly, when there is a simple mixing (stirring) of solution, the Nernst diffusion layer thickness is not known. One of the systems where δ_N can be defined is the rotating disk electrode (RDE), which is shown in Figure 7.6. The main feature of the RDE is that it is equally accessible for the electrochemically active species at any point of the electrode surface. In other words, an equally accessible electrode surface means that the Nernst diffusion layer is the same for any point on the electrode surface.

The theory of the RDE was developed by Benjamin Levich in the 1950s.

Benjamin (Veniamin) Levich (1917–1987) was a Ukraine-born American physical chemist who contributed to the development of the fundamental theory describing the RDE. The famous Levich equation describes the current at an RDE, which is one of the key techniques in electrochemical kinetics.

REFERENCES

A.J. Bard, G. Inzelt, and F. Scholz, *Electrochemical Dictionary*, Springer, Berlin, Germany, 2008.

http://en.wikipedia.org/wiki/Veniamin_Levich (Russian version).

Let us find out how the Nernst diffusion layer can be estimated employing an RDE. The convective–diffusion Equation 7.8 in the case of two-dimensional system in a steady state $[(\partial c/\partial t) = 0]$ in cylindrical coordinates is as follows:

$$D\left(\frac{\partial^2 c}{\partial z^2}\right) = v_r\left(\frac{\partial c}{\partial r}\right) + v_z\left(\frac{\partial c}{\partial z}\right), \tag{7.17}$$

where v_r and v_z are the radial and axial components of the solutions velocity, which depend on the cylindrical coordinates as $v_r = -az^2$ and $v_z = arz$, where a is defined as follows [2,3]:

$$a = 0.51\omega^{3/2}\left(\frac{\eta}{\rho}\right)^{-1/2} D^{-1/3}, \tag{7.18}$$

where ω is the angular velocity with unit of radians per second. After substituting v_r and v_z in Equation 7.17 and solving for limiting current, the concentration gradient at $x = 0$ can be found as follows:

$$\left(\frac{\partial c}{\partial z}\right)_{z=0} = 0.62\omega^{1/2}v^{-1/6}D^{-1/3}c^o, \tag{7.19}$$

where $v = \eta/\rho$ is the kinematic viscosity.

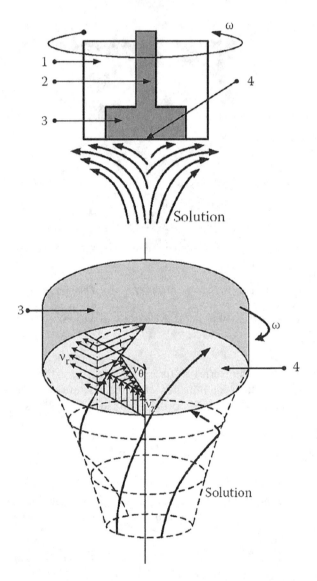

FIGURE 7.6 Rotating disk electrode: (1) outer shaft (nonconducting PTFE), (2) inner electric connection, (3) RDE, and (4) carefully polished disk surface. ω is the angular velocity, v_r is the radial velocity, and v_θ is the axial velocity.

The RDE allows defining the Nernst diffusion layer thickness, δ_N, which depends on (1) the diffusion coefficient of the electrochemically active species, (2) kinematic viscosity of the solution, and (3) rotation rate of the electrode as follows:

$$\delta_N = 1.61 D^{1/3} \nu^{1/6} \omega^{-1/2}, \tag{7.20}$$

which is valid in the presence of a supporting electrolyte.

Also, the limiting current density on the RDE can theoretically be calculated if the bulk concentration of the electrochemically active species of the reactant is known:

$$j_{lim} = nFDc^o/\delta_N = 0.62nFD^{2/3}v^{-1/6}\omega^{1/2}c^o. \tag{7.21}$$

In Equations 7.20 and 7.21, D is an ionic diffusion coefficient in the case of using a supporting electrolyte.

Note that the solution of Equation 7.17 with the partial derivatives is not exact but an approximation. Therefore, some dimensionless coefficients are used in the approximate solutions in Equations 7.18 through 7.21. Equation 7.21 is called the Levich equation.

Figure 7.7 shows how the polarization curves could look using the RDE taken at steady state and demonstrates how they are transformed when the rotation rate is changed.

Also, Equation 7.21 shows that if the limiting current density is plotted vs. square root of the rotation rate, a straight line can be obtained.

This allows calculating (1) the diffusion coefficient from the limiting current data if the kinematic viscosity, v, and the bulk concentration c^0 are known or (2) the bulk concentration if the diffusion coefficient is known.

Another important equation, which was obtained for RDE, allows calculating the current density at infinite rotation rate, j_∞ [2]:

$$j^{-1} = \left(j_\infty^{-1}\right) + \left(j_\infty^{-1}\right)\left[\text{const}/\omega^{1/2}\right]. \tag{7.22}$$

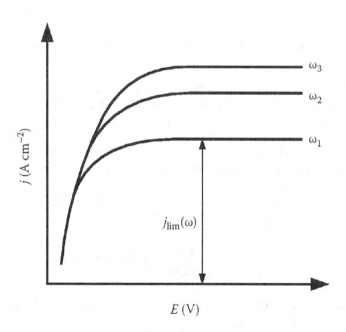

FIGURE 7.7 Polarization curves obtained on the RDE with increasing rotation rate from ω_1 to ω_3.

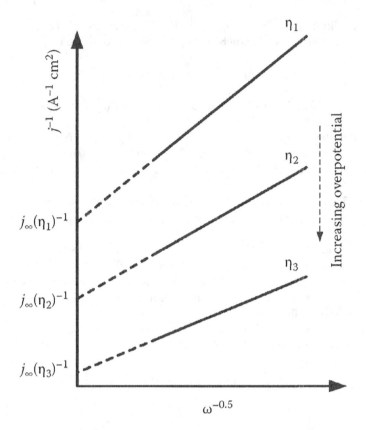

FIGURE 7.8 Determination of the exchange current density at infinite rotation speed using Equation 7.22.

Figure 7.8 and Equation 7.22 show how the current density at infinite rotation rate can be obtained as a function of potential. Taking into account that the infinite rotation rate corresponds to the pure electron transfer limit (when $\delta_N = 0$ and concentration of the electronically active species at the surface is the same as in the bulk), these current densities can be used to obtain the Tafel coefficients and, therefore, calculate the exchange current density and transfer coefficient of the electrochemical reaction taking place on the RDE.

In Figure 7.8, the anodic polarization (positive overpotential) lines are shown, and therefore, an analog of Equation 6.14 can be used:

$$\eta = \frac{RT}{(1-\beta)nF} \ln\left(\frac{j_\infty}{j_o}\right), \tag{7.23}$$

where instead of the experimentally obtained current density, j, the extrapolated value of the current density, j_∞, is used.

At high overpotentials, the limiting current becomes potential independent since it is entirely determined by the transport processes and Equation 7.22 is converted to

$$j^{-1} = \left(j_{lim}^{-1} \right) = \frac{\delta_N}{nFDc^o} = \frac{\text{const}}{\omega^{1/2}}, \tag{7.24}$$

and the j^{-1} vs. $\omega^{-1/2}$ plot should go through the origin. Note that in Equation 7.24, the limiting current can be found using Equation 7.21.

7.7   ROTATING DISK ELECTRODE ELECTROCHEMICAL CELL

A three-electrode cell shown in Figure 7.9 can be used to employ the RDE.

A schematic of the three-electrode RDE cell, a potentiostat to carry out measurements, and a computer to control the system and store the obtained data can be used as shown in Figure 7.10.

The surface of the RDE (Figure 7.11) should be polished to secure the necessary hydrodynamics and get high-quality data. To study the electrochemical kinetics of a redox ferricyanide/ferrocyanide reaction, $Fe(CN)_6^{3-} + e^- = Fe(CN)_6^{4-}$, a platinum RDE can be used while other electrodes can be used instead of platinum if they are of interest.

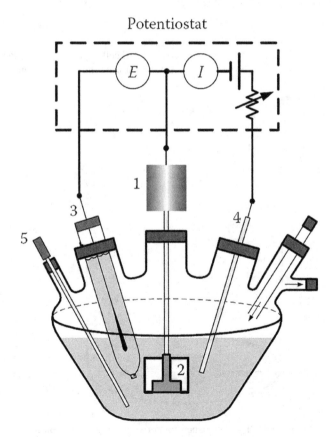

FIGURE 7.9   Three-electrode RDE cell connected to potentiostat: (1) electric motor, (2) RDE, (3) reference electrode, (4) counter electrode, and (5) thermometer.

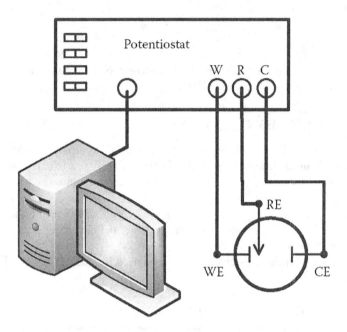

FIGURE 7.10 A schematic of the three-electrode RDE cell, a potentiostat to carry out measurements, and a computer to control the system with storing the obtained data.

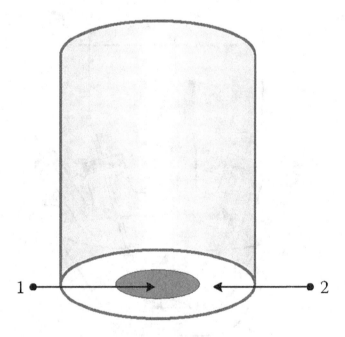

FIGURE 7.11 Platinum RDE: (1) polished mirror smooth surface of platinum and (2) outer shaft made by nonconducting polytetrafluoroethylene (PTFE), which is called Teflon by DuPont Co., who discovered this compound.

7.8 CYCLIC VOLTAMMETRY BACKGROUND

Cyclic voltammetry (CV) is an electrochemical technique in which a three-electrode system is used and the applied potential (between the working and reference electrodes) is changed linearly using a triangular wave form between a negative (cathodic), E_1, and a positive (anodic) turn-around value, E_2, which are usually in an aqueous solution between the hydrogen and oxygen equilibrium electrode potentials (Figure 7.12).

The current flowing between the working and counter electrodes is measured as a function of the applied potential over time. The rate of the potential change or the scan rate is usually above 10 mV s^{-1} and is a variable in this method. The current–potential dependence, which can be obtained in case of a reversible electrochemical reaction, is shown in Figure 7.13. In this case, the magnitudes of the maximum anodic, $j_{a(max)}$, and cathodic, $j_{c(min)}$, current densities are the same, and the equilibrium electrode potential is just in the middle between two potentials, $E_{a(max)}$ and $E_{c(min)}$, that correspond to the maximum anodic and cathodic current densities. The distance between these potentials is well defined and at a constant temperature depends on the number of electrons, n, participating in the electron transfer reaction (theoretically 57 mV n^{-1} at 25°C).

When the applied potential is changing from E_1 to E_2, the current density dependence reflects two coupling processes: (1) electron transfer reaction at the electrode and (2) diffusion of the electrochemically active species. In the beginning, the current is increasing due to (1) increasing of the potential as the Butler–Volmer equation implies and (2) decreasing of the surface concentration and increasing of the concentration gradient at the Nernst diffusion layer. However, the Nernst diffusion layer is also increasing while increasing the potential with simultaneous decrease of the

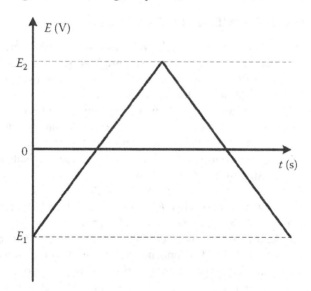

FIGURE 7.12 Potential–time dependences used in CV. E_1 and E_2 are, respectively, negative (cathodic) and positive (anodic) turnaround potentials.

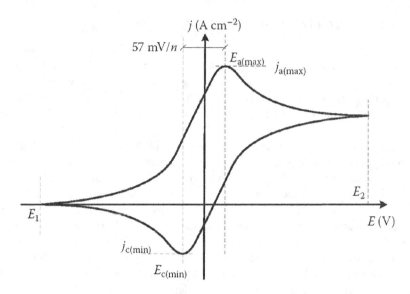

FIGURE 7.13 Current density–potential dependences in the CV of a theoretical reversible electrochemical reaction $Ox + ne^- = Red$.

surface concentration of the electrochemically active species. As soon as the surface concentration reaches zero, the current starts to decrease due to transport limitations. The anodic maximum and cathodic minimum of the current density can be calculated in case of fast and reversible electron transfer reactions [2].

7.9   CYCLIC VOLTAMMETRY OF $Fe(CN)_6^{3-}/Fe(CN)_6^{4-}$ COUPLE

$Fe(CN)_6^{3-} + e^- = Fe(CN)_6^{4-}$ on Pt is a relatively fast and reversible electrochemical half-reaction demonstrating the CV curves (Figure 7.14) similar to those shown in Figure 7.13. Analyzing the CV curves in Figure 7.14, one can make the following conclusions. (1) The equilibrium potential of the reaction vs. SCE (calomel electrode with saturated KCl(aq)) is around 0.2 V, which is comparable to the standard electrode potential of the $Fe(CN)_6^{3-}/Fe(CN)_6^{4-}$ couple (0.356 V) taking into account the nonideality of the solution when the electrode potential is calculated using concentrations and activity coefficients. (2) $E_{a(max)}$ and $E_{c(min)}$ are almost not shifting when the scan rate is increasing from 100 to 300 mV s^{-1}, showing that the electrochemical reaction is quite fast; that is, the exchange current density of the reaction is quite large. (3) The peak potential separation, $E_{a(max)} - E_{c(min)}$, shows a weak dependence on the scan rate and is somewhere around 60 mV. This confirms that the reaction is fast and involves one electron. (4) The peak current ratio is about the same regardless of the scan rate: $j_{a(max)}/|j_{c(min)}| = 1$. This confirms that the reaction is reversible. (5) Both the anodic and cathodic current peaks are proportional to the square root of the scan rate: $j_{c(max)}/E_t^{1/2}$ = constant and $j_{a(max)}/E_t^{1/2}$ = constant, which is consistent with the CV theory of fast and reversible reactions. A reader interested in a variety of the CV studies is encouraged to see other books [2,3,4].

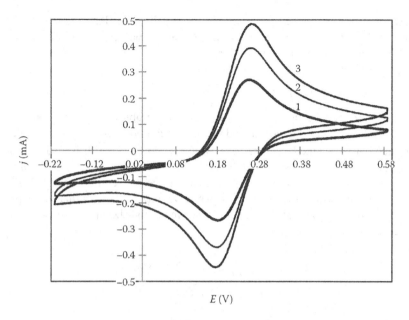

FIGURE 7.14   CV curves for the $Fe(CN)_6^{3-} + e^- = Fe(CN)_6^{4-}$ on platinum (Pt) electrode measured vs. SCE at three scan rates: 1–100, 2–200, 3–300 mV s^{-1}.

7.10   ELECTROCHEMICAL IMPEDANCE SPECTROSCOPY

While Ohm's law, Equation 2.2, is useful for defining the relationship between resistance, current and potential difference (shortened below as potential) for a simple circuit with direct current (dc), almost any real electrochemical cell is more complicated due to time-dependent processes such as charge transfer, diffusion, adsorption, and formation of surface charge. When considering an electrical circuit with time-dependent processes, additional electrical components such as capacitors and inductors typically contribute to the circuit's overall impedance. In this section, impedance represents the ability of a circuit containing resistors, capacitors, and inductors to resist the flow of an alternating current (ac).

We know that electricity flows through a galvanic electrochemical cell as a result of some electrochemical reactions through cell elements such as electrodes, electrolytes, and current collectors. Estimating the impedance of the electrochemical cell helps us to understand the processes occurring at each element mentioned above. When this overall circuit impedance is determined as a function of frequency, we can elucidate the impedance contributions due to specific phenomena taking place inside the cell, such as electrode polarization, species diffusion, and ion migration. To meet this goal, electrochemical impedance spectroscopy (EIS) was developed as a powerful technique to determine the frequency-dependent impedances of elements inside an electrochemical system. EIS is a non-destructive electrochemical technique; it probes an electrochemical system without irreversibly changing it, unlike linear sweep voltammetry or cyclic voltammetry. EIS is performed by introducing an oscillating potential to the electrodes and observing the current response of the cell.

Alternatively, ac can be imposed on the system and the oscillating potential can be observed. Note that in this section, we will slightly depart from the IUPAC convention [5] in the EIS symbols and notations.

7.10.1 COMPLEX VARIABLES

Impedance, Z, can be described using complex variables. A complex variable is defined as the sum of both the real and imaginary components of a system. An imaginary number is assigned the variable i, which is equal to $\sqrt{-1}$. It is also possible to represent the impedance using polar coordinates as graphically shown in Figure 7.15.

For the polar representation of Z, $|Z|\cos\phi$ is the real component of Z, also to be shown in literature as Z_{re} or $\mathrm{Re}\{Z\}$, while $|Z|\sin\phi$ is the imaginary component (also Z_{im} or $\mathrm{Im}\{Z\}$). Therefore, the impedance magnitude $|Z|$ and phase angle, ϕ, are defined as follows:

$$Z = |Z|\left(\cos\phi + i\sin\phi\right), \tag{7.25}$$

$$|Z| = \sqrt{\cos^2\phi + \sin^2\phi}, \tag{7.26}$$

$$\phi = \arctan\frac{\sin\phi}{\cos\phi}. \tag{7.27}$$

Following Euler' theorem, Z can be given in a more convenient exponential form as

$$Z = |Z|\left(\cos\phi + i\sin\phi\right) = |Z|e^{i\phi}. \tag{7.28}$$

Note that in Figure 7.15, the y-axis is $-Z_{im}$ due to negative capacitive impedance as is shown later in Equation 7.37.

7.10.2 RESPONSE OF CURRENT TO APPLIED OSCILLATING POTENTIAL

Figure 3.13 shows the time dependence of an alternating potential oscillating around a steady state value of zero. For the case of small oscillations, the system displays

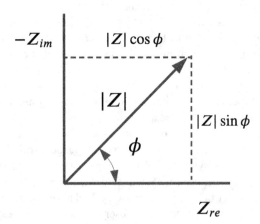

FIGURE 7.15 Relationship among impedance Z, its magnitude |Z| and phase angle ϕ.

linearity when the current signal responds with the same frequency as the potential signal. Because both sine and cosine functions describe the same sustained oscillations, the applied potential can be expressed as either waves. Here, we will start with the cosine function as it was already done in Section 3.8:

$$E(t) = E_m \cos(\omega t) = E_m \cos(2\pi f t), \qquad (7.29)$$

where $E(t)$ is the potential at time t in s, E_m is the amplitude of the electric potential signal in V, ω is the radial frequency in rad s^{-1}, and f is frequency in s^{-1}. For the treatment of impedance, it is important to keep in mind that the potential, $E(t)$, and the current, $I(t)$, show only the oscillating contribution to see how they deviate from a steady state.

The current signal will respond at the same frequency and will be phase-shifted by the phase angle ϕ:

$$I(t) = I_m \cos(\omega t + \phi), \qquad (7.30)$$

where I_m is the amplitude of the current signal.

A comparison of the ac potentials and current responses over different frequencies provides insight into the different time-dependent processes in the electrochemical system. Each process (e.g., adsorption, diffusion, polarization) contributes a unique impedance response over a unique frequency range. As such, the potential–current EIS results, which provide impedance and phase data, can be fit inside an equivalent circuit of the electrochemical system, where the chemical processes are represented by circuit elements such as resistors, capacitors, and inductors.

To better conceptualize the role of impedance within an electrochemical cell and how it defines some of the basic processes, it will be useful to visualize what is taking place in it. Figure 7.16 displays the so-called Randles equivalent circuit, which includes one electric double layer process as a capacitor, C_d, one charge transfer (faradaic) process as a resistance, R_{ct}, which is parallel to the double layer capacitor, and the ohmic resistance of the electrolyte, R_s, oriented in series with respect to both C_d and R_{ct}. Note that R_{ct} in Figure 7.16 is the same as either R_{ct-} or R_{ct+} in Figure 3.14.

The total impedance response is characterized by each element (including parallel and series dependence) inside the equivalent circuit. Therefore, to depict an electrochemical system using an equivalent circuit, we should first assess the current–potential behavior of each circuit element (responding to an EIS scan, for example.)

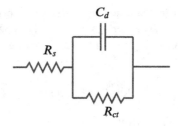

FIGURE 7.16   The Randles equivalent circuit showing three contributions, R_s, R_{ct}, and C_d to an electrode impedance.

The mathematical representation of a capacitor (with capacitance C in F or CV^{-1}) based on the relationship between current and potential is

$$I(t) = C\left(\frac{dE(t)}{dt}\right). \tag{7.31}$$

Recalling the initial representation for an oscillating potential using cosine, we can equivalently use a sine wave:

$$E(t) = E_m \sin(\omega t), \tag{7.32}$$

with either choice resulting in the same current–potential relationship. Because the initial excitation signal from EIS is normally very small in magnitude (1–10 mV), it elicits a pseudo-linear response from the cell. With a linear response, it is possible to apply the principle of superposition and proportionality to simplify the mathematical analysis. Superposition is the ability to take the sum of two inputs and get a combined output equal to the individual outputs that would have occurred from each input. Proportionality enables us to multiply the input by a constant value in which the output responds proportionally. Using these principles and applying them to our linear system, we can add the cosine and sine wave inputs to define an alternative input given in the following equation:

$$E(t) = E_m \cos(\omega t) + iE_m \sin(\omega t). \tag{7.33}$$

This is done so that we can utilize Euler's formula:

$$E(t) = E_m\left[\cos(\omega t) + i\sin(\omega t)\right] = E_m e^{i\omega t}, \tag{7.34}$$

which will reduce the mathematical complexity and allow solving the complex equations for each circuit element.

Now that the equation for a complex input signal is simplified, we can use the current–potential relationship of a capacitor to calculate the complex current response as follows:

$$I(t) = C\left(\frac{dE(t)}{dt}\right) = Ci\omega E_m e^{i\omega t}. \tag{7.35}$$

If we extract the real portion of Equation 7.35, we recover the measurable time-dependent current that responds to the cosine potential input. By doing so, it further simplifies our math by converting the time-dependent derivatives into an algebraic form:

$$I(t) = \mathrm{Re}\{I(t)\} = \mathrm{Re}\{Ci\omega E_m\left[\cos(\omega t) + i\sin(\omega t)\right]\}$$
$$= -C\omega E_m \sin(\omega t). \tag{7.36}$$

Equation 7.36 is a solution of Equation 7.31 showing how the current depends on time as a result of the input potential given in Equation 7.29. This method can be performed on each circuit element to find the relationship between the time dependence of the input potential and output current.

7.10.3   IMPEDANCE OF CIRCUIT COMPONENTS

The impedance of a capacitor can be defined as a ratio of the time-dependent potential given by Equation 7.34 and the current given by Equation 7.35:

$$Z(\omega) = \frac{E(t)}{I(t)} = \frac{1}{i\omega C} = -\frac{i}{\omega C}. \tag{7.37}$$

In general, the impedance of a circuit is defined as

$$Z(\omega) = \frac{E(t)}{I(t)} = \frac{E_m}{I_m} e^{i\phi} = \frac{E_m}{I_m} (\cos\phi + i \sin\phi), \tag{7.38}$$

showing that $Z(\omega)$ is independent of time, contrary to the time-dependent applied potential and corresponding current.

Assigning equivalent circuit elements to cell components helps to model an electrochemical cell impedance and visualize what is taking place inside the cell during an electrochemical test. A widely agreed upon method of constructing a circuit corresponding to an electrochemical cell is to consider three circuit elements such as resistors, capacitors, and inductors. For a resistor, the impedance is independent of frequency and equals resistance

$$Z = R. \tag{7.39}$$

For a capacitor, the impedance depends on frequency and is given by Equation 7.37. For an inductor with inductance, L, in H (henry) or VA^{-1}, the potential–current dependence is

$$E(t) = L\left(\frac{dI(t)}{dt}\right), \tag{7.40}$$

with the frequency-dependent impedance as

$$Z(\omega) = \frac{E(t)}{I(t)} = i\omega L, \tag{7.41}$$

which is again independent of time while both $E(t)$ and $I(t)$ depend on both frequency and time.

To conclude, the impedance of a resistor is equal to its resistance and is independent of frequency, so the current response will be in phase with the applied potential. The impedance of an inductor has an imaginary component, so the current response will be phase shifted $-90°$ from the applied potential. The impedance of a capacitor is the inverse of the impedance of the inductor, meaning the current response will be phase shifted $+90°$ from the applied potential.

These three circuit components and their impedances enable us to model electrochemical cells similarly to electrical circuits such as the Randles circuit. This will be demonstrated in the following section.

7.10.4 IMPEDANCE OF THE RANDLES CIRCUIT

Using the resistors, capacitors, and inductors, we can model the behavior of an electrochemical cell as it pertains to EIS. Here, we will assume the surface of the electrode is charged and the electrode–electrolyte interface is considered an EDL (see Section 3.8). In a cell reaction, an EDL is commonly formed between the electrode and the contacting electrolyte. The EDL is an interface that exists as ions from the electrolyte are introduced to the electrode surface. The EDL capacitance C_d, can vary based on a number of factors including the surface area exposed to the electrolyte, the electrode and electrolyte compositions, and the roughness of the electrode.

An effective equivalent circuit to model this system is the previously mentioned Randles circuit given in Figure 7.16. Note that although the electrical components may not perfectly represent the physicochemical processes occurring at the electrode, this circuit is usually very useful for understanding the cell's behavior. For the Randles circuit, the charge transfer occurs simultaneously with the double-layer formation and, therefore, both of them can be modeled in parallel. During an electrochemical reaction involving kinetics, the EDL can be charged and discharged, creating a capacitive effect on the cell. This is a non-Faradaic process because the current has no electron transfer through the EDL. The component on the circuit that is parallel to the capacitive double layer is a faradaic current because it involves electron transfer at the surface and can be described according to the simplified version of the Butler–Volmer equation (Equation 6.10.) The resistance of the charge transfer reaction, shown in Figure 7.16 as R_{ct}, is treated as a pure resistor and related to the exchange current density, j_o, as

$$R_{ct} = \frac{RT}{nFj_o}, \tag{7.42}$$

where n is the number of electrons involved into the charge transfer reaction.

Modeling the EDL capacitor and the charge transfer resistor in parallel and adding the solution resistor in series, the impedance of the electrical circuit can be easily constructed (taking into account Kirchhoff's law presented in Section 2.11):

$$Z(\omega) = R_s + \left(\frac{1}{\dfrac{1}{R_{ct}} + C_d i\omega} \right). \tag{7.43}$$

7.10.5 Nyquist Plot

A simple manipulation (see Section 7.12) allows us to modify Equation 7.43 to the following format:

$$Z(\omega) = R_s + \left(\frac{R_{ct}}{1 + R_{ct}^2 C_d^2 \omega^2} \right) + i\left(\frac{-R_{ct}^2 C_d \omega}{1 + R_{ct}^2 C_d^2 \omega^2} \right), \tag{7.44}$$

which can now be treated by complex algebra with the real and imaginary parts of the total impedance as follows:

$$Z_{re}(\omega) = R_s + \left(\frac{R_{ct}}{1 + R_{ct}^2 C_d^2 \omega^2} \right), \tag{7.45}$$

and

$$Z_{im}(\omega) = \left(\frac{-R_{ct}^2 C_d \omega}{1 + R_{ct}^2 C_d^2 \omega^2} \right). \tag{7.46}$$

The total impedance of the Randles circuit is then

$$Z(\omega) = Z_{re}(\omega) + iZ_{im}(\omega), \tag{7.47}$$

which again does not depend on time but on frequency.

The relationship among impedance Z, its magnitude $|Z|$, and the phase angle ϕ for the Randles circuit can now be plotted using the format shown in Figure 7.15. This format is called a Nyquist plot. Before constructing a Nyquist plot for the Randles circuit, let us analyze $Z_{re}(\omega)$ and $Z_{im}(\omega)$ at three particular values of ω. When $\omega = 0$, $Z_{im}(\omega) = 0$ and $Z_{re}(\omega) = R_s + R_{ct}$; when $\omega = \infty$, $Z_{im}(\omega) = 0$ and $Z_{re}(\omega) = R_s$; when $\omega = 1/(R_{ct}C_d)$, $Z_{im}(\omega) = -\dfrac{R_{ct}}{2}$ and $Z_{re}(\omega) = R_s + \dfrac{R_{ct}}{2}$. These values of $Z_{re}(\omega)$ and $Z_{im}(\omega)$ at the three values of ω could be used to understand how to plot $-Z_{im}(\omega)$ vs. $Z_{re}(\omega)$ as a perfect semicircle as shown in Figure 7.17, which is the Nyquist plot for the Randles circuit given in Figure 7.17.

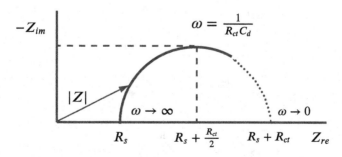

FIGURE 7.17 Nyquist plot of the Randles circuit.

It should be noted that each individual point on a Nyquist plot represents the impedance at one frequency. It is also worth mentioning that while only one semicircle is displayed (characteristic of one time constant, RC in s), an electrochemical cell may display several semicircles from an EIS scan. The point at which the semicircle intersects the real axis denotes that the imaginary component of the impedance equals 0. When creating a Nyquist plot, both axes should be equal in magnitude.

In Figure 7.17, the right part of the semicircle is shown in dots to let reader know that at low frequencies, the EIS test is usually complicated by long duration and, therefore, this part of the semicircle is usually constructed by extrapolating data obtained at high and intermediate frequencies.

7.10.6 BODE PLOT

An alternate display of impedance that allows for the frequency to be identified as a non-implicit variable is the Bode plot. In the Bode plot, $\log[Z(\omega)]$ and $\phi(\omega)$ are presented as functions of $\log(\omega)$. If we consider the Randles circuit described above and use the Nyquist plot shown in Figure 7.17, the corresponding Bode plot will have a format given in Figure 7.18.

Let us first analyze the $\log[Z(\omega)]$ vs. $\log(\omega)$ dependence shown in Figure 7.18. On the left and right sides of the plot, the impedance is independent of frequency, and $\log[Z(\omega)]$ is represented by horizontal lines equal to $\log[R_s + R_{ct}]$ on the left and $\log[R_{ct}]$ on the right. Between these regions $\log[Z(\omega)]$ decreases linearly with $\log(\omega)$, showing a negative slope of -1 ($-45°$). Extrapolating this line to $\log(\omega) = 0$ gives an impedance equals to $\log(C_d^{-1})$. The phase angle ϕ, also shown in Figure 7.18, goes through a maximum $\phi_{max} = (1 + R_{ct}/R_s)^{1/2}/(R_{ct}C_d)$, while on the left and on the right, ϕ approaches zero because the impedance is independent of frequency.

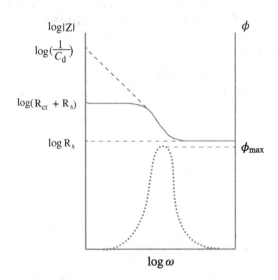

FIGURE 7.18   Bode plot of the Randles circuit.

7.10.7 WARBURG IMPEDANCE

In addition to the charge transfer reaction process, time-dependent mass transfer processes commonly take place at the electrode. This was discussed in Section 6.5 and in other sections of Chapter 6. The equivalent circuit of two electrodes with an electrolyte solution between them is shown in Figure 3.14, where Z_w is an impedance responsible for transfer of electrochemically active species to or from the electrode surface. This circuit element, called the Warburg impedance, may consist of both ohmic and capacitive components. The theory of the Warburg impedance, though rather complicated, arrives to a simple impedance expression as follows:

$$Z(\omega) = Z_{re}(\omega) + iZ_{im}(\omega) = \frac{\sigma}{\omega^{1/2}} + i\frac{\sigma}{\omega^{1/2}}, \quad (7.48)$$

which is represented in a Nyquist plot by a straight line at low frequencies with a positive slope equal to +1 (45°), as shown in Figure 7.19. As frequency becomes larger, the effect of diffusion on impedance is no longer significant, and the Nyquist plot follows the original Randles circuit.

Parameter σ in Equation 7.48 is called the Warburg impedance coefficient, and, in the most simplified form (when diffusion coefficients and concentrations of the oxidized and reduced species are the same,) is as follows:

$$\sigma = \frac{RT2^{1/2}}{n^2F^2c^0D^{1/2}}, \quad (7.49)$$

from which the diffusion coefficient, D, can be extracted.

To complete this section, the author warns that the plots and corresponding models shown above are purely theoretical and serve to help the reader understand the background and principles of EIS. Experimental data from actual EIS scans are usually more complicated and not as easy to interpret. For more information on the EIS theory and applications, the reader is advised to consult additional texts [2,3,5,6].

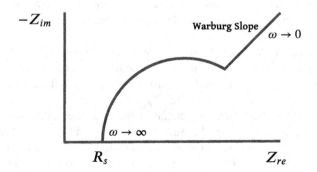

FIGURE 7.19 Nyquist plot of a circuit with Warburg impedance.

7.11   SUMMARY

- Transport processes in electrochemical systems can be treated using the vector analysis, which is part of calculus.
- There are two key equations, which are the basis of the transport process of the electrochemical systems. Simplification of these equations (no convection) leads to Fick's first and second law of diffusion.
- In the potentiostatic regime, current vs. time can be measured at a constant potential. If the charge transfer reaction is very fast, this measurement can be used to obtain the diffusion coefficient or the surface concentration of the electrochemically active species.
- In the galvanostatic regime, potential vs. time can be used (1) to measure the diffusion coefficient if the bulk concentration is known or (2) as an analytical tool to measure the concentration of the electrochemically active species.
- Hydrodynamics (fluid mechanics) can significantly affect the rate of an electrochemical reaction and should be taken into account in accurate electrochemical measurements.
- Electrochemical kinetics significantly depends on the thickness of the Nernst diffusion layer, while hydrodynamics provides the thickness of the Prandtl layer. The ratio of the Nernst diffusion layer to the Prandtl layer can be calculated using the properties of the solution.
- Commonly, when there is a simple mixing (stirring) of solution, the Nernst diffusion layer thickness is not known. One of the systems where δ_N can be defined is the RDE. The main feature of the RDE is that it is equally accessible at any point of the electrode surface.
- The RDE allows defining the Nernst diffusion layer thickness, δ_N, which depends on (1) the diffusion coefficient of the electrochemically active species, (2) the viscosity and density of the solution, and (3) the rotation rate of the electrode.
- The limiting current density on the RDE can be estimated if the Nernst diffusion layer thickness and the bulk concentration of the electrochemically active species are known.
- The current density at infinite rotation rate can be used to treat the data using the Tafel equation and, therefore, calculate the exchange current density and symmetry coefficients of the electrochemical reaction taking place on the RDE.
- A three-electrode cell, potentiostat, computer, and an RDE setup should be used to carry out high-quality electrochemical kinetics measurements. It is important to keep in mind that just stirring is not sufficient to provide a well-known hydrodynamic control.
- CV is an electrochemical technique in which a three-electrode system is used. The current flowing between the working and counter electrodes is measured as a function of the applied potential (time), and the rate of the potential change or the scan rate is usually above 10 mV s^{-1} and is a key variable in this method.

- When the applied potential in CV is changing, the current dependence reflects two coupling processes: (1) diffusion of the electrochemically active species and (2) electron transfer reaction at the electrode. As a result, a maximum or a minimum can appear on the polarization curve.
- A redox ferricyanide/ferrocyanide reaction $Fe(CN)_6^{3-} + e^- = Fe(CN)_6^{4-}$ on Pt is a relatively fast and reversible that can be found analyzing experimental CV curves.
- Background of the EIS is described as a powerful non-destructive technique to estimate characteristics of an electrochemical system without any irreversible changes to it vs. linear sweep voltammetry or cyclic voltammetry.

7.12 EXERCISES

PROBLEM 1

Equation 7.11 can be used to show that in a potentiostatic regime, the Nernst diffusion layer can be approximately calculated as (see Section 7.3) $\delta_N = 3.8 \times Dt^{0.5}$. Using the system presented in Figure 7.1 (Cu^{2+}(aq) in $CuCl_2$(aq) solution with $c^o = 0.05$ mol L^{-1}) with a surface concentration of 0.0 mol L^{-1} and 0.02 mol L^{-1}, calculate $c(x,t)$ as a function of distance from the surface after 5, 25, and 100 seconds. Graphically confirm that (1) δ_N is independent of the surface concentration and that (2) δ_N in the figure matches δ_N calculated using the equation above.

Solution:

The concentration of Cu^{2+}(aq) as a function of distance at surface concentration of 0.02 mol L^{-1} (along with 0.0 mol L^{-1} for comparison) at three time periods of 5, 25, and 100 seconds calculated using Equation 7.11 is shown in figure below.

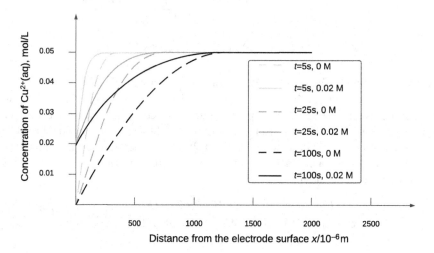

Dependence of the Cu^{2+}(aq) concentration on distance calculated for the conditions shown in Figure 7.1 at surface concentrations of 0.02 and 0.0 mol L^{-1}.

It is clearly seen that the Nernst diffusion layer depends on time and not on the surface concentration. δ_N can be calculated using the given above equation, so at 5 seconds, the Nernst diffusion layer is 301×10^{-6} m, at 25 s it is 674×10^{-6} m, and at 100 s it is 1348×10^{-6} m. These values line up with what can approximately be obtained from the plot. Note that, based on Equation 7.11, $c(x, t)$ is equal to the bulk concentration only at an infinite distance from the electrode, so the Nernst diffusion layer cannot be precisely defined.

PROBLEM 2

Calculate the ratio of the Nernst diffusion layer and Prandtl boundary layer, δ_N/δ_{Pr}, for $CuCl_2$(aq) solution at temperatures 25, 50, 75, and 100°C at ambient pressure for liquid water. Give your conclusion on the temperature dependence of δ_N/δ_{Pr}.

Solution:

In a steady state, the relation between the Prandtl layer thickness, δ_{Pr}, and the Nernst diffusion layer thickness, δ_N, are defined as follows (Equation 7.16):

$$\frac{\delta_N}{\delta_{Pr}} = \left[\frac{\eta}{D\rho}\right]^{-1/3},$$

where η is viscosity in kg m^{-1} s^{-1}, ρ is the density in kg m^{-3}, and D is the diffusion coefficient in m^2 s^{-1} which for a dilute solution can be obtained as (see Section 7.3):

$$D_{CuCl_2(aq)} = \frac{\left(z_{Cu^{2+}(aq)} + \left|z_{Cl^-(aq)}\right|\right) D^0_{Cu^{2+}(aq)} D^0_{Cl^-(aq)}}{z_{Cu^{2+}(aq)} D^0_{Cu^{2+}(aq)} + \left|z_{Cl^-(aq)}\right| D^0_{Cl^-(aq)}}$$

$$= \frac{\left(2 + |-1|\right) \times 0.714 \times 10^{-5} \times 2.032 \times 10^{-5}}{2 \times 0.714 \times 10^{-5} + |-1| \times 2.032 \times 10^{-5}}$$

$$= 1.258 \times 10^{-5}\, cm^2\, s^{-1} = 1.258 \times 10^{-9}\, m^2\, s^{-1},$$

where, according to Table 10.12, $D^0_{Cu^{2+}(aq)}$ and $D^0_{Cl^-(aq)}$ are the limiting diffusion coefficients at 25°C and can be found as 0.714×10^{-5} cm^2 s^{-1} and 2.032×10^{-5} cm^2 s^{-1}, respectively; however, the temperature dependence of the diffusion coefficient should be somehow estimated. Here, we will use the Stokes–Einstein Equation 3.23, first to calculate an effective ionic radius at 25°C and then to use this diameter to calculate the diffusion coefficient assuming that in this temperature range the ionic radius does not depend on temperature.

$$r_{Cu^{2+}(aq)} = \frac{kT}{6\pi D^0_{Cu^{2+}(aq)}\eta} = \frac{1.381 \times 10^{-23} \times 298.15}{6\pi \times 0.714 \times 10^{-9} \times 8.90 \times 10^{-4}} = 0.3437 \times 10^{-9}\, m,$$

$$r_{Cl^-(aq)} = \frac{kT}{6\pi D^0_{Cl^-(aq)}\eta} = \frac{1.381 \times 10^{-23} \times 298.15}{6\pi \times 2.032 \times 10^{-9} \times 8.90 \times 10^{-4}} = 0.1208 \times 10^{-9}\, m.$$

The values of water viscosity and density as a function of temperature can be found in Table 10.22 and are given below.

Temperature (°C)	Viscosity (µPa s)	Density (kg m⁻³)
25	890.02	997.05
50	546.52	988.03
75	377.67	974.81
100	282.75	958.63

Using the diameters of both ions as independent of temperature, their diffusion coefficients as well as the diffusion coefficient of the whole electrolyte can be calculated at elevated temperatures using the equations given above. Finally, the ratio of the Nernst diffusion layer and Prandtl boundary layer can be calculated using Equation 7.16.

Temperature (°C)	$D^0_{Cu^{2+}(aq)}$, $(10^{-9}$ m² s⁻¹$)$	$D^0_{Cl^-(aq)}$, $(10^{-9}$ m² s⁻¹$)$	$D_{CuCl_2(aq)}$, $(10^{-9}$ m² s⁻¹$)$	$\dfrac{\delta_N}{\delta_{Pr}}$
25	0.714	2.032	1.258	0.112
50	1.260	3.587	2.220	0.159
75	1.965	5.592	3.462	0.208
100	2.813	8.005	4.956	0.256

Our calculations show that for a diluted $CuCl_2$(aq) solution, the ratio of the Nernst diffusion layer and Prandtl boundary layer increases with increasing temperature from 25°C to 100°C. While at ambient temperature, the Nernst diffusion layer is ten times smaller, at 100°C it is only four times smaller compared to the Prandtl boundary layer.

PROBLEM 3

Simulated data of a Randles circuit are given below as Nyquist and Bode plots. Estimate the solution resistance (R_s), the charge transfer resistance (R_{ct}), and the capacitance of the EDL (C_d).

Solution:

According to the Nyquist plot, when $\omega = \infty$, $Z_{im}(\omega) = 0$, the solution resistance is obtained from the plot as

$$R_s = Z_{re}(\omega = \infty) = 100 \ \Omega.$$

When $\omega = 0$, $Z_{im}(\omega) = 0$, the charge transfer resistance, R_{ct}, can be calculated as

$$R_{ct} = Z_{re}(\omega = 0) - R_s = 1100 - 100 = 1000 \ \Omega.$$

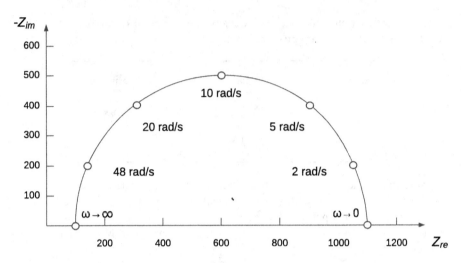

Nyquist plot of the Randles circuit. The radial frequencies, ω, are shown by circles and the corresponding values.

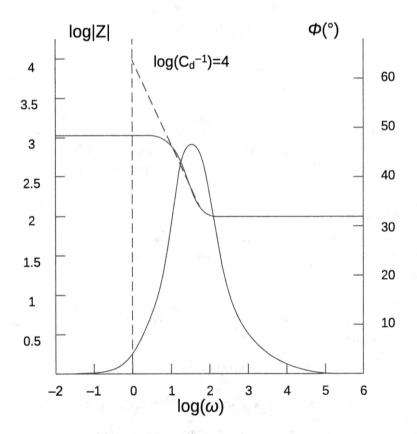

Bode plot of the Randles circuit.

According to the Nyquist plot, the frequency related to the minimum of the imaginary impedance, Z_{im}, (or maximum of $-Z_{im}$) is 1.59 s^{-1}, so the radial frequency can be calculated as

$$\omega(\min Z_{im}) = 2\pi \times 1.59 = 10 \, \text{rad/s} .$$

Therefore, the capacitance of the EDL, C_d, is

$$C_d = \frac{1}{\left[R_{ct} \times \omega(-Z_{im} = \max) \right]} = 10^{-4} \, \text{F.}$$

The values of R_s and R_{ct} can also be estimated from the above Bode plot, which are logarithms of the impedance module, $\log |Z|$, and phase angle ϕ, as a function of logarithm of the radial frequency, $\log \omega$:

$$\log(R_s) = 2, \, R_s = 100\,\Omega,$$

$$\log(R_d + R_s) = 3.04, \, R_d + R_s = 1096.5, \, R_d = 996.5\,\Omega.$$

To verify these calculated values of R_s, R_{ct}, and C_d, let us calculate the radial frequency corresponding to the maximum phase angle using the Bode plot shown above:

$$\omega(\max\phi) = \left(1 + \frac{R_{ct}}{R_s} \right)^{\frac{1}{2}} \bigg/ (R_{ct}C_d) = 33.1 \, \text{rad/s},$$

which gives the value of $\log\left[\omega(\max\phi) \right] = 1.52$.

Extrapolation of $\log|Z|$ to $\log(\omega) = 0$ gives a number equal to $\log(C_d^{-1}) = 4$, so $C_d = 10^{-4}$ F, as it was obtained earlier.

PROBLEM 4

Using Kirchhoff's law, the Randles circuit impedance is

$$Z(\omega) = R_s + \left(\frac{1}{\dfrac{1}{R_{ct}} + C_d i\omega} \right). \tag{7.43}$$

Show how this expression can be transformed to the format of Equation 7.44, which can be treated by complex algebra and used to construct a Nyquist plot.

Solution

First, we modify Equation 7.43 as follows:

$$Z(\omega) = R_s + \left(\frac{1}{\dfrac{1}{R_{ct}} + C_d i\omega} \right) = R_s + \frac{R_{ct}}{1 + iR_{ct}C_d\omega}.$$

Then, we multiply both the numerator and denominator of the second term on the right by $(1 - iR_{ct}C_d\omega)$, with an additional modification:

$$Z(\omega) = R_s + \frac{R_{ct}}{1 + iR_{ct}C_d\omega} = R_s + \frac{R_{ct}}{1 + iR_{ct}C_d\omega} \frac{(1 - iR_{ct}C_d\omega)}{(1 - iR_{ct}C_d\omega)} = R_s + \frac{R_{ct} - iR_{ct}^2 C_d\omega}{1 + \omega^2 R_{ct}^2 C_d^2}$$

$$= \left(R_s + \frac{R_{ct}}{1 + \omega^2 R_{ct}^2 C_d^2} \right) - i \frac{R_{ct}^2 C_d\omega}{1 + \omega^2 R_{ct}^2 C_d^2},$$

which is Equation 7.44.

REFERENCES

1. J. Newman and K.E. Thomas-Alyea, *Electrochemical Systems*, 3rd ed., Wiley, Hoboken, NJ, 2004.
2. C.H. Hamann, A. Hamnett, and W. Vielstich, *Electrochemistry*, 2nd ed., Wiley-VCH, Weinheim, Germany, 2007.
3. A.J. Bard and L.R. Faulkner, *Electrochemical Methods: Fundamentals and Applications*, Wiley, New York, 2001.
4. R.G. Compton, C.E. Banks, *Understanding Voltammetry*, World Scientific, Covent Garden, London, 2007.
5. M. Sluyters-Rehbach, Impedance of electrochemical systems: Terminology, nomenclature, and representation: I. Cells with metal electrodes and liquid solutions. *Pure Appl. Chem.*, **66**, 1994, 1831–1891.
6. M.E. Orazem and B. Tribollet, *Electrochemical Impedance Spectroscopy*, Hoboken, New Jersey: Wiley, 2008.

8 Electrochemical Energy Conversion

8.1 OBJECTIVES

There are a number of well-developed technologies such as aluminum manufacturing, chlorine and caustic soda production, electrodeposition, and the use of electrochemical sensors, which are governed by electrochemical processes. These technologies and their electrochemical science and engineering background are described in a number of published books [1–3] and are not covered in this book. Most recently, electrochemical energy conversion and storage technologies have obtained particular attention [4]. This is due to, for example, a pressing need to develop electric, hybrid, and fuel cell vehicles and by this way to reduce the greenhouse gas emissions, to slow down global warming, to improve energy security, and to develop a more sustainable society. Another example is developing large-scale electric storage (batteries) [5] for a variety of electric grid applications. The main objective of this chapter is to demonstrate how electrochemical thermodynamics and kinetics can be used for analyzing an electrochemical energy conversion system. Fuel cells and corresponding electrolytic cells are used as examples. The materials previously covered in this book are used to analyze the performance (current density–potential dependence) of the electrochemical systems and their efficiency. The heat balance of the electrochemical cells is also briefly discussed.

8.2 MAIN TYPES OF ELECTROCHEMICAL ENERGY CONVERSION SYSTEMS

Electrochemical energy conversion usually means converting the energy of chemicals to electric energy in fuel and flow cells or, vice versa, to produce chemicals using electric energy in electrolytic cells. In rechargeable batteries, both ways of conversion take place. In solar cells, the energy of sunlight is converted to electric energy, and in supercapacitors, the electrochemical energy is stored in an electric double layer at the electrode–electrolyte interface. In this chapter, we will use fuel and electrolytic cells as examples to understand their thermodynamics and kinetics.

8.3 PRINCIPAL DESIGN OF A FUEL CELL

In a fuel cell, electric energy is generated due to continually feeding electrochemically active chemicals, a fuel and an oxidizer. The technology was invented by Christian Friedrich Schönbein in 1838 and developed by William Robert Grove in 1839. Apparently, the first operating fuel cell was the so-called alkaline fuel cell (AFC). A schematic of the AFC is shown in Figure 8.1.

DOI: 10.1201/b21893-8

Christian Friedrich Schönbein (1799–1868) was a German–Swiss chemist who is best known for inventing the hydrogen–oxygen fuel cell.

REFERENCES

A.J. Bard, G. Inzelt, and F. Scholz, *Electrochemical Dictionary*, Springer, Berlin, Germany, 2008.
http://en.wikipedia.org/wiki/Christian_Friedrich_Sch%C3%B6nbein.

William Robert Grove (1811–1896) was a physical scientist (and also a judge) who anticipated the general theory of the conservation of energy and was a pioneer of fuel cell technology. His *gas voltaic battery* consisted of platinum electrodes immersed in acidic solution, with hydrogen over one electrode and oxygen over the other. The *battery* was a prototype of the modern hydrogen–oxygen fuel cell.

REFERENCES

A.J. Bard, G. Inzelt, and F. Scholz, *Electrochemical Dictionary*, Springer, Berlin, Germany, 2008.
http://en.wikipedia.org/wiki/William_Robert_Grove.

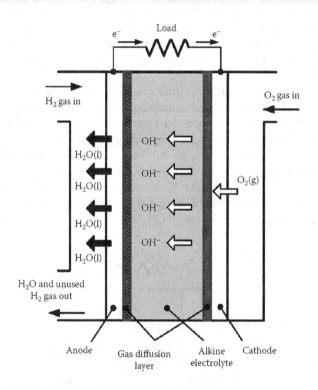

FIGURE 8.1 Schematic of an AFC fueled by hydrogen.

At the moment, the fuel cell technology is under extensive development and is considered as one of the viable options for high-efficiency power generation. As shown in Figure 8.1, four main components of a fuel cell are (1) electrolyte, (2) electrodes (anode and cathode), (3) gas diffusion layers, and (4) chemicals (fuel and oxidizer). The electrolyte (e.g., KOH(aq)) is needed to conduct OH^-(aq) ions from the cathode to the anode. The electrodes are needed to speed up electrochemical reactions and reduce the charge transfer overpotentials. The gas diffusion layers are needed to provide the desirable mass transport of chemicals to (from) the electrodes, reduce the mass transport overpotential, and protect catalytic materials from degradation due to an intensive hydrodynamic flow of chemicals. The fuel is electrochemically oxidized, and oxygen is electrochemically reduced, providing a flow of electrons through the external load and, therefore, generating electric power.

8.4 MAIN KINDS OF FUEL CELLS

There are a number of fuel cells, which can convert the energy of a fuel (usually hydrogen) and an oxidant (usually oxygen) to electricity. We will not cover all fuel cells but consider just a few of them that have some particular historical or technological significance. A detailed description of other fuel cells can be found in Ref. [4].

The AFC, schematically shown in Figure 8.1, uses 30%–40% KOH(aq) as an electrolyte and operates at temperatures up to around 90°C. The electrolyte is selected to optimize its properties such as conductivity and concentration. Temperature is elevated for improving the electrochemical kinetics and reducing the electrolyte resistivity. An AFC does not require using Pt for electrodes, so cheaper materials (e.g., Ni) can be used.

The proton exchange membrane fuel cell (PEMFC), schematically shown in Figure 3.7, uses a polymeric proton exchange membrane (PEM) as an electrolyte at temperatures from ambient up to around 120°C. Nafion is used for PEM and Pt for electrodes. Both of these materials are very expensive. The durability of the PEMFC is still under study for widespread implementation. This fuel cell could be considered as a first candidate for automotive applications.

The direct methanol fuel cell (DMFC) schematically shown in Figure 8.2 is a kind of PEMFC with the main difference being that 5%–10% aqueous methanol (CH_3OH) is used in place of hydrogen. The use of $CH_3OH(aq)$ significantly decreases the fuel

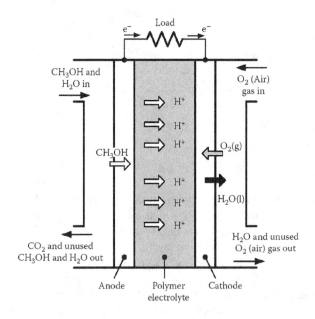

FIGURE 8.2   Schematic of a DMFC.

cell efficiency due to a large charge transfer overpotential at the anode, so another material is used instead of the anode in PEMFC. Crossover of CH_3OH through PEM is another factor in reducing the DMFC efficiency.

The solid oxide fuel cell (SOFC) schematically shown in Figure 3.8 uses anion-conducting ceramic material, yttria-stabilized zirconia (YSZ), as an electrolyte and operates at a high temperature of 800°C–900°C. Due to the high operational temperature, noble metals are not needed for electrodes. The anode is made by a mixture of Ni and porous YSZ, and the cathode is made by dence strontium-doped lanthanum manganite $(La(Sr)MnO_4$ or LSM). This fuel cell is most suitable for stationary power generation systems. In addition to hydrogen, SOFC can use $CO(g)$ or even $CH_4(g)$ as fuels.

The fuel cells mentioned thus far can in principle be used as electrolytic cells to produce chemicals (usually hydrogen and oxygen) using electric energy. However, to run a fuel cell in a reverse mode usually requires some modifications with materials and cell design. If a single electrochemical energy conversion system can efficiently work in both ways as electrolytic and galvanic systems, such a cell is called the regenerative fuel cell.

There are some other fuel cells, such as phosphoric acid (PAFC) or molten carbonate (MCFC) [4], that are commercially available but will probably not find significant implementation due to high cost and durability issues.

8.5   ELECTROCHEMISTRY OF FUEL CELLS

The electrochemical half-reactions taking place at the anode and cathode of a PEMFC, shown in Figure 3.7, are as follows:

$$H_2(g) \rightarrow 2H^+(m) + 2e^- (Pt/C) \quad \text{(Anode)},$$

$$2H^+(m) + (1/2)O_2(g) + 2e^- (Pt/C) \rightarrow H_2O(l) \quad \text{(Cathode)},$$

with the total reaction in the cell as

$$H_2(g) + (1/2)O_2(g) \rightarrow H_2O(l).$$

Note that $H_2O(l)$ is formed in the low-temperature fuel cells when temperature is below 100°C and pressure is 1 bar. If temperature is significantly above 100°C and pressure is ambient, $H_2O(g)$ is formed instead of $H_2O(l)$. A mixture of $H_2O(g)$ and $H_2O(l)$ can also be formed when the temperature is around 100°C. In these half-reactions, Pt/C represents a carbon support slurry with particles of about a micron in size with Pt nanoparticles deposited on the carbon. Nanoparticles are used to increase the surface area of the electrode. In the $H^+(m)$ symbol, "m" represents a proton conductive membrane. While the properties of $H^+(m)$ and $H^+(aq)$ might be different, this fact is usually ignored in most of the studies because the chemical potential (Gibbs energy of formation) of protons in a membrane is not known.

The electrochemical half-reactions taking place at the anode and cathode of a DMFC, shown in Figure 8.2, are as follows:

$$CH_3OH(aq) + H_2O(l) \rightarrow CO_2(g) + 6H^+(m) + 6e^- \left(Pt,Ru/C\right) \quad (Anode),$$

$$6H^+(m) + \left(3/2\right)O_2(g) + 6e^- \left(Pt/C\right) \rightarrow 3H_2O(l) \quad (Cathode),$$

with the total reaction in the cell as follows:

$$CH_3OH(l) + \left(3/2\right)O_2(g) \rightarrow CO_2(g) + 2H_2O(l).$$

Note that in a DMFC, $H_2O(l)$ is consumed at the anode and formed at the cathode. However, the total reaction given earlier does not show this. A better way to represent the total reaction is showing water consumed at the anode, $H_2O(l, a)$, and produced at the cathode, $H_2O(l, c)$:

$$CH_3OH(l) + \left(3/2\right)O_2(g) + H_2O\left(l,a\right) \rightarrow CO_2(g) + 3H_2O\left(l,c\right).$$

The Pt-Ru alloy is used (instead of Pt) in the anodic reaction of DMFC to improve the electrochemical reduction of $CH_3OH(aq)$.

The electrochemical half-reactions taking place at the anode and cathode of an SOFC with hydrogen as fuel (Figure 3.8) are as follows:

$$H_2(g) + O^{2-}(m) \rightarrow H_2O(g) + 2e^-(Ni) \quad (Anode),$$

$$\left(1/2\right)O_2(g) + 2e^-(LSM) \rightarrow O^{2-}(m) \quad (Cathode),$$

with the total reaction in the cell as follows:

$$H_2(g) + \left(1/2\right)O_2(g) \rightarrow H_2O(g).$$

When carbon monoxide is used as a fuel, the SOFC reactions are as follows:

$$CO(g) + O^{2-}(m) \rightarrow CO_2(g) + 2e^-(Ni) \quad (Anode),$$

$$\left(1/2\right)O_2(g) + 2e^-(LSM) \rightarrow O^{2-}(m) \quad (Cathode),$$

$$CO(g) + \left(1/2\right)O_2(g) \rightarrow CO_2(g) \quad (Total).$$

The reader may want to train himself/herself by writing the anodic, cathodic and total reaction when $CH_4(g)$ is used as a fuel in the SOFC.

8.6 THREE-PHASE BOUNDARY ISSUE

A three-phase boundary is required to provide an intimate contact between all necessary phases of an electrochemical half-reaction. In the case of the anodic reaction of a PEMFC, the phases are (1) H_2 gas, (2) H^+ conductive membrane, and (3) e^- conductive Pt/C. If one of the phases is not available, the reaction does not take place and the fuel cell does not work. Note that in the case of the cathodic reaction of a PEMFC, there is a four-phase boundary issue. Therefore, the issue should generally be called the multiphase boundary.

The traditional electrochemical diagram for a PEMFC shown in Figure 3.7

$$Cu(s)|Pt(s)|H_2(g)|H^+(m)|O_2(g),H_2O(l)|Pt(s)|Cu(s),$$

does not illustrate the multiphase boundary problem, and this is a challenge in using the traditional diagrams, which can show only two-phase boundaries (see also Section 2.6). For example, the preceding diagram above does not show that an intimate contact between Pt(s) and H^+(m) should be provided. In contrast, based on the diagram, the phases are electrically separated by a nonconductive $H_2(g)$ phase, so the current should not go through the cell. An additional problem of the traditional diagrams is that in some cases, the cell phases are shown (e.g., Pt(s)) and in other cases, the participating chemicals (e.g., H^+(m)) are presented. This inconsistency makes the traditional diagram quite confusing for a reader who is just starting to learn the subject. The new diagrams described in Section 2.6 eliminate problems of the traditional diagrams.

8.7 NEW ELECTROCHEMICAL DIAGRAMS FOR FUEL CELLS

For the PEMFC described by the half-reactions provided earlier and considered at ambient conditions, the multiphase boundary electrochemical diagram can be presented as follows:

$$\begin{array}{cc} H_2(g) & O_2(g) \\ e^-(Cu)|e^-\left(Pt/C\right) \times H^+(m) \times e^-\left(Pt/C\right)|e^-(Cu), \\ & H_2O(l) \end{array}$$

where the cross on the left shows that three species of the anodic half-reaction ($H_2(g)$, H^+(m), and e^-(Pt/C)) are located, respectively, in three phases (gas, membrane, and metal/carbon) and the phases are in intimate contact. The right-hand cross shows four species (H^+(m), $O_2(g)$, e^-(Pt/C), $H_2O(l)$) of the cathodic half-reaction that are located, respectively, in four phases (membrane, gas, metal/carbon, and liquid). Also, the new diagram shows all electrically conductive species in one row, so the current can go from the left to the right terminals of the cell.

When temperature in the PEMFC is increased significantly above 100°C and $H_2O(l)$ is transformed to $H_2O(g)$, the electrochemical diagram can reflect an important change at the cathode. The cathodic reaction still includes four species but becomes a three-phase electrochemical reaction:

$$H_2(g) \quad O_2(g), H_2O(g)$$

$$e^-(Cu)|e^-\left(Pt/C\right) \times H^+(m) \times e^-\left(Pt/C\right) \,|Cu(s)| \, e^-(Cu).$$

In the electrochemical diagram above, the participants in the electrochemical reactions are clearly shown, while the phases are shown in the brackets immediately following the chemicals. The diagram also shows the intimate contact between the phases that must exist when the reaction occurs at each electrode. The conductive phases are all in contact, and the confusion about what path the charge transfer flows across the phases is eliminated.

8.8 POLARIZATION CURVES OF PEMFC AND SOFC

The dependence of the cell potential difference on current density is called the cell polarization curve. The experimentally obtained open circuit potential (OCP) in a H_2/O_2 PEMFC is always below the theoretically calculated OCP, 1.229 V (at 25°C), due to (1) crossover of chemicals through the membrane and (2) a small internal electron current. This parasitic current density (j_p) can reduce the OCP down to around 1 V for a H_2/O_2 PEMFC and down to around 0.6 V for a DMFC. The PEMFC (Figure 8.3) and SOFC (Figure 8.4) polarization curves at low current densities are quite different due to significant temperature dependence of the charge transfer reaction. At elevated

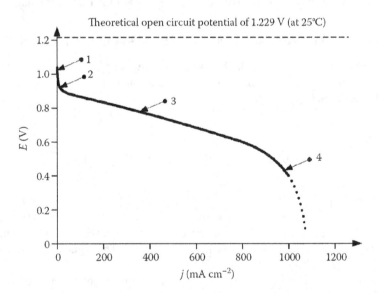

FIGURE 8.3 Polarization curve of a PEMFC: (1) the experimental open circuit potential is less than the theoretical value of 1.229 V (at 25°C) due to permeation of chemicals through membrane, (2) the potential declines due to relatively slow charge transfer cathodic reaction, (3) a linear Ohm's law dependence due to resistance of membrane, (4) the potential drops down due to transport limitation and approaching the limiting current, as shown, of about 1100 mA cm^{-2}.

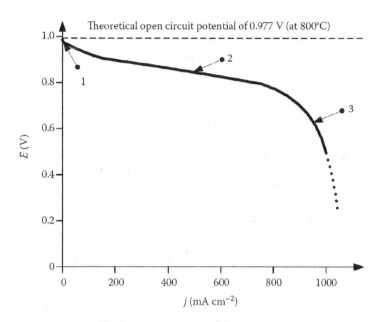

FIGURE 8.4 Polarization curve of an SOFC: (1) the experimental open circuit potential is very close to the theoretical value of about 0.977 V (at 800°C) due to a fast charge transfer reaction, (2) at the moderate current density, the curve is fairly linear due to Ohm's law, and (3) the potential drops down due to transport limitation and approaching the limiting current.

temperatures, the electron transfer reaction is much faster, and therefore, the charge transfer resistance is much smaller. As a result, the experimental and theoretical (0.977 V) OCPs at 800°C are very close.

When the current density is above 0.8 A cm^{-2}, the mass transfer processes are dominating, and the cell potential rapidly decreases down to zero. In the middle part of the curve, the internal ohmic resistance of the electrochemical cell is dominating, so the dependence is almost linear.

Using the electrochemical thermodynamics and kinetics covered in Chapters 4 and 6, the equation

$$E_{FC} = E_{EQ} - A_{FC} \ln\left[\frac{j + j_p}{j_o}\right] + B_{FC} \ln\left(1 - \frac{j}{j_{lim}}\right) - r_{FC} j, \tag{8.1}$$

(suggested in [4]) can be used to describe the cell potential–current density (polarization) curve.

On the right hand, the first term can be calculated using equilibrium electrochemistry of the fuel cell reaction as described in Chapter 4, the second term is actually a modified Tafel equation (Chapter 6), with a parasitic current density correction described earlier, the third term is related to the mass transport of chemicals when the limiting current is approached (Chapter 6), and the last term is simply Ohms' law (Chapter 2). In this equation, A_{FC} and B_{FC} are the semiempirical positive coefficients in V, and r_{FC} is the fuel cell area-specific resistance in Ω cm². Current density is in A cm⁻², and the fuel cell and equilibrium potentials are in V.

The potential–current curve for an electrolyzer can be described using a similar equation with slightly different parameters due to different materials and designs used in fuel and electrolytic cells. Certainly, E_D should be used instead of E_{EQ} when this equation is employed for electrolysis, and a small correction with a sign should be made taking into account the sign convention adapted in this book and described in Chapter 2.

When Equation 8.1 is used for a fuel cell, A_{FC}, B_{FC}, and j are always positive values. Because the last three terms on the right side of Equation 8.1 are always negative and E_{EQ} is always positive, E_{FC} is always less than E_{EQ}, and this corresponds to Figure 2.12.

For an electrolytic cell, Equation 8.1 should be slightly modified as follows:

$$E_{EC} = E_D - A_{EC} \ln\left(\frac{|j| + j_p}{j_o}\right) + B_{EC} \ln\left(1 - \frac{j}{j_{lim}}\right) + r_{EC}j, \qquad (8.2)$$

so that all terms on the right side will be negative, taking into account that the current density and limiting current density for the electrolytic cells are assumed to be negative values. Obviously, the cell potential difference, E_{EC}, of an electrolytic cell will always be more negative than the decomposition potential, E_D, while the magnitude of E_{EC} will always be larger than the magnitude of E_D. For the convenience of a reader, the signs and units of all values in Equations 8.1 and 8.2 are presented in Table 8.1.

TABLE 8.1

Signs and Units of the Values in Equations 8.1 and 8.2

Value and Units	FC	EC
E_{FC} (V)	+	N/A[a]
E_{EC} (V)	N/A	−
E_{EQ} (V)	+	$-E_D$
E_D (V)	$-E_{EQ}$	−
j (A cm⁻²)	+	−
j_{lim} (A cm⁻²)	+	−
j_p (A cm⁻²)	+	+
j_o (A cm⁻²)	+	+
A (V)	+	+
B (V)	+	+
r (Ω cm²)	+	+

[a] N/A stands for not applicable.

8.9 EFFICIENCY OF FUEL CELL VERSUS HEAT ENGINE

The maximum (thermodynamic) efficiency of a heat engine, $\xi_{th, HE}$, is defined by the well-known Carnot's theorem:

$$\xi_{th,HE} = \frac{T_h - T_c}{T_h},$$

where T_h and T_c are thermodynamic temperatures, respectively, of the hot and cold reservoirs.

Assuming that T_h should be about 600°C due to stability of available materials and T_c should be about 100°C due to using a reasonable amount of power for cooling, the absolute maximum theoretical efficiency of a heat engine cannot be above 60%.

The maximum (thermodynamic) efficiency (or thermodynamic ratio) of a fuel cell, $\xi_{th, FC}$, is defined by the ratio of the useful Gibbs energy of the total cell reaction, $\Delta_r G$, to the whole energy of chemicals (fuel and oxidizer), $\Delta_r H$, which is enthalpy:

$$\xi_{th,FC} = \frac{\Delta_r G}{\Delta_r H}. \tag{8.3}$$

The entropy of the fuel cell reaction, $\Delta_r S$, multiplied by cell temperature in K provides the amount of heat to be released or consumed when the fuel cell is in an equilibrium state:

$$T\Delta_r S = \Delta_r H - \Delta_r G, \tag{8.4}$$

and is the difference between the enthalpy and Gibbs energy of the fuel cell reaction. If $\Delta_r S$ is negative, the fuel cell should release heat, and when $\Delta_r S$ is positive, the fuel cell should consume heat when it is in a state around equilibrium.

Comparing the efficiency equations for fuel cell and heat engine, one can see that the efficiency definitions of a heat engine and a fuel cell are completely different. The key difference is that the fuel cell is an isothermal system, and Carnot's theorem cannot be used for defining the fuel cell efficiency. $\Delta_r H$ and $\Delta_r G$ of a fuel cell are usually large negative values, and in most cases, the magnitude of $\Delta_r H$ is larger than the magnitude of $\Delta_r G$. As a result, most fuel cells have a negative value of $T\Delta_r S$, showing that the fuel cell releases heat even in the equilibrium state. However, it might be a case when the magnitude of $\Delta_r H$ can be smaller than the magnitude of $\Delta_r G$, so the thermodynamic efficiency of such a fuel cell will be above 100%. Because an efficiency above 100% sounds unusual, it might be better to call $\Delta_r G/\Delta_r H$ as the ratio of the Gibbs energy to enthalpy.

The relationships between thermodynamic, kinetic, and electrochemical values of a fuel cell are shown in Figure 8.5, where r_{FC} is the internal area-specific resistance and η_{FC} is the sum of the overpotential on both electrodes and a cell potential drop due to crossover of chemicals. The polarization curve shown may correspond to the PEMFC.

If the relationship between the internal resistance R_i and the area-specific resistance r_{FC} is $(R_i I_{FC}) = (r_{FC} j)$, and the overpotential can be defined as $\eta_{FC} = -A_{FC} \ln$

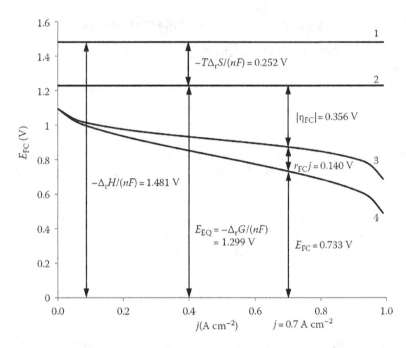

FIGURE 8.5 Polarization curves of a PEMFC with thermodynamic and kinetic explanations of its components: the curves are constructed using Equation 8.1, $E_{FC} = E_{EQ} - A_{FC} \ln[(j + j_p)/j_o] + B_{FC} \ln(1 - (j/j_{lim})) - r_{FC}j = E_{EQ} + \eta_{FC} - r_{FC} j$ with following parameters: $A_{FC} = 0.045$ V, $B_{FC} = 0.05$ V, $j_o = 0.001$ A cm^{-2}, $j_p = 0.02$ A cm^{-2}, $j_{lim} = 1.0$ A cm^{-2}; $r_{FC} = 0.2\ \Omega$ cm^2 for Curve 4 and $r_{FC} = 0\ \Omega$ cm^2 for Curve 3. Line 1 corresponds to the enthalpy of the fuel cell reaction, and Line 2 shows the theoretical equilibrium potential.

$[(j + j_p)/j_o] + B_{FC} \ln (1 - j/j_{lim})$, then Equation 8.1 can be presented in the general form as follows:

$$E_{FC} = E_{EQ} + \eta_{FC} - R_i I_{FC}, \qquad (8.5)$$

which is an extension of Equation 6.1 with overpotentials at the electrodes and cross-over of chemicals through PEM taken into account.

As an example, to illustrate Figure 8.5 and Equation 8.5, let us provide some electrochemical and thermodynamic values calculated for a H_2/O_2 PEMFC operating at ambient conditions with all component activities equal to 1. Using thermodynamic data from Table 10.3 or Table 10.4, the standard thermodynamic values of the fuel cell reaction, $H_2(g) + 0.5\ O_2(g) = H_2O(l)$, can be defined as $\Delta_r G° = -237.141$ kJ mol^{-1}, $\Delta_r H° = -285.83$ kJ mol^{-1}, and $\Delta_r S° = -0.163$ kJ (mol K)$^{-1}$. If all activities are 1 and do not depend on temperature, then $\Delta_r G = \Delta_r G°$, $\Delta_r H = \Delta_r H°$, and $\Delta_r S = \Delta_r S°$. Let us now assume that the parameters of Equation 8.1 are as follows: $A = 0.045$ V, $B = 0.05$ V, $j_o = 0.001$ A cm^{-2}, $j_p = 0.02$ A cm^{-2}, $j_{lim} = 1.0$ A cm^{-2}, $r_{FC} = 0.2\ \Omega$ cm^2, and calculate E_{FC}, η_{FC}, and $R_i I_{FC}$ ($= r_{FC} j$) at two current densities equal to 0 and 0.7 A cm^{-2}. A reader can easily estimate that $E_{EQ} = 1.229$ V, and at zero current density, $\eta_{FC} = -0.135\,V$ and $R_i I_{FC} = 0\,V$, while at 0.7 A cm^{-2}, $\eta_{FC} = -0.356\,V$ and $R_i I_{FC} = r_{FC}\, j = 0.14\,V$. The potential difference of the fuel cell, E_{FC}, at zero current (OCP) is 1.094 V and at, 0.7 A cm^{-2}, is 0.733 V. Note that both $\Delta_r H$ and $T\,\Delta_r S$ can conveniently be presented in V and in the preceding example, respectively, equal to $-\Delta_r H/(nF) = 285{,}830/(2 \times 96{,}485) = 1.481\,V$ and $-T\,\Delta_r S = 298.15 \times 163/(2 \times 96{,}485) = 0.252\,V$.

The efficiency of a fuel cell under load, $\xi_{load,\,FC}$, is always smaller than $\xi_{th,\,FC}$ because of the electrode overpotentials and internal resistance of the fuel cell and therefore $\xi_{load,\,FC}$, can now be defined as follows:

$$\xi_{load,FC} = \frac{\Delta_r G - nF\left(\eta_{FC} - R_i I_{FC}\right)}{\Delta_r H} = \frac{-nFE_{FC}}{\Delta_r H}$$

$$= \left(\frac{\Delta_r G}{\Delta_r H}\right)\left(\frac{E_{FC}}{E_{EQ}}\right) = \xi_{th,FC}\,\xi_{V,FC}, \qquad (8.6)$$

where $\xi_{V,\,FC} = E_{FC}/E_{EQ}$ is the voltage efficiency, which is 0.596 (0.733/1.229) at 0.7 A cm^{-2} in the preceding example. Because $\xi_{th,\,FC}$ in the preceding example is 0.830, the fuel cell efficiency under load, $\xi_{load,\,FC}$, at 0.7 A cm^{-2} is 0.495 or 49.5%. In this estimation of the efficiency under load, we assumed that there are not any other reactions in the PEMFC except the anodic oxidation of hydrogen, which produces protons and electrons, as well as the cathodic reduction of oxygen, which combines with protons and electrons to produce water. Otherwise, the current efficiency should also be taken into account (Section 8.10).

Note that in the PEMFC, both $\xi_{V,\,FC}$ and $\xi_{th,\,FC}$ are commonly less than 1. However for other electrolytic systems, these values could be > 1. Note that when $\xi_{th,\,FC}$ is larger than one, $\Delta_r S$ is positive. An example of a fuel cell with $\Delta_r S > 0$ is given in Section 8.11.

8.10 TOTAL EFFICIENCY OF FUEL AND ELECTROLYTIC CELLS

Generally, the total efficiency, ξ, of an electrochemical energy conversion system is defined as a product of thermodynamic efficiency, ξ_{th} (ratio of the Gibbs energy to enthalpy), voltage efficiency, ξ_V, and current (or Faradaic) efficiency, ξ_i.

$$\xi = \xi_{load}\xi_i = \xi_{th}\xi_V\xi_i. \qquad (8.7)$$

The current efficiency, ξ_i, is defined by Faraday's law. In a fuel cell, the amount of fuel or oxidizer can never be smaller than the theoretically calculated amounts using Faraday's law based on the measured current as a function of time. The ratio of the calculated and measured amount of chemical is the fuel cell current efficiency, $\xi_{i,\,FC}$.

In an electrolytic cell, the amount of chemicals produced can never be larger than the theoretically calculated amounts using Faraday's law based on the measured current as a function of time. The ratio of the measured and calculated amount of chemical is the current efficiency of the electrolytic cell. The current efficiency of either fuel or electrolytic cells is expected to be above 0.95 in the properly operating electrochemical energy conversion system.

The voltage and thermodynamic efficiencies of an electrolytic cell are defined as follows:

$$\xi_{V,EC} = \frac{E_D}{E_{EC}}, \tag{8.8}$$

$$\xi_{th,EC} = \frac{\Delta_r H}{\Delta_r G}. \tag{8.9}$$

In conclusion, the total efficiency of a fuel cell is as follows:

$$\xi_{FC} = \left(\frac{\Delta_r G}{\Delta_r H}\right)\left(\frac{E_{FC}}{E_{EQ}}\right)\xi_{i,FC} = -\left(\frac{nFE_{FC}}{\Delta_r H}\right)\xi_{i,FC}, \tag{8.10}$$

and the total efficiency of an electrolytic cell is as follows:

$$\xi_{EC} = \left(\frac{\Delta_r H}{\Delta_r G}\right)\left(\frac{E_D}{E_{EC}}\right)\xi_{i,EC} = \left(\frac{\Delta_r H}{nFE_{EC}}\right)\xi_{i,EC}. \tag{8.11}$$

In Equations 8.10 and 8.11, $\Delta_r G$, $\Delta_r H$, E_{EQ}, and E_D are theoretically defined values, while E_{FC} and E_{EC} are the measurable values taking a polarization curve, and $\xi_{i,\,FC}$ and $\xi_{i,\,EC}$ are the experimentally obtained values measuring the amount of chemicals consumed or produced versus values calculated from Faraday's law. Note that the definition of the voltage efficiency is applicable if the magnitude of E_D or E_{EQ} is not close to zero, which is an unusual case but still possible. When $E_{EQ} = 0$, the fuel cell voltage efficiency is infinity, and if $E_D = 0$, the electrolytic voltage efficiency is zero.

8.11   HEAT BALANCE IN FUEL AND ELECTROLYTIC CELLS

The negative entropy of the H_2/O_2 PEMFC reaction ($\Delta S° = -163.3$ J mol^{-1} K^{-1} at 25°C [Table 10.4]) shows that about 17% of the total available energy, $\Delta H° = -285.83$ kJ mol^{-1} (Table 10.4), should be released in the form of heat, $T\Delta S°$, even when the fuel cell is in the hypothetical equilibrium state and current is zero.

When the current is larger than zero, additional heat will be produced due to internal resistance of the cell and overpotentials at the electrodes. As a result, when a fuel cell operates at current density, j, the magnitude of the released heat rate, Q_{FC}, in W cm^{-2} is defined as follows:

$$Q_{FC} = j\left[-T\Delta_r S/(nF) - \eta_{FC} + r_{FC}j\right], \tag{8.12}$$

where the two last terms on the right side are always positive and the first one is usually positive but can occasionally be negative. Using the same example of a PEMFC operating at ambient conditions at a current density of 0.7 A cm^{-2}, we can calculate the released heat rate as

$$Q_{FC} = 0.7\left(\text{A cm}^{-2}\right)\left[0.252(\text{V}) + 0.356(\text{V}) + 0.14(\text{V})\right] = 0.524\,\text{W cm}^{-2}.$$

The current density dependence of the released heat rate, Q_{FC}, along with the produced electric power density, $P_{FC} = jE_{FC}$, and potential difference, E_{FC}, of the PEMFC described earlier are presented in Figure 8.6.

Therefore, if the fuel cell is supposed to work at a constant temperature, a heat removal procedure should be considered. Otherwise, the fuel cell can be overheated and destroyed. In a PEMFC at common operations, current densities (0.5–0.7 A cm^{-2}) up to about 50% of the energy of chemicals, $\Delta_r H$, are converted to heat, and the remaining 50% is available for electric energy. In this fuel cell, one-third of the heat comes from the entropy term and two-thirds are from the overpotentials at the electrodes and from the Joule heating, $r_{FC}j^2$. In the preceding example, at 0.7 A cm^{-2}, the generated electric power density is 0.513 W cm^{-2}, the reversible heat rate density due to the entropy of the reaction is 0.176 W cm^{-2}, and the heat rate density generated due to the overpotentials and crossover is 0.524 W cm^{-2}.

Note that for some fuel cells, the entropy of the fuel cell reaction could be positive. An example is $C(s) + (1/2)O_2(g) = CO(g)$ or $C(s) + O_2(g) = CO_2(g)$. These reactions provide a standard thermodynamic efficiency, $\xi_{th}^0\,FC = \Delta_r G^0/\Delta_r H^0$, greater than 100%

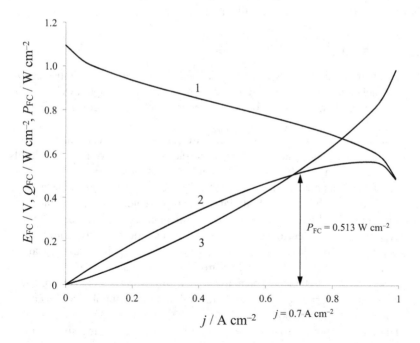

FIGURE 8.6 Dependence of the released heat rate, $Q_{FC} = j\,[-T\Delta_r S/(nF) - \eta_{FC} + r_{FC}j]$, generated electric power density, $P_{FC} = j \times E_{FC}$, and potentials difference, E_{FC}, of the PEMFC described earlier and represented by Figure 8.4: E_{FC}—curve 1; P_{FC}—curve 2; Q_{FC}—curve 3.

(Table 10.4), because the fuel cell must absorb heat in the hypothetical equilibrium state. This conclusion can easily be verified using thermodynamic data from (Table 10.4). As an example, for the reaction of oxidation of C(s) to CO_2(g) at a temperature of around 700°C, $\xi^0_{th} FC = \Delta_r G^0/\Delta_r H^0 = 1.003 = 100.3\%$

Note that the standard thermodynamic efficiency $\xi^0_{th} = \Delta_r G^0/\Delta_r H^0$ and thermodynamic efficiency at a given chemical composition, $\xi_{th} = \Delta_r G/\Delta_r H$, are usually only slightly different, and therefore, in most cases, we could assume that

$$\xi_{th} \approx \xi^0_{th} = \frac{\Delta_r G^0}{\Delta_r H^0}. \tag{8.13}$$

However, in some cases, Equation 8.13 could not be an accurate approximation, particularly in a case when concentrated aqueous solutions are the participating chemicals. An example of a dramatic difference between thermodynamic efficiency and its standard value is given in Ref. [6] for an electrolytic system where highly concentrated aqueous solutions are used.

The heat balance analysis of an electrolytic cell can be carried out in a similar way. For example, in the case of water electrolysis in the reversible process, $T\Delta_r S$ is positive, and therefore, heat is supposed to be consumed by the electrolytic system in the equilibrium state rather than produced like in the fuel cell. However, the overpotentials on the electrodes and the Joule heating will usually compensate this $T\Delta_r S$ heat consumption, and, as a result, the water electrolysis system will not require any significant heat removal procedure like in the fuel cell.

8.12 SUMMARY

- There are a number of electrochemical energy conversion systems such as the fuel cell, electrolyzer, battery, and flow cell. In this chapter, we cover the fundamentals of the electrochemical energy conversion considering only fuel and electrolytic cells.
- Four main components of a fuel cell are (1) electrolyte, (2) electrodes, (3) gas diffusion layers, and (4) chemicals (fuel and oxidizer). Different kinds of fuel cells have different materials of the components.
- The Electrochemistry of the PEMFC, DMFC, and SOFC is considered in detail, and the three-phase boundary issue is presented as a paramount problem in the development of the electrochemical energy conversion systems.
- The traditional electrochemical diagrams cannot properly address the three-phase boundary problem. A new type of electrochemical diagram is proposed to properly describe the three-phase boundary issue.
- Based on the fundamental principles of electrochemical engineering (electrochemical thermodynamics and kinetics), the fuel cell polarization curves can easily be understood.
- The electrolyzer polarization curve can be interpreted using similar principles of electrochemical engineering.
- The theoretical (thermodynamic) efficiency of a fuel cell versus a heat engine is considered. The efficiency of fuel cell under load is also introduced.

- The total efficiency and its components for fuel and electrolytic cells are considered in detail.
- A heat balance approach is introduced to be used for both fuel and electrolytic cells.

8.13 EXERCISES

PROBLEM 1

Experimental polarization curve data for a H_2/O_2 PEMFC obtained at 80°C are given below. The pressures of hydrogen at the anode and oxygen at the cathode are both 1 bar, which is also the total pressure of the fuel cell. The values of j_p, j_{lim}, and j_o, found in advance, are, 0.004, 1.6, and 10^{-6} A cm^{-2}, respectively.

First, calculate E_{EQ} for this cell using thermodynamic data. Then, using software such as Mathematica or MATLAB, fit the data to Equation 8.1 and solve for A_{FC}, B_{FC}, and r_{FC}. On the same plot, compare the experimental polarization data with the fitted data from Equation 8.1 to show the quality of the regression.

j (A cm^{-2})	E_{FC} (V)
0	1.074
0.1	1.004
0.2	0.971
0.3	0.937
0.4	0.911
0.5	0.880
0.6	0.849
0.7	0.825
0.8	0.793
0.9	0.768
1.0	0.738
1.1	0.708
1.2	0.682
1.3	0.647
1.4	0.621
1.5	0.584

Solution:

The PEMFC fueled by $H_2(g)$ half reactions are

$$H_2(g) \rightarrow 2H^+(m) + 2e^- \quad (\text{Anode}),$$

$$2H^+(m) + 0.5O_2(g) + 2e^- \rightarrow H_2O(l) \quad (\text{Cathode}),$$

with the total reaction of the cell as

$$H_2(g) + 0.5O_2(g) \rightarrow H_2O(l).$$

Therefore, the corresponding Nernst equation can be written as

$$E_{EQ} = E^0 + \frac{RT}{2F} \ln \left[\frac{\left(\frac{f_{H_2(g)}}{f^0} \right) \left(\frac{f_{O_2(g)}}{f^0} \right)^{0.5}}{a_{H_2O(l)}} \right] \approx E^0 + \frac{RT}{2F} \ln \left[\left(\frac{p_{H_2(g)}}{p^0} \right) \left(\frac{p_{O_2(g)}^{0.5}}{p^0} \right) \right].$$

The dimensionless activity of water is equal to 1, and the standard fugacity $f^0 = 1$ bar and standard pressure $p^0 = 1$ bar guarantee that the expressions in the squared brackets are dimensionless. Due to the ambient operational pressure, the fugacities of hydrogen and oxygen are very close to their partial pressures [8]. The activity of water, $a_{H_2O(l)}$, equals its mole fraction and is very close to 1 due to a very small solubility of oxygen in water (Table 10.2).

The standard value of the fuel cell potential difference E^0 can be calculated using Equations 4.19 and 4.21:

$$E^0 = -\frac{\Delta_r G_T^0}{nF} = -\frac{\left(\Delta_f G_{H_2O(l)}^0 - \Delta_f G_{H_2(g)}^0 - 0.5 \Delta_f G_{O_2(g)}^0 \right)}{nF}$$

$$= -\frac{-228.3 \times 10^3}{2 \times 96,485} = 1.183 \, V,$$

where the Gibbs energies of formation are interpolated from data in Table 10.4. The equilibrium potential difference to be used in Equation 8.1 is then

$$E_{EQ} = E^0 + \frac{RT}{2F} \ln \left[\left(\frac{p_{H_2(g)}}{p^0} \right) \left(\frac{p_{O_2(g)}^{0.5}}{p^0} \right) \right]$$

$$= 1.183 + \frac{8.314 \times 353.15}{2 \times 96,485} \ln \left(1 \times 1^{0.5} \right) = 1.183 \, V.$$

Now, A_{FC}, B_{FC}, and r_{FC} can be estimated by fitting the experimental data to Equation 8.1:

$$E_{FC} = E_{EQ} - A_{FC} \ln \left[\frac{j + j_p}{j_o} \right] + B_{FC} \ln \left(1 - \frac{j}{j_{lim}} \right) - r_{FC} j,$$

using the values of E_{EQ}, j_p, j_{lim}, and j_o given above. The best-fit calculated values are as follows: $A_{FC} = 0.01321 \, V$, $B_{FC} = 0.01296 \, V$, and $r_{FC} = 0.2502 \, \Omega \, cm^2$ with a 0.9999 R^2 value. The resulting fit, shown below, graphically demonstrates that Equation 8.1 is well suited to fit the experimental data.

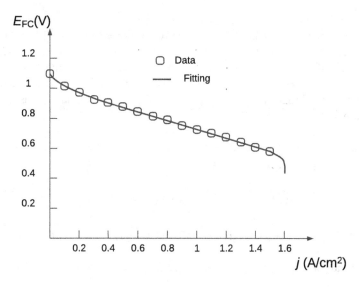

PROBLEM 2

The molten carbonate fuel cell (MCFC) operates at elevated temperatures and pressures and can be fueled by either $H_2(g)$ or $CO(g)$. An MCFC fueled by $CO(g)$ is shown in a simplified version as follows:

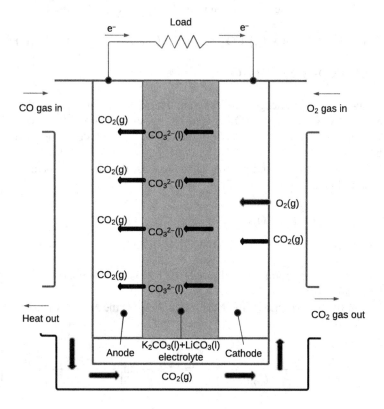

Construct both the traditional and new electrochemical diagrams for the MCFC operating at a temperature of 600°C and a pressure of 9 bar fueled by CO(g). Write the cathodic, anodic, and total reactions, and construct the Nernst equation. Calculate the potential difference assuming that the partial pressures of both CO(g) and O_2(g) are 3 bar and the partial pressure of CO_2(g) is 6 bar at both the cathode and anode. Also, determine if fugacities, rather than partial pressures, should be used in these calculations.

Solution:

The half-reactions of an MCFC fueled by CO(g) are

$$2CO(g) + 2CO_3^{2-}(l) = 4CO_2(g,a) + 4e^- \quad \text{(Anode)},$$

$$O_2(g) + 2CO_2(g,c) + 4e^- = 2CO_3^{2-}(l) \quad \text{(Cathode)},$$

with the total reaction as

$$2CO(g) + O_2(g) + 2CO_2(g,c) = 4CO_2(g,a),$$

so the CO_2(g, a) generated at the anode can be used at the cathode but also should partially be removed from the fuel cell to keep the total pressure constant.

The traditional electrochemical diagram of the MCFC is as follows

$$Cu(s)|Ni(Cr)(s)|CO(g),CO_2(g,a)|CO_3^{2-}(l) \text{ in } K_2CO_3(l) + Li_2CO_3(l)|$$

$$O_2(g),CO_2(g,c)|NiO(s)|Cu(s).$$

In this diagram, it is challenging to represent the electrical conductivity through the cell because CO(g), CO_2(g, a), O_2(g), and CO_2(g, c) are not electrical conductors. Also, the three-phase boundaries at both the cathode and anode are not represented in such a diagram.

Based on the cathodic and anodic half-reactions, the new electrochemical diagram is as follows:

$$CO(g),CO_2(g,a) \qquad\qquad O_2(g),CO_2(g,c)$$
$$Cu(s)\,|\,Ni(Cr)(s)\, \times CO_3^{2-}(l) \text{ in } K_2CO_3(l) + Li_2CO_3(l) \times NiO(s)\,|\,Cu(s).$$

Here, the problem of conductivity through the cell is resolved and the three-phase boundaries are represented.

Based on the half-reactions, the Nernst equation for the MCFC fueled by CO(g) is as follows:

$$E = E^0 + \frac{RT}{4F}\ln\left[\frac{f_{CO(g)}^2 f_{CO_2(g,c)}^2 f_{O_2(g)}}{f_{CO_2(g,a)}^4}\right] \approx E^0 + \frac{RT}{4F}\ln\left[\frac{p_{CO(g)}^2 p_{CO_2(g,c)}^2 p_{O_2(g)}}{p_{CO_2(g,a)}^4}\right],$$

where the standard fugacities, f^0, and standard pressures, p^0, are not shown in the equation for simplicity.

First, we calculate the standard value of the potential difference, E^0, at a temperature of 873.15 K:

$$E^0 = -\frac{\Delta_r G_T^0}{nF} = -\frac{\left(2\Delta_f G_{CO_2(g)}^0 - 2\Delta_f G_{CO(g)}^0 - \Delta_f G_{O_2(g)}^0\right)}{nF}$$

$$= -\frac{\left[2\times(-395.6794) - 2\times(-189.0212) - 0\right]\times 10^3}{4\times 96,485} = 1.071V,$$

where the Gibbs energies of formation can be interpolated from data in Table 10.4. Using the partial pressures of all species in the MCFC reaction, the potential difference can be calculated as follows:

$$E = E^0 + \frac{RT}{4F}\ln\left[\frac{p_{CO(g)}^2 p_{CO_2(g,c)}^2 p_{O_2(g)}}{p_{CO_2(g,a)}^4}\right]$$

$$= 1.071 + \frac{8.314\times 873.15}{4\times 96,485}\ln\left[\frac{3^2\times 6^2\times 3}{6^4}\right] = 1.066 \ V.$$

Let us now verify whether using the partial pressures instead of fugacities is a reasonable approximation at a total pressure of 9 bar. The ratio of fugacity and partial pressure of a gas component can be approximately estimated if the molar volumes of the real, V, and the ideal, V_{id}, gases are available [7]:

$$\ln\frac{f_i}{p_i} \approx \frac{1}{RT}(V - V_{id})(p_i - p^0),$$

where p^0 is the standard pressure of 1 bar.

For calculating the real molar volume of each component, we can use the van der Waals equation:

$$\left(p_i + \frac{n^2 a}{V^2}\right)(V - nb) = nRT,$$

where n is the amount of substance in mol and the values of a and b for the gases can be found in Table 10.24 as follows.

Substance	a (bar L²mol⁻²)	b (L mol⁻¹)
CO_2	3.658	0.0429
CO	1.472	0.0395
O_2	1.382	0.0319

Using CO_2(g, a) as an example, the volume taken up by 4 mol of CO_2(g) using the ideal gas law is

$$V_{CO_2(g,a),id} = \frac{nRT}{p_{CO_2(g,a)}} = \frac{4 \times 0.08314 \times 873.15}{6} = 48.40\,L\,mol^{-1},$$

where n is taken as a stoichiometric coefficient in the anodic half-reaction. Solving the van der Waals equation for CO_2(g, a) with $a = 3.658$ bar L^2mol^{-2} and $b = 0.0429$ L mol^{-1}, $V_{CO_2(g,a)} = 48.37$L. The difference, $V_{id} - V$, is 0.03 L mol^{-1} and the ratio of fugacity and partial pressure at an excess pressure of 5 bar is 0.998. Similar calculations can be carried out for all other gas components:

Substance	V_{id} (L)	V (L)	$(V_{id} - V)$ (L)	$p_i - p^0$ (bar)	f_i/p_i
CO_2(g, a)	48.40	48.37	0.030	5	0.998
CO_2(g, c)	24.20	24.19	0.010	5	0.999
CO	48.40	48.44	−0.040	2	1.001
O_2	24.20	24.21	−0.010	2	1.000

As shown in the table above, the ratio of fugacity to partial pressure for all gasses is all close to 1. Therefore, we confirmed that the partial pressures can be used instead of fugacities in this problem without compromising the result.

REFERENCES

1. V.S. Bagotsky, *Fundamentals of Electrochemistry*, 2nd ed., Wiley, New York, 2006.
2. C.H. Hamann, A. Hamnett, and W. Vielstich, *Electrochemistry*, 2nd ed., Wiley-VCH, Weinheim, Germany, 2007.
3. K.B. Oldham, J.C. Myland, and A.M. Bond, *Electrochemical Science and Technology*, Wiley, Chichester, UK, 2012.
4. J. Larminie and A. Dicks, *Fuel Cell Systems Explained*, 2nd ed., Wiley-VCH, Chichester, UK, 2007.
5. K.W. Beard and T.B. Reddy, *Linden's Handbook of Batteries*, 5th ed., McGraw-Hill, New York, 2019.
6. D.M. Hall, N.N. Akinfiev, E.G. LaRow, R.S. Schatz, and S.N. Lvov, Thermodynamics and efficiency of a CuCl(aq)/HCl(aq) electrolyzer. *Electrochim. Acta*, **143**, 2014, 70–82.
7. I.N. Levine, *Physical Chemistry*, 6th ed., McGraw-Hill, New York, 2009.

9 Electrochemical Corrosion

9.1 OBJECTIVES

The main objective of this chapter is to apply previously studied electrochemical thermodynamics and kinetics briefly explain the nature of an important electrochemical phenomenon—corrosion. Pourbaix diagrams, polarization curves, and the rate of the electrochemical corrosion are considered. Main types of corrosion protection and possible kinds of corrosion are mentioned.

9.2 ORIGIN OF ELECTROCHEMICAL CORROSION

The corrosion of a metal (or any other electron conductor) is an anodic electrochemical half-reaction, $M(s) \rightarrow M^{n+}(aq) + ne^-$, when a metallic element dissolves, generating ions and electrons, as shown in Figure 9.1 [1].

An example of an anodic reaction is the electrochemical oxidation of iron:

$$Fe(s) \rightarrow Fe^{2+}(aq) + 2e^-. \tag{9.1}$$

The electrons generated from the anodic half-reaction are consumed by a secondary process, a cathodic reaction such as

$$2H^+(aq) + 2e^- \rightarrow H_2(g). \tag{9.2}$$

The anodic and cathodic half-reactions should balance their charges; that is, current densities of anodic and cathodic processes should be equal assuming that cathodic

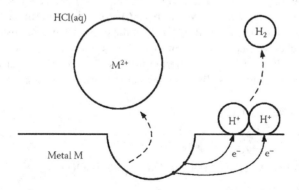

FIGURE 9.1 Schematic illustration of anodic and cathodic reactions during a metal corrosion.

DOI: 10.1201/b21893-9

and anodic surface areas are the same. Sites hosting these two processes can be located close to each other on the metal's surface or far apart depending on circumstances. If surface areas of cathodic and anodic reactions are different, cathodic and anodic currents (not the current densities) should be equal.

The total reaction of the corrosion process is the sum of its half-reactions (9.1) and (9.2):

$$Fe(s) + 2H^+(aq) \rightarrow Fe^{2+}(aq) + H_2(g). \tag{9.3}$$

The cathodic half-reaction depends mainly on pH, dissolved oxygen, and temperature. There are four main cathodic processes, depending on whether the solution is acidic or neutral or basic and aerated or deaerated. They are as follows:

$2H^+(aq) + 2e^- \rightarrow H_2(g)$ in acidic and deaerated solution,

$2H_2O(l) + 2e^- \rightarrow H_2(g) + 2OH^-(aq)$ in neutral or basic and deaerated solution,

$2H^+(aq) + 0.5O_2(g) + 2e^- \rightarrow H_2O(l)$ in acidic and aerated solution,

$H_2O(l) + 0.5O_2(g) + 2e^- \rightarrow 2OH^-(aq)$ in neutral or basic and aerated solution.

These cathodic reactions suggest that corrosion rate may strongly depend on the solution pH and concentration of the dissolved oxygen.

9.3 POURBAIX DIAGRAM IN CORROSION SCIENCE

Generally, the Pourbaix diagram shows the stability of aqueous species and phases depending on the Nernst equation potential, Eh, and pH. As an example, Figure 9.2 [1,2] represents a Pourbaix diagram of Al assuming all activities of aqueous species equal 10^{-6}. In this diagram, $H_2O(l)$ is stable in a region between two dash lines (a) and (b). $O_2(g)$ is stable above (b) in oxygenated aqueous solutions, and $H_2(g)$ is stable below (a) in hydrogenated aqueous solutions. Also, the diagram shows regions of thermodynamic immunity, corrosion, and passivation of Al(s). In the immunity region (e.g., $Eh = -2\,V$ and pH = 4), metal corrosion cannot take place. In the corrosion region (e.g., $Eh = 0.8\,V$ and pH = 2), corrosion will definitely take place. In the passivation region (e.g., $Eh = 0.8\,V$ and pH = 6), corrosion can either take place or not. This depends on the chemical structure of the passivation layer (usually an oxide or a salt film), chemical composition of the aqueous solution, and the rate of the dissolution reaction, including the transport processes. The chemical or electrochemical reactions corresponding to the lines on the diagram are as follows:

a. $2H^+(aq) + 2e^- = H_2(g)$

b. $\left(1/2\right)O_2(g) + 2H^+(aq) + 2e^- = H_2O(l)$

c. $Al^{3+}(aq) + 3e^- = Al(s)$

d. $\left(1/2\right)Al_2O_3(s) + 3H^+(aq) + 3e^- = Al(s) + \left(3/2\right)H_2O(l)$

e. $AlO_2^-(aq) + 4H^+(aq) + 3e^- = Al(s) + 2H_2O(l)$

f. $2Al^{3+}(aq) + 3H_2O(l) = Al_2O_3(s) + 6H^+(aq)$

g. $Al_2O_3(s) + H_2O(l) = 2AlO_2^-(aq) + 2H^+(aq)$

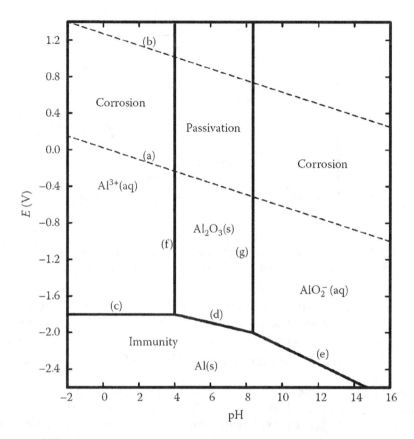

FIGURE 9.2 Pourbaix diagram of Al assuming all activities of aqueous species equal 10^{-6}. (Adapted from D.A. Jones, *Principles and Prevention of Corrosion*, 2nd ed., Prentice Hall, Upper Saddle River, NJ, 1996; M. Pourbaix, *Atlas of Electrochemical Equilibria in Aqueous Solutions*, NACE, Houston, TX, 1974.)

Obviously, corrosion rates can never be obtained from the Pourbaix diagram. Still, Pourbaix diagrams are very useful in corrosion science. For example, Figure 9.2 suggests that it might not be a good idea to use aluminum kitchenware for high acidic food solutions. A substantial collection of Pourbaix diagrams constructed for 25°C and 1.013 bar are given in a well-known reference book [2].

Note that Eh cannot be measured easily so there is a challenge of comparability between calculated and observed Eh values.

9.4 POLARIZATION CURVE OF METAL CORROSION

Electrochemical corrosion of an electron conductive material, such as a metal, can be studied by applying a potential and measuring current using a three-electrode electrochemical cell. The typical current density–applied potential curve in such an experiment is shown in Figure 9.3 [1,3]. The curve consists of three parts: activation, passivation, and transpassivation. $E_{Me^{z+}/Me}$ in Figure 9.3 is the electrode potential of

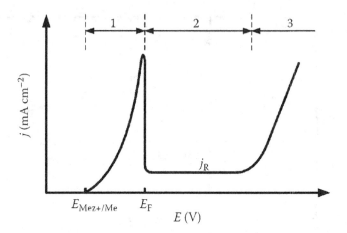

FIGURE 9.3 Typical current density–applied potential curve showing three main processes in corrosion: (1) activation, (2) passivation, and (3) transpassivation.

the metal ion/metal electrochemical couple. Below this potential, no electrochemical metal dissolution can take place. In the case of iron, the (reduction) electrochemical reaction is $Fe^{2+}(aq) + 2e^{-} = Fe(s)$, with electrode potential $E_{Fe^{2+}/Fe} = E^{0}_{Fe^{2+}/Fe} + [RT/(2F)]$ $\ln \left(a_{Fe^{2+}} \right)$ and the standard electrode potential $E^{0}_{Fe^{2+}/Fe} = -0.447\,V$ at 25°C and 1.013 bar.

Note that $E_{Mez+/Me}$ can significantly deviate from $E^{0}_{Mez+/Me}$, and it depends on the corrosion environment, including concentration of species and temperature. Pressure can also affect $E_{Mez+/Me}$, but to a smaller extent if the pressure is not too high, say, below 100 bar. However, the pressure effect can significantly increase when temperature is elevated above 200°C.

Because metal dissolution is an anodic process, for example, $Fe(s) \rightarrow Fe^{2+}(aq) + 2e^{-}$, the current of the process is assumed to be positive. When potential increases from $E_{Mez+/Me}$ to E_F (passivation or Flade potential), the current is increasing exponentially due to the charge transfer reaction, for example, $Fe(s) \rightarrow Fe^{2+}(aq) + 2e^{-}$, and can be described using Tafel's equation. At E_F, the formation of an oxide layer (passive film) starts. When the metal surface is covered by a metal oxide passive film (an insulator or a semiconductor), the resistivity is sharply increasing, and the current density drops down to the rest current density, j_R. This low current corresponds to a slow growth of the oxide layer and possible dissolution of the metal oxide into solution. In the region of transpassivation, another electrochemical reaction can take place, for example, $H_2O(l) \rightarrow (1/2)O_2(g) + 2H^{+}(aq) + 2e^{-}$, or the passive film can be broken down due to a chemical interaction with environment and mechanical instability. Clearly, a three-electrode cell and a potentiostat should be used to obtain the current density–potential curve shown in Figure 9.3.

9.5 CORROSION POTENTIAL AND CURRENT DENSITY

If more than one electron transfer reaction takes place at the same electrode, the observed potential is called the mixed potential. A typical example of the mixed potential is the corrosion potential. In Figure 9.4, two polarization curves are shown. The first one, $j_{Fe}(E)$, corresponds to the anodic dissolution of Fe(s), Fe(s) → $Fe^{2+}(aq) + 2e^-$, and the second one, $j_{H_2}(E)$, relates to the cathodic evolution of hydrogen, $2H^+(aq) + 2e^- \rightarrow H_2(g)$, in an acidic and deaerated environment. The directions of the half-reactions are defined by thermodynamics assuming that E_{H^+/H_2} is more positive than $E_{Fe^{2+}/Fe}$. Note that $E^0_{H^+/H_2}$ is always more positive than $E^0_{Fe^{2+}/Fe}$, but electrode potentials, E_{H^+/H_2} and $E_{Fe^{2+}/Fe}$, also depend on activities of $Fe^{2+}(aq)$ and $H^+(aq)$.

If both reactions taking place on the same surface, the cathodic current density of the $H_2(g)$ evolution should be equal to the anodic current density of the anodic dissolution of Fe(s), and both of them are equal to the corrosion current density, j_{corr}. The potential corresponding to the corrosion current density is called the corrosion potential and is shown in Figure 9.4, too. The corrosion potential can easily be measured using a high-resistance electrometer and a reference electrode. In contrast, the corrosion current cannot be directly measured but can be calculated using an electrochemical corrosion theory. Also, the corrosion potential is not Eh in the Pourbaix diagram, but these two can be close in some special cases.

One of the most famous theories to estimate the corrosion current was developed by Stern and Geary in 1957 [1]. The theory allows calculating the corrosion current density when a three-electrode cell is used and a small current density, j, occurs between the working and counter electrodes. If j is a function of overpotential, $\eta = E - E_{corr}$, then the area-specific resistance, r_{corr}, can immediately be calculated as follows:

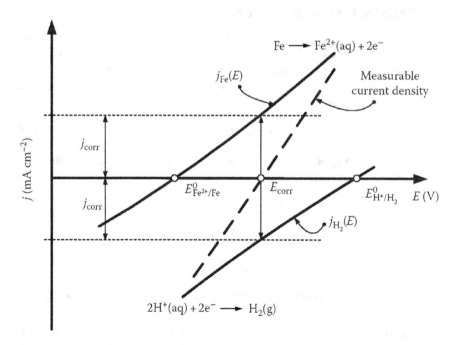

FIGURE 9.4 Illustration of electrochemical corrosion of Fe in an acidic and deaerated aqueous solution showing the corrosion potential E_{corr} and corrosion current density j_{corr} assuming that activities of the iron and hydrogen ions equal 1.

$$\frac{j}{\eta} = \left(r_{corr}\right)^{-1} = j_{corr} B',$$ (9.4)

where B' is a simple combination of the cathodic, B_c', and anodic, B_a', Tafel slopes

$$B' = \ln 10 \frac{B_c' + B_a'}{B_c' B_a'}.$$ (9.5)

The cathodic and anodic Tafel slopes are defined as

$$B_c' = \ln 10 \frac{RT}{\beta_c nF},$$ (9.6)

$$B_a' = \ln 10 \frac{RT}{\beta_a nF},$$ (9.7)

where β_c and β_a are, respectively, the cathodic and anodic symmetry factors. Note that Equations 9.4 through 9.7 are theoretically suitable for the elementary anodic and cathodic reactions and, therefore, cannot be applicable for some real corrosion processes.

Note that in such corrosion studies, when $j=0$, $\eta=0$; that is, the working electrode is at the corrosion potential, E_{corr}.

9.6 RATE OF ELECTROCHEMICAL CORROSION

If the corrosion current is estimated, and we believe that the assumed electrochemical half-reactions, for example, (9.1) and (9.2), as well as the total reaction, for example, (9.3), take place on the surface of an electrode (or a corroding metal), the rate of the electrochemical corrosion (CR) (as depth of the corrosion penetration per time) can be calculated based on Faraday's law as follows:

$$CR = j_{corr} M_M / \left(\rho_M zF\right),$$ (9.8)

where ρ_M is the density of the corroding metal and M_M is its molar mass (Table 10.2). For example, for iron with the anodic reaction (9.1) and measured corrosion current density of 90 µA cm^{-2}, the corrosion rate can be calculated as follows:

$$CR = 55.85\left[g\,mol^{-1}\right] \times 90 \times 10^{-6}\left[A\,cm^{-2}\right] /$$
$$\left(7.87\left[g\,cm^{-3}\right] \times 2 \times 96,485\left[A \times s\,mol^{-1}\right]\right)$$
$$= 3.31 \times 10^{-9}\left[cm\,s^{-1}\right] = 1.98 \times 10^{-6}\left[mm\,min^{-1}\right] = 1.04\left[mm\,year^{-1}\right].$$

Therefore, the corrosion rate of Fe(s) was calculated as 1 mm year^{-1}, which based on Table 9.1 [4] is considered as quite high, showing a poor corrosion resistance of the metal in the solution used for obtaining the current density of 90 µA cm^{-2}.

TABLE 9.1
Conventional Relative Corrosion Resistances Based on the Corrosion Rates of Typical Ferrous- and Nickel-Based Alloys

Relative Corrosion Resistance of Typical Ferrous- and Nickel-Based Alloys	Corrosion Rate (mm year^{-1})
Outstanding	<0.02
Excellent	0.02–0.1
Good	0.1–0.5
Fair	0.5–1
Poor	1–5
Unacceptable	>5

Source: M.G. Fontana, *Corrosion Engineering*, 3rd ed., McGraw-Hill, New York, 1986.

Note that corrosion measurements and calculations of corrosion rate can be used to speculate about the corrosion mechanism. For example, if a study shows that in Equation 9.8, z should be 3 rather than 2, then the anodic dissolution reaction should be

$$Fe(s) \rightarrow Fe^{3+}(aq) + 3e^-,$$

rather than the half-reaction (9.1). This can take place in an aerated aqueous solution where the cathodic reaction will be a different half-reaction than (9.2).

It would be prudent to mention that Equations 9.4 through 9.8 are valid for active corrosion or charge transfer–controlled processes. However, many corrosion processes can be limited by mass transport for both the anodic and cathodic reactions, especially if O_2 is a reacting species. Thus, the active corrosion relations expressed by Equations 9.4 through 9.8 may not always be accurate and should be used with caution.

9.7 CORROSION PROTECTION

There are a number of methods to protect or at least mitigate metal corrosion [1]. We will briefly describe only three of them.

1. A polymer coating (or paint) can provide long-term protection from corrosion by isolating the metal surface from an aggressive environment as long as this coating is not damaged.
2. The inhibition of corrosion can be provided by a chemically formed protective film. For example, iron corrosion can significantly be mitigated by treating the surface with phosphoric acid H_3PO_4 due to the following reaction:

$$2H_3PO_4(aq) + Fe(s) \rightarrow Fe(H_2PO_4)_2(s) + H_2(aq),$$

where $Fe(H_2PO_4)_2$ is a thin electrically nonconductive film isolating the metal surface from oxidizing environments.

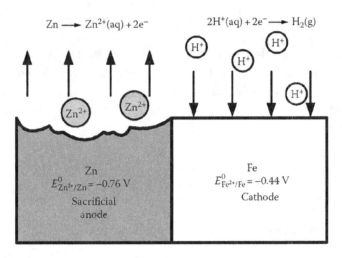

FIGURE 9.5 Illustration of electrochemical cathodic protection: the cathode consists of a metal (e.g., Fe) to be protected, and the anode is a metal (e.g., Zn) to be consumed.

3. Electrochemical (cathodic) protection can be used to protect a metal of interest (e.g., gas pipeline, ship body, etc.) by the dissolution of another metal using a galvanic couple (Figure 9.5). The other metal is referred to as the sacrificial anode, which is consumed in order to preserve the metal of interest. In such a galvanic couple, the metal dissolution reaction with a lower (more negative) electrode potential on the electrochemical series will be the anode, assuming that activities of ions will not significantly move electrode potentials from their standard values.

In Figure 9.5, the electrochemical cathodic protection of Fe ($E^0_{Fe^{2+}/Fe} = -0.447\,V$) with Zn ($E^0_{Zn^{2+}/Zn} = -0.762\,V$) is shown. Until Zn is completely dissolved, Fe will not be corroded because Fe is the cathode and the electrochemical corrosion is an anodic reaction, which takes place at the Zn surface.

Cathodic protection could also be provided by applying an appropriate potential, and this is common in some industries such as petroleum and natural gas engineering. Another approach is to design a metal alloy and adjust the conditions to push the corrosion potential into the passive region, where the corrosion rate is often at a much more acceptable rate. However, these regions are often metastable, and failure to properly maintain the potential could place the metal back into the active area or even in the transpassive state.

9.8 KINDS OF CORROSION

There are different forms of corrosion. General (uniform) and galvanic corrosion mechanisms were mainly considered in this chapter. The other forms of possible corrosion are crevice, pitting, environmentally induced cracking, hydrogen-induced cracking, intergranular, dealloying, and erosion-induced corrosion [1]. These forms of corrosion are not covered in this book but can be very damaging and, therefore, should seriously be considered. A number of excellent texts, such as Ref. [1], are available for a reader interested in corrosion. Note that all kinds of corrosion can take place in the electrochemical energy conversion systems and, therefore, should be taking into account in studying such systems.

9.9 SUMMARY

- The corrosion of a metal (or any other electron conductor) is an anodic electrochemical half-reaction, which is coupled with a cathodic reaction. The cathodic reaction depends on pH and the concentration of oxygen.
- Potential–pH (Pourbaix) diagram shows the thermodynamic stability of a metal. A region of corrosion immunity can be defined using this diagram. The Pourbaix diagram of Al is discussed in detail.
- Electrochemistry of metal corrosion (dissolution) under an applied potential is presented in the form of a typical polarization curve. The origins of three common regions (activation, passivation, and transpassivation) are explained.
- Using the common mixed potential definition, the background of corrosion potential is presented. An example of the corrosion potential for Fe(s) is discussed.
- When corrosion occurs, the cathodic current density is equal to the anodic current density, and both of them are equal to the corrosion current, j_{corr}. The corrosion current cannot be experimentally measured but can be calculated.
- On the contrary, the corrosion potential can be measured and its value is somewhere between equilibrium potentials of the cathodic and anodic reactions.
- Faraday's law allows calculating the corrosion rate if the anodic half-reaction and density of the corroding metal are known. The most common unit for the rate of corrosion penetration is mm year^{-1}.
- Three methods of corrosion protection are discussed: (1) isolation of metal surface from an aggressive environment by polymer coating (or paint), (2) inhibition of corrosion by a chemical reaction to provide a protective film, and (3) cathodic protection with using another metal and forming a galvanic couple.
- There are different forms of corrosion. General (uniform) and galvanic corrosion mechanisms were mainly considered in this chapter. Other kinds of corrosion such as pitting, stress cracking, and hydrogen-induced cracking are not covered in this introductory book but can be learned from a number of excellent books such as Ref. [1].

9.10 EXERCISES

PROBLEM 1

The overpotential–current density relationship inside the activation region of corrosion (dashed curve in Figure 9.4) can be represented by the following equation:

$$j = j_{corr}\left[\exp\left(\frac{\beta_a nF\eta}{RT}\right) - \exp\left(\frac{-\beta_c nF\eta}{RT}\right)\right].$$

Though similar to the Butler–Volmer Equation 6.6, the equation above represents two different half reactions, rather than one. Here, overpotential $(\eta = E - E_{corr})$ represents a departure from corrosion potential, rather than the equilibrium potential, and $(\beta_a + \beta_c)$ is not supposed to be 1. The $\eta - j$ dependence of the anodic half-reaction (e.g., $Fe \rightarrow Fe^{2+}(aq) + 2e^-$) is given in the first exponential term, and the $\eta - j$ dependence for cathodic half-reaction (e.g., $2H^+(aq) + 2e^- \rightarrow H_2(g)$) is given in the second exponential term. Starting with this equation, derive the Stern and Geary Equation 9.4.

Solution:

Assuming that the overpotential, η, is sufficiently small, both exponential terms can be expanded into a series and limiting the sum to the first two terms:

$$\exp\left(\frac{\beta_a nF\eta}{RT}\right) = 1 + \frac{\beta_a nF\eta}{RT},$$

$$\exp\left(\frac{-\beta_c nF\eta}{RT}\right) = 1 - \frac{\beta_c nF\eta}{RT},$$

then

$$j = j_{corr}\left[\exp\left(\frac{\beta_a nF\eta}{RT}\right) - \exp\left(\frac{-\beta_c nF\eta}{RT}\right)\right] = j_{corr}\eta\left(\frac{\beta_a nF}{RT} + \frac{\beta_c nF}{RT}\right).$$

Using the Tafel slopes as defined by Equations 9.6 and 9.7, the above equation can be further modified to

$$\frac{j}{\eta} = j_{corr}2.303\left(\frac{1}{B_a'} + \frac{1}{B_c'}\right) = j_{corr}2.303\left(\frac{B_c' + B_a'}{B_a'B_c'}\right) = j_{corr}B',$$

which, using the definition of B' (Equation 9.5), gives the Stern and Geary Equation 9.4.

PROBLEM 2

The simulated polarization data of a carbon steel in a deaerated aqueous solution is given below.

Anodic		Cathodic	
η/V	j/A cm^{-2}	η/V	j/A cm^{-2}
0	0.00	0	0.00
0.0005	1.94×10^{-6}	−0.0005	-1.95×10^{-6}
0.001	3.88×10^{-6}	−0.001	-3.91×10^{-6}

(Continued)

Anodic		Cathodic	
η/V	j/A cm^{-2}	η/V	j/A cm^{-2}
0.005	1.91×10^{-5}	-0.005	-1.99×10^{-5}
0.01	3.77×10^{-5}	-0.01	-4.07×10^{-5}
0.015	5.59×10^{-5}	-0.015	-6.28×10^{-5}
0.02	7.39×10^{-5}	-0.02	-8.63×10^{-5}
0.025	9.19×10^{-5}	-0.025	-1.12×10^{-4}
0.03	1.10×10^{-4}	-0.03	-1.39×10^{-4}
0.04	1.47×10^{-4}	-0.04	-2.01×10^{-4}
0.05	1.87×10^{-4}	-0.05	-2.76×10^{-4}
0.06	2.30×10^{-4}	-0.06	-3.67×10^{-4}
0.07	2.78×10^{-4}	-0.07	-4.80×10^{-4}
0.08	3.32×10^{-4}	-0.08	-6.20×10^{-4}
0.09	3.94×10^{-4}	-0.09	-7.94×10^{-4}
0.1	4.65×10^{-4}	-0.1	-1.01×10^{-3}
0.11	5.47×10^{-4}	-0.11	-1.29×10^{-3}
0.12	6.42×10^{-4}	-0.12	-1.64×10^{-3}
0.13	7.53×10^{-4}	-0.13	-2.07×10^{-3}
0.14	8.81×10^{-4}	-0.14	-2.62×10^{-3}
0.15	1.03×10^{-3}	-0.15	-3.32×10^{-3}
0.16	1.21×10^{-3}	-0.16	-4.19×10^{-3}
0.17	1.41×10^{-3}	-0.17	-5.30×10^{-3}
0.18	1.65×10^{-3}	-0.18	-6.70×10^{-3}
0.19	1.93×10^{-3}	-0.19	-8.47×10^{-3}
0.2	2.25×10^{-3}	-0.2	-1.07×10^{-2}
0.21	2.63×10^{-3}	-0.21	-1.35×10^{-2}

The anodic and cathodic data sets correspond to the half-reactions of Equations 9.1 and 9.2, respectively.

1. Plot the current density $|j|$ (log scale, y-axis) vs. overpotential η (x-axis) using the polarization data. Write the linear portions of the cathodic and anodic polarization curves in the form of the Tafel Equations 6.13 and 6.14. Calculate j_{corr} by finding the intersection of these two fits (around $\eta = 0$ V).
2. Use the Tafel Equations 6.13 through 6.15 to calculate the Tafel slopes B_c' and B_a'. Calculate the symmetry factors β_c and β_a using Equations 9.6 and 9.7.
3. Calculate the Stern–Geary parameter B' using Equation 9.5. Calculate the inverse of the area specific resistance, r_{corr}^{-1}. Compare r_{corr}^{-1} with the slope of $j(\eta)$ vs. η around $\eta = 0$ V (Equation 9.4).
4. Calculate the corrosion rate CR using Equation 9.8, using data for carbon steel (Fe is a good approximation), and make you conclusion on the relative corrosion resistance using Table 9.1.

Solution:

1. The plot of $\log_{10}|j|$ vs. η is shown below. In this problem, the data sets with $|\eta| \geq 0.1\,V$ are used for curve fitting the cathodic and anodic polarizations to the linear Tafel Equations (6.15):

$$\eta_c = -0.393 - 0.0979\log_{10}|j_c|,$$

$$\eta_a = 0.587 + 0.146\log_{10}|j_a|.$$

By setting $\eta_c = \eta_a = 0$, we solve for the corrosion current at the intersection $j_c = j_a = j_{corr}$, giving $j_{corr} = 10^{-4}\,A\,cm^2$.

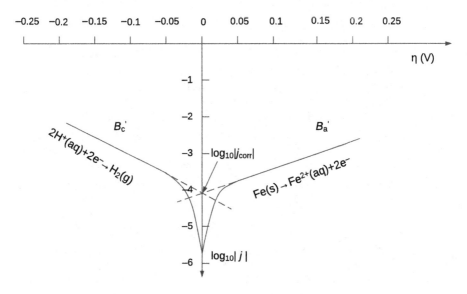

2. According to the results from (1), $B_c' = 0.0957$ $B_a' = 0.142$. The symmetry factors, β_c and β_a, are calculated using Equations 9.6 and 9.7:

$$\beta_c = \frac{(\ln 10)RT}{B_c'nF} = \frac{(\ln 10)\times 8.314 \times 298.15}{0.0979 \times 2 \times 96,485} = 0.302,$$

$$\beta_a = \frac{(\ln 10)RT}{B_a'nF} = \frac{(\ln 10)\times 8.314 \times 298.15}{0.146 \times 2 \times 96,485} = 0.202.$$

3. Using Equation 9.5, the Stern–Geary parameter, B', can be obtained as

$$B' = (\ln 10)\frac{B_c' + B_a'}{B_c'B_a'} = (\ln\ 10)\times \frac{(0.0979 + 0.146)}{0.0979 \times 0.146} = 39.3\ V^{-1}.$$

Then, the inverse area-specific resistance, r_{corr}^{-1}, can be obtained with Equation 9.5:

$$r_{corr}^{-1} = j_{corr} B' = 3.93 \text{ m}\Omega^{-1}\text{cm}^{-2} \, .$$

The slope $dj/d\eta$ surrounding zero overpotential (using polarization data from -10 to $10\,$mV) is approximately $3.92\,\text{m}\Omega^{-1}\text{cm}^{-2}$. The difference between the two calculated values (Tafel analysis and linear polarization analysis) is negligible.

4. The corrosion rate can be calculated as

$$CR = \frac{j_{corr} M_M}{\rho_M z F} = \frac{10^{-4} \times 55.845}{7.87 \times 2 \times 96,485} = 3.68 \times 10^{-9} \text{ cm s}^{-1} = 0.0193\,\text{mm year}^{-1},$$

where CR is the corrosion rate, M_M is the molar mass of the metal, $55.845\,$g mol^{-1} (Table 10.2), and ρ_M is the density of the metal, $7.87\,$g cm^{-3} (Table 10.2). Comparing the calculated corrosion rate with the reference in Table 9.1, the relative corrosion resistance of carbon steel is outstanding.

REFERENCES

1. D.A. Jones, *Principles and Prevention of Corrosion*, 2nd ed., Prentice Hall, Upper Saddle River, NJ, 1996.
2. M. Pourbaix, *Atlas of Electrochemical Equilibria in Aqueous Solutions*, NACE, Houston, TX, 1974.
3. C.H. Hamann, A. Hamnett, and W. Vielstich, *Electrochemistry*, 2nd ed., Wiley-VCH, Weinheim, Germany, 2007.
4. M.G. Fontana, *Corrosion Engineering*, 3rd ed., McGraw-Hill, New York, 1986.

10 Data Section[1]

10.1 CODATA VALUES OF THE FUNDAMENTAL PHYSICAL CONSTANTS[2]

This table gives the 2010 self-consistent set of values of the fundamental physical constants of physics and chemistry recommended by the Committee on Data for Science and Technology (CODATA) for international use.

Quantity	Symbol	Numerical Value	Unit	Relative Std. Uncert. u_r
Speed of light in vacuum	c, c_0	299 792 458	m s^{-1}	Exact
Magnetic constant	μ_0	$4\pi \times 10^{-7}$ $= 12.566\ 370\ 614 \times 10^{-7}$	N A^{-2}	Exact
Electric constant $1/\mu_0 c^2$	ϵ_0	$8.854\ 187\ 817 \times 10^{-12}$	F m^{-1}	Exact
Elementary charge	e	$1.602\ 176\ 565(35) \times 10^{-19}$	C	2.2×10^{-8}
Electron mass	m_e	$9.109\ 382\ 91(40) \times 10^{-31}$	kg	4.4×10^{-8}
Proton mass	m_p	$1.672\ 621\ 777(74) \times 10^{-27}$	kg	4.4×10^{-8}
Avogadro constant	N_A, L	$6.022\ 141\ 29(27) \times 10^{23}$	mol^{-1}	4.4×10^{-8}
Faraday constant $N_A e$	F	96 485.3365(21)	C mol^{-1}	2.2×10^{-8}
Molar gas constant	R	8.314 4621(75)	J mol^{-1} K^{-1}	9.1×10^{-7}
Boltzmann constant R/N_A	k	$1.380\ 6488(13) \times 10^{-23}$	J K^{-1}	9.1×10^{-7}
Electron volt (e/C) J	eV	$1.602\ 176\ 565(35) \times 10^{-19}$	J	2.2×10^{-8}
(Unified) atomic mass unit $\frac{1}{12} m\left(^{12}C\right)$	u	$1.660\ 538\ 921(73) \times 10^{-27}$	kg	4.4×10^{-8}

---

[1] The symbols in Chapter 10 were electronically transformed from W.M. Haynes (Editor-in-Chief), *CRC Handbook of Chemistry and Physics,* 95th edition, CRC Press, Boca Raton, 2014 and are not always compatible with symbols in Chapters 1 through 9. In a case of uncertainty, see the *CRC Handbook of Chemistry and Physics* for clarifications. Note that the data in the *CRC Handbook of Chemistry and Physics* does not significantly change from one edition to another, so that this chapter is almost identical to Chapter 10 of the 1st edition of the book.

[2] Adapted from W.M. Haynes (Editor-in-Chief), *CRC Handbook of Chemistry and Physics*, 95th Edition, CRC Press, Boca Raton, 2014, pp. 1–1.

10.2 PHYSICAL CONSTANTS OF INORGANIC COMPOUNDS[3]

The following data fields appear in the table:

- **Name**: Systematic name for the substance. The valence state of a metallic element is indicated by a Roman numeral, e.g., copper in the +1 state is written as copper(I) rather than cuprous, iron in the +3 state is iron(III) rather than ferric.
- **Formula**: The simplest descriptive formula is given, but this does not necessarily specify the actual structure of the compound. For example, aluminum chloride is designated as $AlCl_3$, even though a more accurate representation of the structure in the solid phase (and, under some conditions, in the gas phase) is Al_2Cl_6. A few exceptions are made, such as the use of Hg_2^{+2} for the mercury(I) ion.
- **CAS Registry Number**: Chemical Abstracts Service Registry Number. An asterisk* following the CAS RN for a hydrate indicates that the number refers to the anhydrous compound. In most cases the generic CAS RN for the compound is given rather than the number for a specific crystalline form or mineral.
- **Mol. Weight (Mass)**: Molecular weight (relative molar mass) as calculated with the 2005 IUPAC Recommended Atomic Weights. The number of decimal places corresponds to the number of places in the atomic weight of the least accurately known element (e.g., one place for lead compounds, two places for compounds of selenium, germanium, etc.); a maximum of three places is given. For compounds of radioactive elements for which IUPAC makes no recommendation, the mass number of the isotope with longest half-life is used.
- **Physical Form**: The crystal system is given, when available, for compounds that are solid at room temperature, together with color and other descriptive features. Abbreviations are listed below.
- *mp*: Normal melting point in °C. The notation *tp* indicates the temperature where solid, liquid, and gas are in equilibrium at a pressure greater than one atmosphere (i.e., the normal melting point does not exist). When available, the triple point pressure is listed.
- *bp*: Normal boiling point in °C (referred to 101.325 kPa or 760 mmHg pressure). The notation *sp* following the number indicates the temperature where the pressure of the vapor in equilibrium with the solid reaches 101.325 kPa. A notation "sublimes" without a temperature being given indicates that there is a perceptible sublimation pressure above the solid at ambient temperatures.
- **Density**: Density values for solids and liquids are always in unit of grams per cubic centimeter and can be assumed to refer to temperatures near room

---

[3] Adapted from W.M. Haynes (Editor-in-Chief), *CRC Handbook of Chemistry and Physics*, 95th Edition, CRC Press, Boca Raton, 2014, pp. 4–43.

temperature unless otherwise stated. Values for gases are the calculated ideal gas densities in grams per liter at 25°C and 101.325 kPa; the unit is always specified for a gas value.

- **Aqueous Solubility**: Solubility is expressed as the number of grams of the compound (excluding any water of hydration) that will dissolve in 100 g of water. The temperature in °C is given as a superscript.
- **Qualitative Solubility**: Qualitative information on the solubility in other solvents (and in water, if quantitative data are unavailable) is given here. The abbreviations are:

i	insoluble
sl	slightly soluble
s	soluble
vs	very soluble
reac	reacts with the solvent

ABBREVIATIONS

Ac	acetyl
ace	acetone
acid	acid solutions
alk	alkaline solutions
amorp	amorphous
anh	anhydrous
aq	aqueous
blk	black
brn	brown
bz	benzene
chl	chloroform
col	colorless
conc	concentrated
cry	crystals, crystalline
cub	cubic
cyhex	cyclohexane
dec	decomposes
dil	dilute
diox	dioxane
eth	ethyl ether
EtOH	ethanol
exp	explodes, explosive
extrap	extrapolated
flam	flammable
gl	glass, glassy

(Continued)

grn	green
hc	hydrocarbon solvents
hex	hexagonal, hexane
hp	heptane
HT	high temperature
hyd	hydrate
hyg	hygroscopic
i	insoluble in
liq	liquid
LT	low temperature
MeOH	methanol
monocl	monoclinic
octahed	octahedral
oran	orange
orth	orthorhombic
os	organic solvents
peth	petroleum ether
pow	powder
prec	precipitate
pur	purple
py	pyridine
reac	reacts with
refrac	refractory
rhom	rhombohedral
r.t.	room temperature
s	soluble in
silv	silvery
sl	slightly soluble in
soln	solution
sp	sublimation point
stab	stable
subl	sublimes
temp	temperature
tetr	tetragonal
thf	tetrahydrofuran
tol	toluene
tp	triple point
trans	transition, transformation
tricl	triclinic
trig	trigonal
unstab	unstable
viol	violet
visc	viscous
vs	very soluble in
wh	white
xyl	xylene
yel	yellow

No.	Name	Formula	CAS Reg No.	Mol. Weight	Physical Form	mp (°C)	bp (°C)	Density (g cm^{-3})	Solubility g/100 g H$_2$O	Qualitative Solubility
1	Aluminum	Al	7429-90-5	26.982	silv-wh metal; cub cry	660.323	2519	2.70		i H$_2$O; s acid, alk
2	Aluminum chloride	AlCl$_3$	7446-70-0	133.341	wh hex cry or powder; hyg	192.6	180 sp	2.48	45.1^{25}	s bz, ctc, chl
3	Aluminum hydroxide	Al(OH)$_3$	21645-51-2	78.004	wh amorp powder			2.42		i H$_2$O; s alk, acid
4	Aluminum oxide (α)	Al$_2$O$_3$	1344-28-1	101.961	wh powder; hex	2054	2977	3.99		i H$_2$O; os; sl alk
5	Aluminum oxide (γ)	Al$_2$O$_3$	1344-28-1	101.961	soft wh pow	trans to corundum 1200		3.97		i H$_2$O; s acid; sl alk
6	Aluminum oxyhydroxide (boehmite)	AlO(OH)	1318-23-6	59.989	wh orth cry	trans to diasphore 227		3.07		i H$_2$O; s hot acid, alk
7	Aluminum oxyhydroxide (diaspore)	AlO(OH)	14457-84-2	59.989	orth cry	dec 450		3.38		i H$_2$O; s acid, alk
8	Ammonia	NH$_3$	7664-41-7	17.031	col gas	−77.73	−33.33	0.696 g L^{-1}		vs H$_2$O; s EtOH, eth
9	Ammonium chloride	NH$_4$Cl	12125-02-9	53.492	col cub cry	520.1 tp (dec)	338 sp	1.519	39.5^{25}	
10	Ammonium hydroxide	NH$_4$OH	1336-21-6	35.046	exists only in soln					
11	Argon	Ar	7440-37-1	39.948	col gas	−189.34	−185.847	1.633 g L^{-1}		sl H$_2$O
12	Barium chloride	BaCl$_2$	10361-37-2	208.233	wh orth cry; hyg	961	1560	3.9	37.0^{25}	

(Continued)

No.	Name	Formula	CAS Reg No.	Mol. Weight	Physical Form	mp (°C)	bp (°C)	Density (g cm⁻³)	Solubility g/100 g H₂O	Qualitative Solubility
13	Boric acid	H_3BO_3	10043-35-3	61.833	col tricl cry	170.9		1.5	5.80[25]	sl EtOH
14	Calcium chloride	$CaCl_2$	10043-52-4	110.984	wh cub cry or powder; hyg	775	1935	2.15	81.3[25]	vs EtOH
15	Carbon (diamond)	C	7782-40-3	12.011	col cub cry	4440 (12.4 GPa)		3.513		i H_2O
16	Carbon (graphite)	C	7782-42-5	12.011	soft blk hex cry	4489 tp (10.3 MPa)	3825 sp	2.2		i H_2O
17	Carbon monoxide	CO	630-08-0	28.010	col gas	−205.02	−191.5	1.145 g L⁻¹		sl H_2O; s chl, EtOH
18	Carbon dioxide	CO_2	124-38-9	44.010	col gas	−56.558 tp	−78.464 sp	1.799 g L⁻¹		s H_2O
19	Chlorine	Cl_2	7782-50-5	70.90	grn-yel gas	−101.5	−34.04	2.898 g L⁻¹		sl H_2O
20	Chromium	Cr	7440-47-3	51.996	blue-wh metal; cub	1907	2671	7.15		reac dil acid
21	Copper	Cu	7440-50-8	63.546	red metal; cub	1084.62	2560	8.96		sl dil acid
22	Copper(I) oxide	Cu_2O	1317-39-1	143.091	red-brn cub cry	1244	1800 dec	6.0		i H_2O
23	Copper(II) hydroxide	$Cu(OH)_2$	20427-59-2	97.561	blue-grn powder			3.37		i H_2O; s acid, conc alk
24	Copper(II) oxide	CuO	1317-38-0	79.545	blk powder or monocl cry	1227		6.31		i H_2O, EtOH; s dil acid
25	Copper(II) sulfate	$CuSO_4$	7758-98-7	159.609	wh-grn amorp powder or rhomb cry	560 dec		3.60	22.0[25]	i EtOH
26	Gold	Au	7440-57-5	196.967	soft yel metal	1064.18	2836	19.3		s aqua regia
27	Hydrazine	N_2H_4	302-01-2	32.045	col oily liq	1.54	113.55	1.0036		vs H_2O, EtOH, MeOH

(Continued)

No.	Name	Formula	CAS Reg No.	Mol. Weight	Physical Form	mp (°C)	bp (°C)	Density (g cm⁻³)	Solubility g/100 g H₂O	Qualitative Solubility
28	Hydrogen	H_2	1333-74-0	2.016	col gas; flam	−259.16	−252.762	0.082 g L⁻¹		sl H_2O
29	Hydrogen chloride	HCl	7647-01-0	36.461	col gas	−114.17	−85	1.490 g L⁻¹		vs H_2O
30	Hydrogen sulfide	H_2S	7783-06-4	34.081	col gas; flam	−85.5	−59.55	1.393 g L⁻¹		s H_2O
31	Iron	Fe	7439-89-6	55.845	silv-wh or gray met	1538	2861	7.87		s dil acid
32	Iron(II) carbonate	$FeCO_3$	563-71-3	115.854	gray-brn hex cry			3.944	0.000062²⁰	
33	Iron(II) chloride	$FeCl_2$	7758-94-3	126.751	wh hex cry; hyg	677	1023	3.16	65.0²⁵	vs EtOH, ace; sl bz
34	Iron(II) oxide	FeO	1345-25-1	71.844	blk cub cry	1377		6.0		i H_2O, alk; s acid
35	Iron(II) sulfate	$FeSO_4$	7720-78-7	151.908	wh orth cry; hyg			3.65	29.5²⁵	
36	Iron(II) sulfide	FeS	1317-37-9	87.910	col hex or tetr cry; hyg	1188	dec	4.7		i H_2O; reac acid
37	Iron(II,III) oxide	Fe_3O_4	1317-61-9	231.533	blk cub cry or amorp powder	1597		5.17		i H_2O; s acid
38	Iron(III) chloride	$FeCl_3$	7705-08-0	162.204	grn hex cry; hyg	307.6	≈316	2.90	91.2²⁵	s EtOH, eth, ace
39	Iron(III) oxide	Fe_2O_3	1309-37-1	159.688	red-brn hex cry	1539		5.25		i H_2O; s acid
40	Iron(III) sulfate	$Fe_2(SO_4)_3$	10028-22-5	399.878	gray-wh rhomb cry; hyg			3.10	440²⁰	sl EtOH; i ace
41	Lanthanum chloride	$LaCl_3$	10099-58-8	245.264	wh hex cry; hyg	858		3.84	95.7²⁵	
42	Lead	Pb	7439-92-1	207.2	soft silv-gray metal; cub	327.462	1749	11.3		s conc acid

(Continued)

No.	Name	Formula	CAS Reg No.	Mol. Weight	Physical Form	mp (°C)	bp (°C)	Density (g cm^{-3})	Solubility g/100 g H$_2$O	Qualitative Solubility
43	Lithium chloride	LiCl	7447-41-8	42.394	wh cub cry or powder; hyg	610	1383	2.07	84.5[25]	s EtOH, ace, py
44	Mercury	Hg	7439-97-6	200.59	heavy silv liq	−38.829	356.619	13.5336		i H$_2$O
45	Mercury(I) chloride	Hg$_2$Cl$_2$	10112-91-1	472.09	wh tetr cry	525 tp	383 sp	7.16	0.0004[25]	i EtOH, eth
46	Mercury(I) oxide	Hg$_2$O	15829-53-5	417.18	prob mixture of HgO+Hg	100 dec		9.8		i H$_2$O; s HNO$_3$
47	Nickel	Ni	7440-02-0	58.693	wh metal; cub	1455	2913	8.90		i H$_2$O; sl dil acid
48	Nitrogen	N$_2$	7727-37-9	28.014	col gas	−210.0	−195.798	1.145 g L^{-1}		sl H$_2$O; i EtOH
49	Oxygen	O$_2$	7782-44-7	31.998	col gas	−218.79	−182.953	1.308 g L		sl H$_2$O, EtOH, os
50	Palladium	Pd	7440-05-3	106.42	silv-wh metal; cub	1554.8	2963	12.0		s aqua regia
51	Phosphoric acid	H$_3$PO$_4$	7664-38-2	97.995	col visc liq	42.4	407		548[20]	s EtOH
52	Platinum	Pt	7440-06-4	195.084	silv-gray metal; cub	1768.2	3825	21.5		i acid; s aqua regia
53	Potassium chloride	KCl	7447-40-7	74.551	wh cub cry	771		1.988	35.5[25]	i eth, ace
54	Potassium hydroxide	KOH	1310-58-3	56.105	wh rhomb cry; hyg	406	1327	2.044	121[25]	s EtOH; s MeOH
55	Silicon	Si	7440-21-3	28.085	gray cry or brn amorp solid	1414	3265	2.3296		i H$_2$O, acid; s alk
56	Silicon dioxide (α-quartz)	SiO$_2$	14808-60-7	60.085	col hex cry	trans to beta quartz 573	2950	2.648		i H$_2$O, acid; s HF

(Continued)

No.	Name	Formula	CAS Reg No.	Mol. Weight	Physical Form	mp (°C)	bp (°C)	Density (g cm⁻³)	Solubility g/100 g H₂O	Qualitative Solubility
57	Silicon dioxide (β-quartz)	SiO_2	14808-60-7	60.085	col hex cry	trans to tridymite 867	2950	2.533^{600}		i H_2O, acid; s HF
58	Silicon dioxide (tridymite)	SiO_2	15468-32-3	60.085	col hex cry	trans cristobalite 1470	2950	2.265		i H_2O, acid; s HF
59	Silicon dioxide (cristobalite)	SiO_2	14464-46-1	60.085	col hex cry	1722	2950	2.334		i H_2O, acid; s HF
60	Silicon dioxide (vitreous)	SiO_2	60676-86-0	60.085	col amorp solid	1713	2950	2.196		i H_2O, acid; s HF
61	Silver	Ag	7440-22-4	107.868	silv metal; cub	961.78	2162	10.5		i EtOH; s acid, alk
62	Silver(I) chloride	AgCl	7783-90-6	143.321	wh cub cry	455	1547	5.56	0.00019^{25}	sl EtOH
63	Silver(I) oxide	Ag_2O	20667-12-3	231.735	brn-blk cub cry	≈200 dec		7.2	0.0025	vs H_2O
64	Sodium chloride	NaCl	7647-14-5	58.443	col cub cry	800.7	1465	2.17	36.0^{25}	s dil acid
65	Sulfuric acid	H_2SO_4	7664-93-9	98.079	col oily liq	10.31	337	1.8302		s acid, alk
66	Ytterbium(III) oxide	Yb_2O_3	1314-37-0	394.08	col cub cry	2355	4070	9.2		
67	Zinc	Zn	7440-66-6	65.38	blue-wh metal; hex	419.527	907	7.134		
68	Zinc hydroxide	$Zn(OH)_2$	20427-58-1	99.424	col orth cry	125 dec		3.05	0.000042^{20}	i H_2O; s dil acid
69	Zinc oxide	ZnO	1314-13-2	81.408	wh powder; hex	1974		5.6		
70	Zinc sulfate	$ZnSO_4$	7733-02-0	161.472	col orth cry	680 dec		3.8	57.7^{25}	i H_2O; sl acid
71	Zirconium(IV) oxide	ZrO_2	1314-23-4	123.223	wh amorp powder	2710	4300	5.68		

10.3   STANDARD THERMODYNAMIC PROPERTIES
OF CHEMICAL SUBSTANCES[4]

This table gives the standard state thermodynamic properties of a number of substances in the crystalline, liquid, and gaseous states. Substances are listed by molecular formula. The properties tabulated are:

$\Delta_f H^\circ$	Standard molar enthalpy of formation at 298.15 K in kJ/mol
$\Delta_f G^\circ$	Standard molar Gibbs energy of formation at 298.15 K in kJ/mol
S°	Standard molar entropy at 298.15 K in J/mol K
C_p	Molar heat capacity at constant pressure at 298.15 K in J/mol K

The standard state pressure is 100 kPa (1 bar). The standard states are defined for different phases by:

- The standard state of a pure gaseous substance is that of the substance as a (hypothetical) ideal gas at the standard state pressure.
- The standard state of a pure liquid substance is that of the liquid under the standard state pressure.
- The standard state of a pure crystalline substance is that of the crystalline substance under the standard state pressure.

An entry of 0.0 for $\Delta_f H^\circ$ for an element indicates the reference state of that element. A blank means no value is available.

---

[4] Adapted from W.M. Haynes (Editor-in-Chief), CRC Handbook of Chemistry and Physics, 95th Edition, CRC Press, Boca Raton, 2014, pp. 5–4.

	Name	Crystal				Liquid				Gas			
		$\Delta_f H°$ (kJ mol^{-1})	$\Delta_f G°$ (kJ mol^{-1})	$S°$ (J mol^{-1} K^{-1})	C_p (J mol^{-1} K^{-1})	$\Delta_f H°$ (kJ mol^{-1})	$\Delta_f G°$ (kJ mol^{-1})	$S°$ (J mol^{-1} K^{-1})	C_p (J mol^{-1} K^{-1})	$\Delta_f H°$ (kJ mol^{-1})	$\Delta_f G°$ (kJ mol^{-1})	$S°$ (J mol^{-1} K^{-1})	C_p (J mol^{-1} K^{-1})
					Substances Not Containing Carbon:								
Ag	Silver	0.0		42.6	25.4					284.9	246.0	173.0	20.8
AgBr	Silver(I) bromide	−100.4	−96.9	107.1	52.4								
AgCl	Silver(I) chloride	−127.0	−109.8	96.3	50.8								
Al	Aluminum	0.0		28.3	24.2					330.0	289.4	164.6	21.4
Al$_2$O$_3$	Aluminum oxide (corundum)	−1675.7	−1582.3	50.9	79.0								
Au	Gold	0.0		47.4	25.4					366.1	326.3	180.5	20.8
BaCl$_2$	Barium chloride	−855.0	−806.7	123.7	75.1								
CaCl$_2$	Calcium chloride	−795.4	−748.8	108.4	72.9								
ClK	Potassium chloride	−436.5	−408.5	82.6	51.3					−214.6	−233.3	239.1	36.5
ClNa	Sodium chloride	−411.2	−384.1	72.1	50.5								
Cl$_2$	Chlorine									0.0		223.1	33.9
Cl$_2$Cu	Copper(II) chloride	−220.1	−175.7	108.1	71.9								
Cl$_2$Hg$_2$	Mercury(I) chloride	−265.4	−210.7	191.6									
Cl$_3$La	Lanthanum chloride	−1072.2			108.8								
Cu	Copper	0.0		33.2	24.4					337.4	297.7	166.4	20.8

(Continued)

	Name	Crystal				Liquid				Gas			
		$\Delta_f H^\circ$ (kJ mol^{-1})	$\Delta_f G^\circ$ (kJ mol^{-1})	S° (J mol^{-1} K^{-1})	C_p (J mol^{-1} K^{-1})	$\Delta_f H^\circ$ (kJ mol^{-1})	$\Delta_f G^\circ$ (kJ mol^{-1})	S° (J mol^{-1} K^{-1})	C_p (J mol^{-1} K^{-1})	$\Delta_f H^\circ$ (kJ mol^{-1})	$\Delta_f G^\circ$ (kJ mol^{-1})	S° (J mol^{-1} K^{-1})	C_p (J mol^{-1} K^{-1})
CuO	Copper(II) oxide	−157.3	−129.7	42.6	42.3								
Cu$_2$O	Copper(I) oxide	−168.6	−146.0	93.1	63.6								
Fe	Iron	0.0		27.3	25.1					416.3	370.7	180.5	25.7
FeO	Iron(II) oxide	−272.0											
Fe$_2$O$_3$	Iron(III) oxide	−824.2	−742.2	87.4	103.9								
Fe$_3$O$_4$	Iron(II,III) oxide	−1118.4	−1015.4	146.4	143.4								
H$_2$	Hydrogen									0.0		130.7	28.8
H$_2$O	Water					−285.8	−237.1	70.0	75.3	−241.8	−228.6	188.8	33.6
H$_2$O$_2$Zn	Zinc hydroxide	−691.9	−553.5	81.2									
H$_2$S	Hydrogen sulfide									−20.6	−33.4	205.8	34.2
H$_3$N	Ammonia									−45.9	−16.4	192.8	35.1
H$_4$N$_2$	Hydrazine					50.6	149.3	121.2	98.9	95.4	159.4	238.5	48.4
Hg	Mercury					0.0		75.9	28.0	61.4	31.8	175.0	20.8
HgO	Mercury(II) oxide	−90.8	−58.5	70.3	44.1								
O$_2$	Oxygen									0.0		205.2	29.4
O$_2$Zr	Zirconium(IV) oxide	−1100.6	−1042.8	50.4	56.2								
O$_3$Y$_2$	Yttrium oxide	−1905.3	−1816.6	99.1	102.5								
Pt	Platinum	0.0		41.6	25.9					565.3	520.5	192.4	25.5
Zn	Zinc	0.0		41.6	25.4					130.4	94.8	161.0	20.8

(Continued)

Substances Containing Carbon:

Name	Crystal $\Delta_f H°$ (kJ mol⁻¹)	$\Delta_f G°$ (kJ mol⁻¹)	$S°$ (J mol⁻¹ K⁻¹)	C_p (J mol⁻¹ K⁻¹)	Liquid $\Delta_f H°$ (kJ mol⁻¹)	$\Delta_f G°$ (kJ mol⁻¹)	$S°$ (J mol⁻¹ K⁻¹)	C_p (J mol⁻¹ K⁻¹)	Gas $\Delta_f H°$ (kJ mol⁻¹)	$\Delta_f G°$ (kJ mol⁻¹)	$S°$ (J mol⁻¹ K⁻¹)	C_p (J mol⁻¹ K⁻¹)
C Carbon (graphite)	0.0		5.7	8.5					716.7	671.3	158.1	20.8
C Carbon (diamond)	1.9	2.9	2.4	6.1								
CH_4 Methane									−74.6	−50.5	186.3	35.7
CH_4O Methanol					−239.2	−166.6	126.8	81.1	−201.0	−162.3	239.9	44.1
CO Carbon monoxide									−110.5	−137.2	197.7	29.1
CO_2 Carbon dioxide									−393.5	−394.4	213.8	37.1
$C_2H_4O_2$ Acetic acid					−484.3	−389.9	159.8	123.3	−432.2	−374.2	283.5	63.4
C_2H_6O Ethanol					−277.6	−174.8	160.7	112.3	−234.8	−167.9	281.6	65.6

10.4 THERMODYNAMIC PROPERTIES AS A FUNCTION OF TEMPERATURE[5]

The thermodynamic properties $C_p^\circ(T)$, $S^\circ(T)$, $H^\circ(T)-H^\circ(T_r)$, $-[G^\circ(T)-H^\circ(T_r)]/T$ and formation properties $\Delta_f H^\circ(T)$, $\Delta_f G^\circ(T)$, $\log K_f(T)$ are tabulated as functions of temperature in the range 298.15–1500 K for a number of substances in the standard state. The reference temperature, T_r, is equal to 298.15 K. The standard state pressure is taken as 1 bar (100,000 Pa).

	J K⁻¹mol⁻¹				kJ mol⁻¹		
T (K)	C_p°	S°	$-(G^\circ-H^\circ(Tr))/T$	$H^\circ-H^\circ$ (Tr)	ΔfH°	ΔfG°	Log K_f
1. Carbon (Graphite) C (cr; Graphite)							
298.15	8.536	5.740	5.740	0.000	0.000	0.000	0.000
300	8.610	5.793	5.740	0.016	0.000	0.000	0.000
400	11.974	8.757	6.122	1.054	0.000	0.000	0.000
500	14.537	11.715	6.946	2.385	0.000	0.000	0.000
600	16.607	14.555	7.979	3.945	0.000	0.000	0.000
700	18.306	17.247	9.113	5.694	0.000	0.000	0.000
800	19.699	19.785	10.290	7.596	0.000	0.000	0.000
900	20.832	22.173	11.479	9.625	0.000	0.000	0.000
1000	21.739	24.417	12.662	11.755	0.000	0.000	0.000
1100	22.452	26.524	13.827	13.966	0.000	0.000	0.000
1200	23.000	28.502	14.968	16.240	0.000	0.000	0.000
1300	23.409	30.360	16.082	18.562	0.000	0.000	0.000
1400	23.707	32.106	17.164	20.918	0.000	0.000	0.000
1500	23.919	33.749	18.216	23.300	0.000	0.000	0.000
2. Carbon Oxide CO (g)							
298.15	29.141	197.658	197.658	0.000	−110.530	−137.168	24.031
300	29.142	197.838	197.659	0.054	−110.519	−137.333	23.912
400	29.340	206.243	198.803	2.976	−110.121	−146.341	19.110
500	29.792	212.834	200.973	5.930	−110.027	−155.412	16.236
600	30.440	218.321	203.419	8.941	−110.157	−164.480	14.319
700	31.170	223.067	205.895	12.021	−110.453	−173.513	12.948
800	31.898	227.277	208.309	15.175	−110.870	−182.494	11.915
900	32.573	231.074	210.631	18.399	−111.378	−191.417	11.109
1000	33.178	234.538	212.851	21.687	−111.952	−200.281	10.461
1100	33.709	237.726	214.969	25.032	−112.573	−209.084	9.928
1200	34.169	240.679	216.990	28.426	−113.228	−217.829	9.482
1300	34.568	243.430	218.920	31.864	−113.904	−226.518	9.101

(*Continued*)

[5] Adapted from W.M. Haynes (Editor-in-Chief), *CRC Handbook of Chemistry and Physics*, 95th Edition, CRC Press, Boca Raton, 2014, pp. 5–43.

T (K)	C_p°	S°	$-(G^\circ - H^\circ(Tr))/T$	$H^\circ - H^\circ$ (Tr)	ΔfH°	ΔfG°	Log K_f
	J K⁻¹mol⁻¹				kJ mol⁻¹		
1400	34.914	246.005	220.763	35.338	−114.594	−235.155	8.774
1500	35.213	248.424	222.527	38.845	−115.291	−243.742	8.488

3. Carbon Dioxide CO₂ (g)

T (K)	C_p°	S°	$-(G^\circ - H^\circ(Tr))/T$	$H^\circ - H^\circ$ (Tr)	ΔfH°	ΔfG°	Log K_f
298.15	37.135	213.783	213.783	0.000	−393.510	−394.373	69.092
300	37.220	214.013	213.784	0.069	−393.511	−394.379	68.667
400	41.328	225.305	215.296	4.004	−393.586	−394.656	51.536
500	44.627	234.895	218.280	8.307	−393.672	−394.914	41.256
600	47.327	243.278	221.762	12.909	−393.791	−395.152	34.401
700	49.569	250.747	225.379	17.758	−393.946	−395.367	29.502
800	51.442	257.492	228.978	22.811	−394.133	−395.558	25.827
900	53.008	263.644	232.493	28.036	−394.343	−395.724	22.967
1000	54.320	269.299	235.895	33.404	−394.568	−395.865	20.678
1100	55.423	274.529	239.172	38.893	−394.801	−395.984	18.803
1200	56.354	279.393	242.324	44.483	−395.035	−396.081	17.241
1300	57.144	283.936	245.352	50.159	−395.265	−396.159	15.918
1400	57.818	288.196	248.261	55.908	−395.488	−396.219	14.783
1500	58.397	292.205	251.059	61.719	−395.702	−396.264	13.799

4. Methane CH₄ (g)

T (K)	C_p°	S°	$-(G^\circ - H^\circ(Tr))/T$	$H^\circ - H^\circ$ (Tr)	ΔfH°	ΔfG°	Log K_f
298.15	35.695	186.369	186.369	0.000	−74.600	−50.530	8.853
300	35.765	186.590	186.370	0.066	−74.656	−50.381	8.772
400	40.631	197.501	187.825	3.871	−77.703	−41.827	5.462
500	46.627	207.202	190.744	8.229	−80.520	−32.525	3.398
600	52.742	216.246	194.248	13.199	−82.969	−22.690	1.975
700	58.603	224.821	198.008	18.769	−85.023	−12.476	0.931
800	64.084	233.008	201.875	24.907	−86.693	−1.993	0.130
900	69.137	240.852	205.773	31.571	−88.006	8.677	−0.504
1000	73.746	248.379	209.660	38.719	−88.996	19.475	−1.017
1100	77.919	255.607	213.511	46.306	−89.698	30.358	−1.442
1200	81.682	262.551	217.310	54.289	−90.145	41.294	−1.797
1300	85.067	269.225	221.048	62.630	−90.367	52.258	−2.100
1400	88.112	275.643	224.720	71.291	−90.390	63.231	−2.359
1500	90.856	281.817	228.322	80.242	−90.237	74.200	−2.584

5. Ethane C₂H₆ (g)

T (K)	C_p°	S°	$-(G^\circ - H^\circ(Tr))/T$	$H^\circ - H^\circ$ (Tr)	ΔfH°	ΔfG°	Log K_f
298.15	52.487	229.161	229.161	0.000	−84.000	−32.015	5.609
300	52.711	229.487	229.162	0.097	−84.094	−31.692	5.518
400	65.459	246.378	231.379	5.999	−88.988	−13.473	1.759
500	77.941	262.344	235.989	13.177	−93.238	5.912	−0.618
600	89.188	277.568	241.660	21.545	−96.779	26.086	−2.271
700	99.136	292.080	247.835	30.972	−99.663	46.800	−3.492

(Continued)

	J K⁻¹mol⁻¹			kJ mol⁻¹		
T (K)	$C_p^°$	$S°$	$-(G°-H°(Tr))/T$	$H°-H°$ (Tr)	$\Delta fH°$	$\Delta fG°$	Log K_f
800	107.936	305.904	254.236	41.334	−101.963	67.887	−4.433
900	115.709	319.075	260.715	52.525	−103.754	89.231	−5.179
1000	122.552	331.628	267.183	64.445	−105.105	110.750	−5.785
1100	128.553	343.597	273.590	77.007	−106.082	132.385	−6.286
1200	133.804	355.012	279.904	90.131	−106.741	154.096	−6.708
1300	138.391	365.908	286.103	103.746	−107.131	175.850	−7.066
1400	142.399	376.314	292.178	117.790	−107.292	197.625	−7.373
1500	145.905	386.260	298.121	132.209	−107.260	219.404	−7.640
6. Propane C₃H₈ (g)							
298.15	73.597	270.313	270.313	0.000	−103.847	−23.458	4.110
300	73.931	270.769	270.314	0.136	−103.972	−22.959	3.997
400	94.014	294.739	273.447	8.517	−110.33	15.029	−0.657
500	112.591	317.768	280.025	18.872	−115.658	34.507	−3.605
600	128.700	339.753	288.162	30.955	−119.973	64.961	−5.655
700	142.674	360.668	297.039	44.540	−123.384	96.065	−7.168
800	154.766	380.528	306.245	59.427	−126.016	127.603	−8.331
900	165.352	399.381	315.555	75.444	−127.982	159.430	−9.253
1000	174.598	417.293	324.841	92.452	−129.380	191.444	−10.000
1100	182.673	434.321	334.026	110.325	−130.296	223.574	−10.617
1200	189.745	450.526	343.064	128.954	−130.802	255.770	−11.133
1300	195.853	465.961	351.929	148.241	−130.961	287.993	−11.572
1400	201.209	480.675	360.604	168.100	−130.829	320.217	−11.947
1500	205.895	494.721	369.080	188.460	−130.445	352.422	−12.272
7. Methanol CH₃OH (g)							
298.15	44.101	239.865	239.865	0.000	−201.000	−162.298	28.434
300	44.219	240.139	239.866	0.082	−201.068	−162.057	28.216
400	51.713	253.845	241.685	4.864	−204.622	−148.509	19.393
500	59.800	266.257	245.374	10.442	−207.750	−134.109	14.010
600	67.294	277.835	249.830	16.803	−210.387	−119.125	10.371
700	73.958	288.719	254.616	23.873	−212.570	−103.737	7.741
800	79.838	298.987	259.526	31.569	−214.350	−88.063	5.750
900	85.025	308.696	264.455	39.817	−215.782	−72.188	4.190
1000	89.597	317.896	269.343	48.553	−216.916	−56.170	2.934
1100	93.624	326.629	274.158	57.718	−217.794	−40.050	1.902
1200	97.165	334.930	278.879	67.262	−218.457	−23.861	1.039
1300	100.277	342.833	283.497	77.137	−218.936	−7.624	0.306
1400	103.014	350.367	288.007	87.304	−219.261	8.644	−0.322
1500	105.422	357.558	292.405	97.729	−219.456	24.930	−0.868

(*Continued*)

	J K⁻¹mol⁻¹			kJ mol⁻¹			
T (K)	C_p°	S°	$-(G^\circ - H^\circ(Tr))/T$	$H^\circ - H^\circ$ (Tr)	ΔfH°	ΔfG°	Log K_f
8. Ethanol C₂H₅OH (g)							
298.15	65.652	281.622	281.622	0.000	−234.800	−167.874	29.410
300	65.926	282.029	281.623	0.122	−234.897	−167.458	29.157
400	81.169	303.076	284.390	7.474	−239.826	−144.216	18.832
500	95.400	322.750	290.115	16.318	−243.940	−119.820	12.517
600	107.656	341.257	297.112	26.487	−247.260	−94.672	8.242
700	118.129	358.659	304.674	37.790	−249.895	−69.023	5.151
800	127.171	375.038	312.456	50.065	−251.951	−43.038	2.810
900	135.049	390.482	320.276	63.185	−253.515	−16.825	0.976
1000	141.934	405.075	328.033	77.042	−254.662	9.539	−0.498
1100	147.958	418.892	335.670	91.543	−255.454	36.000	−1.709
1200	153.232	431.997	343.156	106.609	−255.947	62.520	−2.721
1300	157.849	444.448	350.473	122.168	−256.184	89.070	−3.579
1400	161.896	456.298	357.612	138.160	−256.206	115.630	−4.314
1500	165.447	467.591	364.571	154.531	−256.044	142.185	−4.951
9. Acetic Acid C₂H₄O₂ (g)							
298.15	63.438	283.470	283.470	0.000	−432.249	−374.254	65.567
300	63.739	283.863	283.471	0.118	−432.324	−373.893	65.100
400	79.665	304.404	286.164	7.296	−436.006	−353.840	46.206
500	93.926	323.751	291.765	15.993	−438.875	−332.950	34.783
600	106.181	341.988	298.631	26.014	−440.993	−311.554	27.123
700	116.627	359.162	306.064	37.169	−442.466	−289.856	21.629
800	125.501	375.331	313.722	49.287	−443.395	−267.985	17.497
900	132.989	390.558	321.422	62.223	−443.873	−246.026	14.279
1000	139.257	404.904	329.060	75.844	−443.982	−224.034	11.702
1100	144.462	418.429	336.576	90.039	−443.798	−202.046	9.594
1200	148.760	431.189	343.933	104.707	−443.385	−180.086	7.839
1300	152.302	443.240	351.113	119.765	−442.795	−158.167	6.355
1400	155.220	454.637	358.105	135.146	−442.071	−136.299	5.085
1500	157.631	465.432	364.903	150.793	−441.247	−114.486	3.987
10. Dichlorine Cl₂ (g)							
298.15	33.949	223.079	223.079	0.000	0.000	0.000	0.000
300	33.981	223.290	223.080	0.063	0.000	0.000	0.000
400	35.296	233.263	224.431	3.533	0.000	0.000	0.000
500	36.064	241.229	227.021	7.104	0.000	0.000	0.000
600	36.547	247.850	229.956	10.736	0.000	0.000	0.000
700	36.874	253.510	232.926	14.408	0.000	0.000	0.000
800	37.111	258.450	235.815	18.108	0.000	0.000	0.000

(Continued)

T (K)	C_p°	S°	$-(G^\circ - H^\circ(Tr))/T$	$H^\circ - H^\circ$ (Tr)	ΔfH°	ΔfG°	Log K_f
	J K⁻¹mol⁻¹			kJ mol⁻¹			
900	37.294	262.832	238.578	21.829	0.000	0.000	0.000
1000	37.442	266.769	241.203	25.566	0.000	0.000	0.000
1100	37.567	270.343	243.692	29.316	0.000	0.000	0.000
1200	37.678	273.617	246.052	33.079	0.000	0.000	0.000
1300	37.778	276.637	248.290	36.851	0.000	0.000	0.000
1400	37.872	279.440	250.416	40.634	0.000	0.000	0.000
1500	37.961	282.056	252.439	44.426	0.000	0.000	0.000

11. Dihydrogen H₂ (g)

T (K)	C_p°	S°					
298.15	28.836	130.680	130.680	0.000	0.000	0.000	0.000
300	28.849	130.858	130.680	0.053	0.000	0.000	0.000
400	29.181	139.217	131.818	2.960	0.000	0.000	0.000
500	29.260	145.738	133.974	5.882	0.000	0.000	0.000
600	29.327	151.078	136.393	8.811	0.000	0.000	0.000
700	29.440	155.607	138.822	11.749	0.000	0.000	0.000
800	29.623	159.549	141.172	14.702	0.000	0.000	0.000
900	29.880	163.052	143.412	17.676	0.000	0.000	0.000
1000	30.204	166.217	145.537	20.680	0.000	0.000	0.000
1100	30.580	169.113	147.550	23.719	0.000	0.000	0.000
1200	30.991	171.791	149.460	26.797	0.000	0.000	0.000
1300	31.422	174.288	151.275	29.918	0.000	0.000	0.000
1400	31.860	176.633	153.003	33.082	0.000	0.000	0.000
1500	32.296	178.846	154.653	36.290	0.000	0.000	0.000

12. Water H₂O (l)

298.15	75.300	69.950	69.950	0.000	−285.830	−237.141	41.546
300	75.281	70.416	69.951	0.139	−285.771	−236.839	41.237
373.21	76.079	86.896	71.715	5.666	−283.454	−225.160	31.513

13. Water H₂O (g)

298.15	33.598	188.832	188.832	0.000	−241.826	−228.582	40.046
300	33.606	189.040	188.833	0.062	−241.844	−228.500	39.785
400	34.283	198.791	190.158	3.453	−242.845	−223.900	29.238
500	35.259	206.542	192.685	6.929	−243.822	−219.050	22.884
600	36.371	213.067	195.552	10.509	−244.751	−214.008	18.631
700	37.557	218.762	198.469	14.205	−245.620	−208.814	15.582
800	38.800	223.858	201.329	18.023	−246.424	−203.501	13.287
900	40.084	228.501	204.094	21.966	−247.158	−198.091	11.497
1000	41.385	232.792	206.752	26.040	−247.820	−192.603	10.060
1100	42.675	236.797	209.303	30.243	−248.410	−187.052	8.882
1200	43.932	240.565	211.753	34.574	−248.933	−181.450	7.898
1300	45.138	244.129	214.108	39.028	−249.392	−175.807	7.064
1400	46.281	247.516	216.374	43.599	−249.792	−170.132	6.348
1500	47.356	250.746	218.559	48.282	−250.139	−164.429	5.726

(*Continued*)

	J K⁻¹mol⁻¹				kJ mol⁻¹		
T (K)	C_p°	S°	$-(G^\circ-H^\circ(Tr))/T$	$H^\circ-H^\circ\ (Tr)$	ΔfH°	ΔfG°	Log K_f
			14. Dioxygen O_2 (g)				
298.15	29.378	205.148	205.148	0.000	0.000	0.000	0.000
300	29.387	205.330	205.148	0.054	0.000	0.000	0.000
400	30.109	213.873	206.308	3.026	0.000	0.000	0.000
500	31.094	220.695	208.525	6.085	0.000	0.000	0.000
600	32.095	226.454	211.045	9.245	0.000	0.000	0.000
700	32.987	231.470	213.612	12.500	0.000	0.000	0.000
800	33.741	235.925	216.128	15.838	0.000	0.000	0.000
900	34.365	239.937	218.554	19.244	0.000	0.000	0.000
1000	34.881	243.585	220.878	22.707	0.000	0.000	0.000
1100	35.314	246.930	223.096	26.217	0.000	0.000	0.000
1200	35.683	250.019	225.213	29.768	0.000	0.000	0.000
1300	36.006	252.888	227.233	33.352	0.000	0.000	0.000
1400	36.297	255.568	229.162	36.968	0.000	0.000	0.000
1500	36.567	258.081	231.007	40.611	0.000	0.000	0.000

10.5 THERMODYNAMIC PROPERTIES OF AQUEOUS IONS[6]

This table contains standard state thermodynamic properties of positive and negative ions in aqueous solution. It includes enthalpy and Gibbs energy of formation, partial molar entropy, and partial molar heat capacity. The standard state is the hypothetical ideal solution with molality equals 1 mol kg⁻¹.

All values refer to standard conditions of 298.15 K and 100 kPa pressure.

Ions	$\Delta_f H^\circ$ (kJ mol⁻¹)	$\Delta_f G^\circ$ (kJ mol⁻¹)	S° (J mol⁻¹K⁻¹)	C_p (J mol⁻¹K⁻¹)
		Cations		
Ag^+	105.6	77.1	72.7	21.8
Al^{+3}	−531.0	−485.0	−321.7	
$AlOH^{+2}$		−694.1		
Ba^{+2}	−537.6	−560.8	9.6	
$BaOH^+$		−730.5		
Be^{+2}	−382.8	−379.7	−129.7	
Bi^{+3}		82.8		
$BiOH^{+2}$		−146.4		
Ca^{+2}	−542.8	−553.6	−53.1	
$CaOH^+$		−718.4		
Cd^{+2}	−75.9	−77.6	−73.2	

(Continued)

[6] Adapted from W.M. Haynes (Editor-in-Chief), *CRC Handbook of Chemistry and Physics*, 95th Edition, CRC Press, Boca Raton, 2014, pp. 5–66.

Ions	$\Delta_f H°$ (kJ mol^{-1})	$\Delta_f G°$ (kJ mol^{-1})	$S°$ (J mol^{-1}K^{-1})	C_p (J mol^{-1}K^{-1})
CdOH$^+$		−261.1		
Ce^{+3}	−696.2	−672.0	−205.0	
Ce^{+4}	−537.2	−503.8	−301.0	
Co^{+2}	−58.2	−54.4	−113.0	
Co^{+3}	92.0	134.0	−305.0	
Cr^{+2}	−143.5			
Cs$^+$	−258.3	−292.0	133.1	−10.5
Cu$^+$	71.7	50.0	40.6	
Cu^{+2}	64.8	65.5	−99.6	
Dy^{+3}	−699.0	−665.0	−231.0	21.0
Er^{+3}	−705.4	−669.1	−244.3	21.0
Eu^{+2}	−527.0	−540.2	−8.0	
Eu^{+3}	−605.0	−574.1	−222.0	8.0
Fe^{+2}	−89.1	−78.9	−137.7	
Fe^{+3}	−48.5	−4.7	−315.9	
FeOH$^+$	−324.7	−277.4	−29.0	
FeOH^{+2}	−290.8	−229.4	−142.0	
Fe(OH)$_2^+$		−438.0		
Ga^{+2}		−88.0		
Ga^{+3}	−211.7	−159.0	−331.0	
GaOH^{+2}		−380.3		
Ga(OH)$_2{}^+$		−597.4		
Gd^{+3}	−686.0	−661.0	−205.9	
H$^+$	0	0	0	0
Hg^{+2}	171.1	164.4	−32.2	
Hg$_2{}^{+2}$	172.4	153.5	84.5	
HgOH$^+$	−84.5	−52.3	71.0	
Ho^{+3}	−705.0	−673.7	−226.8	17.0
In$^+$		−12.1		
In^{+2}		−50.7		
In^{+3}	−105.0	−98.0	−151.0	
InOH^{+2}	−370.3	−313.0	−88.0	
In(OH)$_2^+$	−619.0	−525.0	25.0	
K$^+$	−252.4	−283.3	102.5	21.8
La^{+3}	−707.1	−683.7	−217.6	−13.0
Li$^+$	−278.5	−293.3	13.4	68.6
Lu^{+3}	−665.0	−628.0	−264.0	25.0
LuF^{+2}		−931.4		
Mg^{+2}	−466.9	−454.8	−138.1	
MgOH$^+$		−626.7		
Mn^{+2}	−220.8	−228.1	−73.6	50.0
MnOH$^+$	−450.6	−405.0	−17.0	

(Continued)

Ions	$\Delta_f H°$ (kJ mol^{-1})	$\Delta_f G°$ (kJ mol^{-1})	$S°$ (J mol^{-1}K^{-1})	C_p (J mol^{-1}K^{-1})
NH_4^+	−132.5	−79.3	113.4	79.9
$N_2H_5^+$	−7.5	82.5	151.0	70.3
Na^+	−240.1	−261.9	59.0	46.4
Nd^{+3}	−696.2	−671.6	−206.7	−21.0
Ni^{+2}	−54.0	−45.6	−128.9	
$NiOH^+$	−287.9	−227.6	−71.0	
PH_4^+		92.1		
Pa^{+4}	−619.0			
Pb^{+2}	−1.7	−24.4	10.5	
$PbOH^+$		−226.3		
Pd^{+2}	149.0	176.5	−184.0	
Po^{+2}		71.0		
Po^{+4}		293.0		
Pr^{+3}	−704.6	−679.1	−209.0	−29.0
Pt^{+2}		254.8		
Ra^{+2}	−527.6	−561.5	54.0	
Rb^+	−251.2	−284.0	121.5	
Re^+		−33.0		
Sc^{+3}	−614.2	−586.6	−255.0	
$ScOH^{+2}$	−861.5	−801.2	−134.0	
Sm^{+2}		−497.5		
Sm^{+3}	−691.6	−666.6	−211.7	−21.0
Sn^{+2}	−8.8	−27.2	−17.0	
$SnOH^+$	−286.2	−254.8	50.0	
Sr^{+2}	−545.8	−559.5	−32.6	
$SrOH^+$		−721.3		
Tb^{+3}	−682.8	−651.9	−226.0	17.0
$Te(OH)_3^+$	−608.4	−496.1	111.7	
Th^{+4}	−769.0	−705.1	−422.6	
$Th(OH)^{+3}$	−1030.1	−920.5	−343.0	
$Th(OH)_2^{+2}$	−1282.4	−1140.9	−218.0	
Tl^+	5.4	−32.4	125.5	
Tl^{+3}	196.6	214.6	−192.0	
$TlOH^{+2}$		−15.9		
$Tl(OH)_2^+$		−244.7		
Tm^{+3}	−697.9	−662.0	−243.0	25.0
U^{+3}	−489.1	−476.2	−188.0	
U^{+4}	−591.2	−531.9	−410.0	
Y^{+3}	−723.4	−693.8	−251.0	
$Y_2(OH)_2^{+4}$		−1780.3		
Yb^{+2}		−527.0		

(*Continued*)

Ions	$\Delta_f H°$ (kJ mol^{-1})	$\Delta_f G°$ (kJ mol^{-1})	$S°$ (J mol^{-1}K^{-1})	C_p (J mol^{-1}K^{-1})
Yb^{+3}	−674.5	−644.0	−238.0	25.0
Y(OH)$^{+2}$		−879.1		
Zn^{+2}	−153.9	−147.1	−112.1	46.0
ZnOH$^+$		−330.1		
Anions				
AlO$_2^-$	−930.9	−830.9	−36.8	
Al(OH)$_4^-$	−1502.5	−1305.3	102.9	
As(O)$_2^-$	−429.0	−350.0	40.6	
As(O)$_4^{-3}$	−888.1	−648.4	−162.8	
BF$_4^-$	−1574.9	−1486.9	180.0	
BH$_4^-$	48.2	114.4	110.5	
BO$_2^-$	−772.4	−678.9	−37.2	
B$_4$O$_7^{-2}$		−2604.8		
BeO$_2^{-2}$	−790.8	−640.1	−159.0	
Br$^-$	−121.6	−104.0	82.4	−141.8
BrO$^-$	−94.1	−33.4	42.0	
BrO$_3^-$	−67.1	18.6	161.7	
BrO$_4^-$	13.0	118.1	199.6	
CHOO$^-$	−425.6	−351.0	92.0	−87.9
CH$_3$COO$^-$	−486.0	−369.3	86.6	−6.3
C$_2$O$_4^{-2}$	−825.1	−673.9	45.6	
C$_2$O$_4$H$^-$	−818.4	−698.3	149.4	
Cl$^-$	−167.2	−131.2	56.5	−136.4
ClO$^-$	−107.1	−36.8	42.0	
ClO$_2^-$	−66.5	17.2	101.3	
ClO$_3^-$	−104.0	−8.0	162.3	
ClO$_4^-$	−129.3	−8.5	182.0	
CN$^-$	150.6	172.4	94.1	
CO$_3^{-2}$	−677.1	−527.8	−56.9	
CrO$_4^{-2}$	−881.2	−727.8	50.2	
Cr$_2$O$_7^{-2}$	−1490.3	−1301.1	261.9	
F$^-$	−332.6	−278.8	−13.8	−106.7
Fe(CN)$_6^{-3}$	561.9	729.4	270.3	
Fe(CN)$_6^{-4}$	455.6	695.1	95.0	
HB$_4$O$_7^-$		−2685.1		
HCO$_3^-$	−692.0	−586.8	91.2	
HF$_2^-$	−649.9	−578.1	92.5	
HPO$_3$F$^-$		−1198.2		
HPO$_4^{-2}$	−1292.1	−1089.2	−33.5	
HP$_2$O$_7^{-3}$	−2274.8	−1972.2	46.0	

(Continued)

Ions	$\Delta_f H°$ (kJ mol⁻¹)	$\Delta_f G°$ (kJ mol⁻¹)	$S°$ (J mol⁻¹K⁻¹)	C_p (J mol⁻¹K⁻¹)
HS^-	−17.6	12.1	62.8	
HSO_3^-	−626.2	−527.7	139.7	
HSO_4^-	−887.3	−755.9	131.8	−84.0
$HS_2O_4^-$		−614.5		
HSe^-	15.9	44.0	79.0	
$HSeO_3^-$	−514.6	−411.5	135.1	
$HSeO_4^-$	−581.6	−452.2	149.4	
$H_2ASO_3^-$	−714.8	−587.1	110.5	
$H_2ASO_4^-$	−909.6	−753.2	117.0	
$H_2PO_4^-$	−1296.3	−1130.2	90.4	
$H_2P_2O_7^{-2}$	−2278.6	−2010.2	163.0	
I^-	−55.2	−51.6	111.3	−142.3
IO^-	−107.5	−38.5	−5.4	
IO_3^-	−221.3	−128.0	118.4	
IO_4^-	−151.5	−58.5	222.0	
MnO_4^-	−541.4	−447.2	191.2	−82.0
MnO_4^{-2}	−653.0	−500.7	59.0	
MoO_4^{-2}	−997.9	−836.3	27.2	
NO_2^-	−104.6	−32.2	123.0	−97.5
NO_3^-	−207.4	−111.3	146.4	−86.6
N_3^-	275.1	348.2	107.9	
OCN^-	−146.0	−97.4	106.7	
OH^-	−230.0	−157.2	−10.8	−148.5
PO_4^{-3}	−1277.4	−1018.7	−220.5	
$P_2O_7^{-4}$	−2271.1	−1919.0	−117.0	
Re^-	46.0	10.1	230.0	
S^{-2}	33.1	85.8	−14.6	
SCN^-	76.4	92.7	144.3	−40.2
SO_3^{-2}	−635.5	−486.5	−29.0	
SO_4^{-2}	−909.3	−744.5	20.1	−293.0
S_2^{-2}	30.1	79.5	28.5	
$S_2O_3^{-2}$	−652.3	−522.5	67.0	
$S_2O_4^{-2}$	−753.5	−600.3	92.0	
$S_2O_8^{-2}$	−1344.7	−1114.9	244.3	
Se^{-2}		129.3		
SeO_3^{-2}	−509.2	−369.8	13.0	
SeO_4^{-2}	−599.1	−441.3	54.0	
VO_3^-	−888.3	−783.6	50.0	
VO_4^{-3}		−899.0		
WO_4^{-2}	−1075.7			

10.6 IONIZATION CONSTANT OF WATER AS A FUNCTION OF TEMPERATURE AND PRESSURE[7]

This table gives values of $pK_w = -\log_{10}(K_w)$, where K_w is the equilibrium constant of the reaction $2H_2O(l) = H_3O^+(aq) + OH^-(aq)$. K_w is defined as $K_w = a_{H_3O^+} a_{OH^-}/a_{H_2O}^2$, where a_i is the dimensionless activity of species i. The activities are on the molality basis for ions and mole fraction basis for water molecules. It is assumed that the activity of $H_3O^+(aq)$ is the same as the activity of $H^+(aq)$, so that K_w is numerically equal to a $a_{H^+} a_{OH^-}/a_{H_2O}$, the equilibrium constant for the ionization reaction of water, $H_2O(l) = H^+(aq) + OH^-(aq)$, that is most commonly used in the literature.

	Temperature, °C								
Pressure, MPa	0	25	50	75	100	150	200	250	300
0.1 or p_s	14.946	13.995	13.264	12.696	12.252	11.641	11.310	11.205	11.339
25	14.848	13.908	13.181	12.613	12.165	11.543	11.189	11.050	11.125
50	14.754	13.824	13.102	12.533	12.084	11.450	11.076	10.898	10.893
75	14.665	13.745	13.026	12.458	12.006	11.364	10.974	10.769	10.715
100	14.580	13.668	12.953	12.385	11.933	11.283	10.880	10.655	10.568
150	14.422	13.524	12.815	12.249	11.795	11.135	10.713	10.458	10.327
200	14.278	13.390	12.687	12.123	11.668	11.000	10.564	10.289	10.131
250	14.145	13.265	12.567	12.004	11.549	10.876	10.430	10.140	9.963
300	14.021	13.148	12.453	11.892	11.437	10.760	10.306	10.005	9.814

Note: Pressure for first row is 0.1 MPa at $t < 100°C$ or p_s (saturated liquid) for $100°C \le t \le 300°C$.

[7] Adapted from W.M. Haynes (Editor-in-Chief), *CRC Handbook of Chemistry and Physics*, 95th Edition, CRC Press, Boca Raton, 2014, pp. 5–70.

10.7   ELECTRIC CONDUCTIVITY OF WATER[8]

This table gives the electric conductivity of highly purified water over a range of temperature and pressure. The first column of conductivity data refers to water at its own vapor pressure.

	Conductivity in μS cm^{-1} at the Indicated Pressure			
t (°C)	Sat. Vapor	50 MPa	100 MPa	200 MPa
0	0.0115	0.0150	0.0189	0.0275
25	0.0550	0.0686	0.0836	0.117
100	0.765	0.942	1.13	1.53
200	2.99	4.08	5.22	7.65
300	2.41	4.87	7.80	14.1

10.8   ELECTRIC CONDUCTIVITY OF AQUEOUS SOLUTIONS[9]

The following table gives the electric conductivity of aqueous solutions of some acids, bases, and salts as a function of concentration. All values refer to 293.15 K.

---

[8] Adapted from W.M. Haynes (Editor-in-Chief), *CRC Handbook of Chemistry and Physics*, 95th Edition, CRC Press, Boca Raton, 2014, pp. 5–72.
[9] Adapted from W.M. Haynes (Editor-in-Chief), *CRC Handbook of Chemistry and Physics*, 95th Edition, CRC Press, Boca Raton, 2014, pp. 5–73.

Electric Conductivity κ in mS cm^{-1} for the Indicated Concentration in Mass Percent (pph)

Name	Formula	0.5%	1%	2%	5%	10%	15%	20%	25%	30%	40%	50%
Acetic acid	CH_3COOH	0.3	0.6	0.8	1.2	1.5	1.7	1.7	1.6	1.4	1.1	0.8
Ammonia	NH_3	0.5	0.7	1.0	1.1	1.0	0.7	0.5	0.4			
Ammonium chloride	NH_4Cl	10.5	20.4	40.3	95.3	180						
Ammonium sulfate	$(NH_4)_2SO_4$	7.4	14.2	25.7	57.4	105	147	185	215			
Barium chloride	$BaCl_2$	4.7	9.1	17.4	40.4	76.7	109.0	137.0				
Calcium chloride	$CaCl_2$	8.1	15.7	29.4	67.0	117	157	177	183	172	106	
Cesium chloride	$CsCl$	3.8	7.4	13.8	32.9	65.8	102	142				
Citric acid	$H_3C(OH)(COO)_3$	1.2	2.1	3.0	4.7	6.2	7.0	7.2	7.1			
Copper(II) sulfate	$CuSO_4$	2.9	5.4	9.3	19.0	32.2	42.3					
Formic acid	$HCOOH$	1.4	2.4	3.5	5.6	7.8	9.0	9.9	10.4	10.5	9.9	8.6
Hydrogen chloride	HCl	45.1	92.9	183								
Lithium chloride	$LiCl$	10.1	19.0	34.9	76.4	127	155	170	165	146		
Magnesium chloride	$MgCl_2$	8.6	16.6	31.2	66.9	108	129	134	122	98		
Magnesium sulfate	$MgSO_4$	4.1	7.6	13.3	27.4	42.7	54.2	51.1	44.1			
Manganese(II) sulfate	$MnSO_4$		6.2	10.6	21.6	34.5	43.7	47.6				
Nitric acid	HNO_3	28.4	56.1	108								
Oxalic acid	$H_2C_2O_4$	14.0	21.8	35.3	65.6							
Phosphoric acid	H_3PO_4	5.5	10.1	16.2	31.5	59.4	88.4	118	146	173	209	
Potassium bromide	KBr	5.2	10.2	19.5	47.7	95.6	144	194				
Potassium carbonate	K_2CO_3	7.0	13.6	25.4	58.0	109	152	188	223			
Potassium chloride	KCl	8.2	15.7	29.5	71.9	143	208					
Potassium dihydrogen phosphate	KH_2PO_4	3.0	5.9	11.0	25.0	44.6						
Potassium hydrogen carbonate	$KHCO_3$	4.6	8.9	17.0	38.8	72.4	101	128				
Name	Formula	0.5%	1%	2%	5%	10%	15%	20%	25%	30%	40%	50%
Potassium hydrogen phosphate	K_2HPO_4	5.2	9.9	18.3	40.3							

(Continued)

Electric Conductivity κ in mS cm^{-1} for the Indicated Concentration in Mass Percent (pph)

Name	Formula	0.5%	1%	2%	5%	10%	15%	20%	25%	30%	40%	50%
Potassium hydroxide	KOH	20.0	38.5	75.0	178							
Potassium iodide	KI	3.8	7.5	14.2	35.2	71.8	110	188	224			
Potassium nitrate	KNO$_3$	5.5	10.7	20.1	47.0	87.3	124	157	182			
Potassium permanganate	KMnO$_4$	3.5	6.9	13.0	30.5							
Potassium sulfate	K$_2$SO$_4$	5.8	11.2	21.0	48.0	88.6						
Silver(I) nitrate	AgNO$_3$	3.1	6.1	12.0	26.7	49.8	72.0	92.8	112	129	162	
Sodium acetate	NaCH$_3$COO	3.9	7.6	14.4	30.9	53.4	64.1	69.3	69.2	64.3		
Sodium bromide	NaBr	5.0	9.7	18.4	44.0	84.6	122	157	191	216		
Sodium carbonate	Na$_2$CO$_3$	7.0	13.1	23.3	47.0	74.4	88.6					
Sodium chloride	NaCl	8.2	16.0	30.2	70.1	126	171	204	222			
Sodium citrate	Na$_3$C$_6$H$_5$O$_7$		7.4	12.8	26.2	42.1	52.0	57.1	57.3	53.5		
Sodium dihydrogen phosphate	NaH$_2$PO$_4$	2.2	4.4	9.1	21.0	33.2	43.3	49.6	53.1	54.0	46.1	
Sodium hydrogen carbonate	NaHCO$_3$	4.2	8.2	15.0	31.4							
Sodium hydrogen phosphate	Na$_2$HPO$_4$	4.6	8.7	15.6	31.4							
Sodium hydroxide	NaOH	24.8	48.6	93.1	206							
Sodium nitrate	NaNO$_3$	5.4	10.6	20.4	46.2	82.6	111	134	152	165	178	
Sodium phosphate	Na$_3$PO$_4$	7.3	14.1	22.7	43.5							
Sodium sulfate	Na$_2$SO$_4$	5.9	11.2	19.8	42.7	71.3	91.1	109				
Sodium thiosulfate	Na$_2$S$_2$O$_3$	5.7	10.7	19.5	43.3	76.7	104	123	134	136	118	
Strontium chloride	SrCl$_2$	5.9	11.4	22.0	49.1	91.5	127	153	168	178		
Sulfuric acid	H$_2$SO$_4$	24.3	47.8	92	211							
Trichloroacetic acid	CCl$_3$COOH	10.3	19.6	37.2	84.7	148	193	221				
Zinc sulfate	ZnSO$_4$	2.8	5.4	10.0	20.5	33.7	43.3					

10.9 STANDARD KCL SOLUTIONS FOR CALIBRATING CONDUCTIVITY CELLS[10]

This table presents recommended electric conductivity (κ) values for aqueous potassium chloride solutions with molalities of 0.01–1 mol kg^{-1} at temperatures from 0°C to 50°C. The conductivity of water saturated with atmospheric CO_2 is given in the last column.

	$10^4 \kappa$ (S m^{-1})			
t (°C)	0.01 m KCl	0.1 m KCl	1.0 m KCl	H_2O (CO_2 sat.)
0	772.92	7 116.85	63 488	0.58
5	890.96	8 183.70	72 030	0.68
10	1 013.95	9 291.72	80 844	0.79
15	1 141.45	10 437.1	89 900	0.89
18	1 219.93	11 140.6	—	0.95
20	1 273.03	11 615.9	99 170	0.99
25	1 408.23	12 824.6	108 620	1.10
30	1 546.63	14 059.2	118 240	1.20
35	1 687.79	15 316.0	127 970	1.30
40	1 831.27	16 591.0	137 810	1.40
45	1 976.62	17 880.6	147 720	1.51
50	2 123.43	19 180.9	157 670	1.61

10.10 MOLAR CONDUCTIVITY OF AQUEOUS HCL[11]

The molar conductivity Λ of an electrolyte solution is defined as the electric conductivity divided by amount-of-substance concentration. The customary unit is S cm^{2}mol^{-1} (i.e., Ω^{-1}cm^{2}mol^{-1}). The first part of this table gives the molar conductivity of the hydrochloric acid at 298.15 K as a function of the concentration in mol L^{-1}. The second part gives the temperature dependence of Λ for HCl (aq).

c (mol L^{-1})	HCl
Inf. dil.	426.1
0.0001	424.5
0.0005	422.6
0.001	421.2
0.005	415.7

(Continued)

[10] Adapted from W.M. Haynes (Editor-in-Chief), *CRC Handbook of Chemistry and Physics,* 95th edition, CRC Press, Boca Raton, 2014, pp. 5–74.

[11] Adapted from W.M. Haynes (Editor-in-Chief), *CRC Handbook of Chemistry and Physics,* 95th edition, CRC Press, Boca Raton, 2014, pp. 5–75.

c (mol L^{-1})	HCl
0.01	411.9
0.05	398.9
0.10	391.1
0.5	360.7
1.0	332.2
1.5	305.8
2.0	281.4
2.5	258.9
3.0	237.6
3.5	218.3
4.0	200.0
4.5	183.1
5.0	167.4
5.5	152.9
6.0	139.7
6.5	127.7
7.0	116.9
7.5	107.0
8.0	98.2
8.5	90.3
9.0	83.1
9.5	76.6
10.0	70.7

c (mol L^{-1})	253.15 K	263.15 K	273.15 K	283.15 K	293.15 K	303.15 K	313.15 K	323.15 K
				HCl				
0.5			228.7	283.0	336.4	386.8	436.9	482.4
1.0			211.7	261.6	312.2	359.0	402.9	445.3
1.5			196.2	241.5	287.5	331.1	371.6	410.8
2.0			182.0	222.7	262.9	303.3	342.4	378.2
2.5		131.7	168.5	205.1	239.8	277.0	315.2	347.6
3.0		120.8	154.6	188.5	219.3	253.3	289.3	319.0
3.5	85.5	111.3	139.6	172.2	201.6	232.9	263.9	292.1
4.0	79.3	102.7	129.2	158.1	185.6	214.2	242.2	268.2
4.5	73.7	94.9	119.5	145.4	170.6	196.6	222.5	246.7
5.0	68.5	87.8	110.3	133.5	156.6	180.2	204.1	226.5
5.5	63.6	81.1	101.7	122.5	143.6	165.0	187.1	207.7
6.0	58.9	74.9	93.7	112.3	131.5	151.0	171.3	190.3
6.5	54.4	69.1	86.2	103.0	120.4	138.2	156.9	174.3
7.0	50.2	63.7	79.3	94.4	110.2	126.4	143.3	159.7
7.5	46.3	58.6	73.0	86.5	100.9	115.7	131.6	146.2

(Continued)

c (mol L⁻¹)	253.15 K	263.15 K	273.15 K	283.15 K	293.15 K	303.15 K	313.15 K	323.15 K
8.0	42.7	54.0	67.1	79.4	92.4	106.1	120.6	134.0
8.5	39.4	49.8	61.7	72.9	84.7	97.3	110.7	123.0
9.0	36.4	45.9	56.8	67.1	77.8	89.4	101.7	112.9
9.5	33.6	42.3	52.3	61.8	71.5	82.3	93.6	103.9
10.0	31.2	39.1	48.2	57.0	65.8	75.9	86.3	95.7
10.5	28.9	36.1	44.5	52.7	60.7	70.1	79.6	88.4
11.0	26.8	33.4	41.1	48.8	56.1	64.9	73.6	81.7
11.5	24.9	31.0	38.0	45.3	51.9	60.1	68.0	75.6
12.0	23.1	28.7	35.3	42.0	48.0	55.6	62.8	70.0
12.5	21.4	26.7	32.7	39.0	44.4	51.4	57.9	64.8

10.11 MOLAR CONDUCTIVITY OF ELECTROLYTES[12]

This table gives the limiting ionic conductivity λ_i^0 for common ions at infinite dilution. All values refer to aqueous solutions at 298.15 K. It also lists the limiting diffusion coefficient D_i^0 of the ion in infinitely dilute aqueous solution, which is related to λ_i^0 through the equation

$$D_i^0 = \left(\frac{RT}{F^2} \right)\left(\frac{\lambda_i^0}{|z_i|^2} \right)$$

where R is the molar gas constant, T the thermodynamic temperature, F the Faraday constant, and z_i the charge on the ion. Note that the fraction in the left column in this table should be used only for estimating λ_i^0. For example, $\lambda_i^0(Ca^{2+}) = 118.94 \times 10^{-4} m^2 s\ mol^{-1}$ and $D_i^0 = 0.792 \times 10^{-5} cm^2 s^{-1}$.

Compound	Infinite Dilution $\Lambda°$	Concentration (mol L⁻¹)						
		0.0005	0.001	0.005	0.01	0.02	0.05	0.1
		Λ (10⁻⁴ m² S mol⁻¹)						
$AgNO_3$	133.29	131.29	130.45	127.14	124.70	121.35	115.18	109.09
$1/2BaCl_2$	139.91	135.89	134.27	127.96	123.88	119.03	111.42	105.14
$1/2CaCl_2$	135.77	131.86	130.30	124.19	120.30	115.59	108.42	102.41
$1/2Ca(OH)_2$	258	—	—	233	226	214	—	—
$CoSO_4^a$	133.6	121.6	115.20	94.02	83.08	72.16	59.02	50.55
HCl	425.95	422.53	421.15	415.59	411.80	407.04	398.89	391.13
KBr	151.9	149.8	148.9	146.02	143.36	140.41	135.61	131.32
KCl	149.79	147.74	146.88	143.48	141.20	138.27	133.30	128.90
$KClO_4$	139.97	138.69	137.80	134.09	131.39	127.86	121.56	115.14
$1/3K_3Fe(CN)_6$	174.5	166.4	163.1	150.7	—	—	—	—
$1/4K_4Fe(CN)_6$	184	—	167.16	146.02	134.76	122.76	107.65	97.82

(Continued)

[12] Adapted from W.M. Haynes (Editor-in-Chief), *CRC Handbook of Chemistry and Physics*, 95th edition, CRC Press, Boca Raton, 2014, pp. 5–76.

		Concentration (mol L^{-1})						
	Infinite Dilution	0.0005	0.001	0.005	0.01	0.02	0.05	0.1
Compound	$\Lambda°$	Λ (10^{-4}m^2 S mol^{-1})						
KHCO$_3$	117.94	116.04	115.28	112.18	110.03	107.17	—	—
KI	150.31	148.2	143.32	144.30	142.11	139.38	134.90	131.05
KIO$_4$	127.86	125.74	124.88	121.18	118.45	114.08	106.67	98.2
KNO$_3$	144.89	142.70	141.77	138.41	132.75	132.34	126.25	120.34
KMnO$_4$	134.8	132.7	131.9	—	126.5	—	—	113
KOH	271.5	—	234	230	228	—	219	213
KReO$_4$	128.20	126.03	125.12	121.31	118.49	114.49	106.40	97.40
1/3LaCl$_3$	145.9	139.6	137.0	127.5	121.8	115.3	106.2	99.1
LiCl	114.97	113.09	112.34	109.35	107.27	104.60	100.06	95.81
LiClO$_4$	105.93	104.13	103.39	100.52	98.56	96.13	92.15	88.52
1/2MgCl$_2$	129.34	125.55	124.15	118.25	114.49	109.99	103.03	97.05
NH$_4$Cl	149.6	147.5	146.7	143.9	141.21	138.25	133.22	128.69
NaCl	126.39	124.44	123.68	120.59	118.45	115.70	111.01	106.69
NaClO$_4$	117.42	115.58	114.82	111.70	109.54	106.91	102.35	98.38
NaI	126.88	125.30	124.19	121.19	119.18	116.64	112.73	108.73
NaOOCCH$_3$	91.0	89.2	88.5	85.68	83.72	81.20	76.88	72.76
NaOH	247.7	245.5	244.6	240.7	237.9	—	—	—
Na picrate	80.45	78.7	78.6	75.7	73.7	—	66.3	61.8
1/2Na$_2$SO$_4$	129.8	125.68	124.09	117.09	112.38	106.73	97.70	89.94
1/2SrCl2	135.73	131.84	130.27	124.18	120.23	115.48	108.20	102.14
ZnSO4	132.7	121.3	114.47	95.44	84.87	74.20	61.17	52.61

[a] For 2:2 electrolytes (e.g., CuSO$_4$) given in the table, values are equivalent conductivities. The molar conductivities are two times larger.

10.12 IONIC CONDUCTIVITY AND DIFFUSION COEFFICIENT AT INFINITE DILUTION[13]

This table gives the limiting ionic conductivity λ_i^0 for common ions at infinite dilution. All values refer to aqueous solutions at 298.15 K. It also lists the limiting diffusion coefficient D_i^0 of the ion in infinitely dilute aqueous solution, which is related to λ_i^0 through the equation

$$D_i^0 = \left(\frac{RT}{F^2}\right)\left(\frac{\lambda_i^0}{|z_i|^2}\right)$$

Where R is the molar gas constant, T the thermodynamic temperature, F the Faraday constant, and z_i the charge on the ion.

Note that the fraction in the left column in this table should be used only for estimating λ_i^0. For example, $\lambda_i^0(Ca^{2+}) = 118.94 \times 10^{-4}$m^{2}s mol^{-1} and $D_i^0 = 0.792 \times 10^{-5}$cm^{2}s^{-1}.

[13] Adapted from W.M. Haynes (Editor-in-Chief), *CRC Handbook of Chemistry and Physics*, 95th edition, CRC Press, Boca Raton, 2014, pp. 5–77.

Ion	$\lambda_i \times 10^{-4} m^2\ S\ mol^{-1}$	$D_i \times 10^{-5} cm^2\ S^{-1}$
Inorganic Cations		
Ag^+	61.9	1.648
$1/3 Al^{3+}$	61	0.541
$1/2 Ba^{2+}$	63.6	0.847
$1/2 Be^{2+}$	45	0.599
$1/2 Ca^{2+}$	59.47	0.792
$1/2 Cd^{2+}$	54	0.719
$1/3 Ce^{3+}$	69.8	0.620
$1/2 Co^{2+}$	55	0.732
$1/3 [Co(NH_3)_6]^{3+}$	101.9	0.904
$1/3 [Co(en)_3]^{3+}$	74.7	0.663
$1/6 [Co_2(trien)_3]^{6+}$	69	0.306
$1/3 Cr^{3+}$	67	0.595
Cs^+	77.2	2.056
$1/2 Cu^{2+}$	53.6	0.714
D^+	249.9	6.655
$1/3 Dy^{3+}$	65.6	0.582
$1/3 Er^{3+}$	65.9	0.585
$1/3 Eu^{3+}$	67.8	0.602
$1/2 Fe^{2+}$	54	0.719
$1/3 Fe^{3+}$	68	0.604
$1/3 Gd^{3+}$	67.3	0.597
H^+	349.65	9.311
$1/2 Hg^{2+}$	68.6	0.913
$1/2 Hg^{2+}$	63.6	0.847
$1/3 Ho^{3+}$	66.3	0.589
K^+	73.48	1.957
$1/3 La^{3+}$	69.7	0.619
Li^+	38.66	1.029
$1/2 Mg^{2+}$	53.0	0.706
$1/2 Mn^{2+}$	53.5	0.712
NH_4^+	73.5	1.957
$N_2H_5^+$	59	1.571
Na^+	50.08	1.334
$1/3 Nd^{3+}$	69.4	0.616
$1/2 Ni^{2+}$	49.6	0.661
$1/4 [Ni_2(trien)_3]^{4+}$	52	0.346
$1/2 Pb^{2+}$	71	0.945
$1/3 Pr^{3+}$	69.5	0.617
$1/2 Ra^{2+}$	66.8	0.889
Rb^+	77.8	2.072
$1/3 Sc^{3+}$	64.7	0.574
$1/3 Sm^{3+}$	68.5	0.608
$1/2 Sr^{2+}$	59.4	0.791
Tl^+	74.7	1.989
$1/3 Tm^{3+}$	65.4	0.581

(Continued)

Ion	$\lambda_i \times 10^{-4} m^2\ S\ mol^{-1}$	$Di \times 10^{-5} cm^2\ S^{-1}$
$1/2UO_2^{2+}$	32	0.426
$1/3Y^{3+}$	62	0.550
$1/3Yb^{3+}$	65.6	0.582
$1/2Zn^{2+}$	52.8	0.703
Inorganic Anions		
$Au(CN)_2^-$	50	1.331
$Au(CN)_4^-$	36	0.959
$B(C_6H_5)_4^-$	21	0.559
Br^-	78.1	2.080
Br_3^-	43	1.145
BrO_3^-	55.7	1.483
CN^-	78	2.077
CNO^-	64.6	1.720
$1/2CO_3^{2-}$	69.3	0.923
Cl^-	76.31	2.032
ClO_2^-	52	1.385
ClO_3^-	64.6	1.720
ClO_4^-	67.3	1.792
$1/3[Co(CN)_6]^{3-}$	98.9	0.878
$1/2CrO_4^{2-}$	85	1.132
F^-	55.4	1.475
$1/4[Fe(CN)_6]^{4-}$	110.4	0.735
$1/3[Fe(CN)_6]^{3-}$	100.9	0.896
$H_2AsO_4^-$	34	0.905
HCO_3^-	44.5	1.185
HF_2^-	75	1.997
$1/2HPO_4^{2-}$	57	0.759
$H_2PO_4^-$	36	0.959
$H_2PO_2^-$	46	1.225
HS^-	65	1.731
HSO_3^-	58	1.545
HSO_4^-	52	1.385
$H_2SbO_4^-$	31	0.825
I^-	76.8	2.045
IO_3^-	40.5	1.078
IO_4^-	54.5	1.451
MnO_4^-	61.3	1.632
$1/2MoO_4^{2-}$	74.5	1.984
$N(CN)_2^-$	54.5	1.451
NO_2^-	71.8	1.912
NO_3^-	71.42	1.902
$NH_2SO_3^-$	48.3	1.286
N_3^-	69	1.837
OCN^-	64.6	1.720
OD^-	119	3.169
OH^-	198	5.273

(Continued)

Ion	$\lambda_i \times 10^{-4} m^2\ S\ mol^{-1}$	$D_i \times 10^{-5} cm^2\ S^{-1}$
PF_6^-	56.9	1.515
$1/2PO_3F^{2-}$	63.3	0.843
$1/3PO_4^{3-}$	92.8	0.824
$1/4P_2O_7^{4-}$	96	0.639
$1/3P_3O_9^{3-}$	83.6	0.742
$1/5P_3O_{10}^{5-}$	109	0.581
ReO_4^-	54.9	1.462
SCN^-	66	1.758
$1/2SO_3^{2-}$	72	0.959
$1/2SO_4^{2-}$	80.0	1.065
$1/2S_2O_3^{2-}$	85.0	1.132
$1/2S_2O_4^{2-}$	66.5	0.885
$1/2S_2O_6^{2-}$	93	1.238
$1/2S_2O_8^{2-}$	86	1.145
$Sb(OH)_6^-$	31.9	0.849
$SeCN^-$	64.7	1.723
$1/2SeO_4^{2-}$	75.7	1.008
$1/2WO_4^{2-}$	69	0.919
Acetate$^-$	40.9	1.089
Formate$^-$	54.6	1.454

10.13   ELECTROCHEMICAL SERIES[14]

The table lists standard electrode potentials, $E°$ values, of the reduction half-reactions at 298.15 K, and at a pressure of 101.325 kPa (1 atm) (not the standard pressure of 1 bar). All ions are aqueous and other species are in a state which is stable at 1 atm and 298.15 K.

Reaction	$E°$ (V)
$Ac^{3+} + 3e \rightleftharpoons Ac$	−2.20
$Ag^+ + e \rightleftharpoons Ag$	0.7996
$Ag^{2+} + e \rightleftharpoons Ag^+$	1.980
$Ag(ac) + e \rightleftharpoons Ag + (ac)^-$	0.643
$AgBr + e \rightleftharpoons Ag + Br^-$	0.07133
$AgBrO_3 + e \rightleftharpoons Ag + BrO_3^-$	0.546
$Ag_2C_2O_4 + 2e \rightleftharpoons 2Ag + C_2O_4^{2-}$	0.4647
$AgCl + e \rightleftharpoons Ag + Cl^-$	0.22233
$AgCN + e \rightleftharpoons Ag + CN^-$	−0.017
$Ag_2CO_3 + 2e \rightleftharpoons 2Ag + CO_3^{2-}$	0.47
$Ag_2CrO_4 + 2e \rightleftharpoons 2Ag + CrO_4^{2-}$	0.4470

(Continued)

[14] Adapted from W.M. Haynes (Editor-in-Chief), *CRC Handbook of Chemistry and Physics,* 95th edition, CRC Press, Boca Raton, 2014, pp. 5–80.

Reaction	$E°$ (V)
$AgF + e \rightleftharpoons Ag + F^-$	0.779
$Ag_4[Fe(CN)_6] + 4e \rightleftharpoons 4Ag + [Fe(CN)_6]^{4-}$	0.1478
$AgI + e \rightleftharpoons Ag + I^-$	-0.15224
$AgIO_3 + e \rightleftharpoons Ag + IO_3^-$	0.354
$Ag_2MoO_4 + 2e \rightleftharpoons 2Ag + MoO_4^{2-}$	0.4573
$AgNO_2 + e \rightleftharpoons Ag + 2NO_2^-$	0.564
$Ag_2O + H_2O + 2e \rightleftharpoons 2Ag + 2OH^-$	0.342
$Ag_2O_3 + H_2O + 2e \rightleftharpoons 2AgO + 2OH^-$	0.739
$Ag^{3+} + 2e \rightleftharpoons Ag^+$	1.9
$Ag^{3+} + e \rightleftharpoons Ag^{2+}$	1.8
$Ag_2O_2 + 4H^+ + e \rightleftharpoons 2Ag + 2H_2O$	1.802
$2AgO + H_2O + 2e \rightleftharpoons Ag_2O + 2OH^-$	0.607
$AgOCN + e \rightleftharpoons Ag + OCN^-$	0.41
$Ag_2S + 2e \rightleftharpoons 2Ag + S^{2-}$	-0.691
$Ag_2S + 2H^+ + 2e \rightleftharpoons 2Ag + H_2S$	-0.0366
$AgSCN + e \rightleftharpoons Ag + SCN^-$	0.08951
$Ag_2SeO_3 + 2e \rightleftharpoons 2Ag + SeO_4^{2-}$	0.3629
$Ag_2SO_4 + 2e \rightleftharpoons 2Ag + SO_4^{2-}$	0.654
$Ag_2WO_4 + 2e \rightleftharpoons 2Ag + WO_4^{2-}$	0.4660
$Al^{3+} + 3e \rightleftharpoons Al$	-1.676
$Al(OH)_3 + 3e \rightleftharpoons Al + 3OH^-$	-2.30
$Al(OH)_4^- + 3e \rightleftharpoons Al + 4OH^-$	-2.310
$H_2AlO_3^- + H_2O + 3e \rightleftharpoons Al + 4OH^-$	-2.33
$AlF_6^{3-} + 3e \rightleftharpoons Al + 6F^-$	-2.069
$Am^{4+} + e \rightleftharpoons Am^{3+}$	2.60
$Am^{2+} + 2e \rightleftharpoons Am$	-1.9
$Am^{3+} + 3e \rightleftharpoons Am$	-2.048
$Am^{3+} + e \rightleftharpoons Am^{2+}$	-2.3
$As + 3H^+ + 3e \rightleftharpoons AsH_3$	-0.608
$As_2O_3 + 6H^+ + 6e \rightleftharpoons 2As + 3H_2O$	0.234
$HAsO_2 + 3H^+ + 3e \rightleftharpoons As + 2H_2O$	0.248
$AsO_2^- + 2H_2O + 3e \rightleftharpoons As + 4OH^-$	-0.68
$H_3AsO_4 + 2H^+ + 2e^- \rightleftharpoons HAsO_2 + 2H_2O$	0.560
$AsO_4^{3-} + 2H_2O + 2e \rightleftharpoons AsO_2^- + 4OH^-$	-0.71
$At_2 + 2e \rightleftharpoons 2At^-$	0.2
$Au^+ + e \rightleftharpoons Au$	1.692
$Au^{3+} + 2e \rightleftharpoons Au^+$	1.401
$Au^{3+} + 3e \rightleftharpoons Au$	1.498
$Au^{2+} + e - \rightleftharpoons Au^+$	1.8
$AuOH^{2+} + H^+ + 2e \rightleftharpoons Au^+ + H_2O$	1.32
$AuBr_2^- + e \rightleftharpoons Au + 2Br^-$	0.959
$AuBr_4^- + 3e \rightleftharpoons Au + 4Br^-$	0.854
$AuCl_4^- + 3e \rightleftharpoons Au + 4Cl^-$	1.002

(Continued)

Reaction	$E°$ (V)
$Au(OH)_3 + 3H^+ + 3e \rightleftharpoons Au + 3H_2O$	1.45
$H_2BO_3^- + 5H_2O + 8e \rightleftharpoons BH_4^- + 8OH^-$	−1.24
$H_2BO_3^- + H_2O + 3e \rightleftharpoons B + 4OH^-$	−1.79
$H_3BO_3 + 3H^+ + 3e \rightleftharpoons B + 3H_2O$	−0.8698
$B(OH)_3 + 7H^+ + 8e \rightleftharpoons BH_4^- + 3H_2O$	−0.481
$Ba^{2+} + 2e \rightleftharpoons Ba$	−2.912
$Ba^{2+} + 2e \rightleftharpoons Ba(Hg)$	−1.570
$Ba(OH)_2 + 2e \rightleftharpoons Ba + 2OH^-$	−2.99
$Be^{2+} + 2e \rightleftharpoons Be$	−1.847
$Be_2O_3^{2-} + 3H_2O + 4e \rightleftharpoons 2Be + 6OH^-$	−2.63
p–benzoquinone $+ 2H^+ + 2e \rightleftharpoons$ hydroquinone	0.6992
$Bi^+ + e \rightleftharpoons Bi$	0.5
$Bi^{3+} + 3e \rightleftharpoons Bi$	0.308
$Bi^{3+} + 2e \rightleftharpoons Bi^+$	0.2
$Bi + 3H^+ + 3e \rightleftharpoons BiH_3$	−0.8
$BiCl_4^- + 3e \rightleftharpoons Bi + 4Cl^-$	0.16
$Bi_2O_3 + 3H_2O + 6e \rightleftharpoons 2 Bi + 6OH^-$	−0.46
$Bi_2O_4 + 4H^+ + 2e \rightleftharpoons 2 BiO^+ + 2H_2O$	1.593
$BiO^+ + 2H^+ + 3e \rightleftharpoons Bi + H_2O$	0.320
$BiOCl + 2H^+ + 3e \rightleftharpoons Bi + Cl^- + H_2O$	0.1583
$Bk^{4+} + e \rightleftharpoons Bk^{3+}$	1.67
$Bk^{2+} + 2e \rightleftharpoons Bk$	−1.6
$Bk^{3+} + e \rightleftharpoons Bk^{2+}$	−2.8
$Br_2(aq) + 2e \rightleftharpoons 2Br^-$	1.0873
$Br_2(l) + 2e \rightleftharpoons 2Br^-$	1.066
$HBrO + H^+ + 2e \rightleftharpoons Br^- + H_2O$	1.331
$HBrO + H^+ + e \rightleftharpoons 1/2Br_2(aq) + H_2O$	1.574
$HBrO + H^+ + e \rightleftharpoons 1/2Br_2(l) + H_2O$	1.596
$BrO^- + H_2O + 2e \rightleftharpoons Br^- + 2OH^-$	0.761
$BrO_3^- + 6H^+ + 5e \rightleftharpoons 1/2 Br_2 + 3H_2O$	1.482
$BrO_3^- + 6H^+ + 6e \rightleftharpoons Br^- + 3H_2O$	1.423
$BrO_3^- + 3H_2O + 6e \rightleftharpoons Br^- + 6OH^-$	0.61
$(CN)_2 + 2H^+ + 2e \rightleftharpoons 2HCN$	0.373
$2HCNO + 2H^+ + 2e \rightleftharpoons (CN)_2 + 2H_2O$	0.330
$(CNS)_2 + 2e \rightleftharpoons 2CNS^-$	0.77
$CO_2 + 2H^+ + 2e \rightleftharpoons HCOOH$	−0.199
$Ca^+ + e \rightleftharpoons Ca$	−3.80
$Ca^{2+} + 2e \rightleftharpoons Ca$	−2.868
$Ca(OH)_2 + 2e \rightleftharpoons Ca + 2OH^-$	−3.02
Calomel electrode, 1 molal KCl	0.2800
Calomel electrode, 1 molar KCl (NCE)	0.2801
Calomel electrode, 0.1 molar KCl	0.3337
Calomel electrode, saturated KCl (SCE)	0.2412

(Continued)

Reaction	$E°$ (V)
Calomel electrode, saturated NaCl (SSCE)	0.2360
$Cd^{2+} + 2e \rightleftharpoons Cd$	−0.4030
$Cd^{2+} + 2e \rightleftharpoons Cd(Hg)$	−0.3521
$Cd(OH)_2 + 2e \rightleftharpoons Cd(Hg) + 2OH^-$	−0.809
$CdSO_4 + 2e \rightleftharpoons Cd + SO_4^{2-}$	−0.246
$Cd(OH)_4^{2-} + 2e \rightleftharpoons Cd + 4OH^-$	−0.658
$CdO + H_2O + 2e \rightleftharpoons Cd + 2OH^-$	−0.783
$Ce^{3+} + 3e \rightleftharpoons Ce$	−2.336
$Ce^{3+} + 3e \rightleftharpoons Ce(Hg)$	−1.4373
$Ce^{4+} + e \rightleftharpoons Ce^{3+}$	1.72
$CeOH^{3+} + H^+ + e \rightleftharpoons Ce^{3+} + H_2O$	1.715
$Cf^{4+} + e \rightleftharpoons Cf^{3+}$	3.3
$Cf^{3+} + e \rightleftharpoons Cf^{2+}$	−1.6
$Cf^{3+} + 3e \rightleftharpoons Cf$	−1.94
$Cf^{2+} + 2e \rightleftharpoons Cf$	−2.12
$Cl_2(g) + 2e \rightleftharpoons 2Cl^-$	1.35827
$HClO + H + + e \rightleftharpoons 1/2Cl_2 + H_2O$	1.611
$HClO + H^+ + 2e \rightleftharpoons Cl^- + H_2O$	1.482
$ClO^- + H_2O + 2e \rightleftharpoons Cl^- + 2OH^-$	0.81
$ClO_2 + H^+ + e \rightleftharpoons HClO_2$	1.277
$HClO_2 + 2H^+ + 2e \rightleftharpoons HClO + H_2O$	1.645
$HClO_2 + 3H^+ + 3e \rightleftharpoons 1/2Cl_2 + 2H_2O$	1.628
$HClO_2 + 3H^+ + 4e \rightleftharpoons Cl^- + 2H_2O$	1.570
$ClO_2^- + H_2O + 2e \rightleftharpoons ClO^- + 2OH^-$	0.66
$ClO_2^- + 2H_2O + 4e \rightleftharpoons Cl^- + 4OH^-$	0.76
$ClO_2(aq) + e \rightleftharpoons ClO_2^-$	0.954
$ClO_3^- + 2H^+ + e \rightleftharpoons ClO_2 + H_2O$	1.152
$ClO_3^- + 3H^+ + 2e \rightleftharpoons HClO_2 + H_2O$	1.214
$ClO_3^- + 6H^+ + 5e \rightleftharpoons 1/2Cl_2 + 3H_2O$	1.47
$ClO_3^- + 6H^+ + 6e \rightleftharpoons Cl^- + 3H_2O$	1.451
$ClO_3^- + H_2O + 2e \rightleftharpoons ClO_2^- + 2OH^-$	0.33
$ClO_3^- + 3H_2O + 6e \rightleftharpoons Cl^- + 6OH^-$	0.62
$ClO_4^- + 2H^+ + 2e \rightleftharpoons ClO_3^- \ H_2O$	1.189
$ClO_4^- + 8H^+ + 7e \rightleftharpoons 1/2Cl_2 + 4H_2O$	1.39
$ClO_4^- + 8H^+ + 8e \rightleftharpoons Cl^- + 4H_2O$	1.389
$ClO_4^- + H_2O + 2e \rightleftharpoons ClO_3^- + 2OH^-$	0.36
$Cm^{4+} + e \rightleftharpoons Cm^{3+}$	3.0
$Cm^{3+} + 3e \rightleftharpoons Cm$	−2.04
$Co^{2+} + 2e \rightleftharpoons Co$	−0.28
$Co^{3+} + e \rightleftharpoons Co^{2+}$	1.92
$[Co(NH_3)_6]^{3+} + e \rightleftharpoons [Co(NH_3)_6]^{2+}$	0.108
$Co(OH)_2 + 2e \rightleftharpoons Co + 2OH^-$	−0.73
$Co(OH)_3 + e \rightleftharpoons Co(OH)_2 + OH^-$	0.17

(*Continued*)

Reaction	$E°$ (V)
$Cr^{2+} + 2e \rightleftharpoons Cr$	−0.913
$Cr^{3+} + e \rightleftharpoons Cr^{2+}$	−0.407
$Cr^{3+} + 3e \rightleftharpoons Cr$	−0.744
$Cr_2O_7^{2-} + 14H^+ + 6e \rightleftharpoons 2Cr^{3+} + 7H_2O$	1.36
$CrO_2^- + 2H_2O + 3e \rightleftharpoons Cr + 4OH^-$	−1.2
$HCrO_4^- + 7H^+ + 3e \rightleftharpoons Cr^{3+} + 4H_2O$	1.350
$CrO_2 + 4H^+ + e \rightleftharpoons Cr^{3+} + 2H_2O$	1.48
$Cr(V) + e \rightleftharpoons Cr(IV)$	1.34
$CrO_4^{2-} + 4H_2O + 3e \rightleftharpoons Cr(OH)_3 + 5OH^-$	−0.13
$Cr(OH)_3 + 3e \rightleftharpoons Cr + 3OH^-$	−1.48
$Cs^+ + e \rightleftharpoons Cs$	−3.026
$Cu^+ + e \rightleftharpoons Cu$	0.521
$Cu^{2+} + e \rightleftharpoons Cu^+$	0.153
$Cu^{2+} + 2e \rightleftharpoons Cu$	0.3419
$Cu^{2+} + 2e \rightleftharpoons Cu(Hg)$	0.345
$Cu^{3+} + e \rightleftharpoons Cu^{2+}$	2.4
$Cu_2O_3 + 6H^+ + 2e \rightleftharpoons 2Cu^{2+} + 3H_2O$	2.0
$Cu^{2+} + 2\ CN^- + e \rightleftharpoons [Cu(CN)_2]^-$	1.103
$CuI_2^- + e \rightleftharpoons Cu + 2I^-$	0.00
$Cu_2O + H_2O + 2e \rightleftharpoons 2Cu + 2OH^-$	−0.360
$Cu(OH)_2 + 2e \rightleftharpoons Cu + 2OH^-$	−0.222
$2\ Cu(OH)_2 + 2e \rightleftharpoons Cu_2O + 2OH^- + H_2O$	−0.080
$2\ D^+ + 2e \rightleftharpoons D_2$	−0.013
$Dy^{2+} + 2e \rightleftharpoons Dy$	−2.2
$Dy^{3+} + 3e \rightleftharpoons Dy$	−2.295
$Dy^{3+} + e \rightleftharpoons Dy^{2+}$	−2.6
$Er^{2+} + 2e \rightleftharpoons Er$	−2.0
$Er^{3+} + 3e \rightleftharpoons Er$	−2.331
$Er^{3+} + e \rightleftharpoons Er^{2+}$	−3.0
$Es^{3+} + e \rightleftharpoons Es^{2+}$	−1.3
$Es^{3+} + 3e \rightleftharpoons Es$	−1.91
$Es^{2+} + 2e \rightleftharpoons Es$	−2.23
$Eu^{2+} + 2e \rightleftharpoons Eu$	−2.812
$Eu^{3+} + 3e \rightleftharpoons Eu$	−1.991
$Eu^{3+} + e \rightleftharpoons Eu^{2+}$	−0.36
$F_2 + 2H^+ + 2e \rightleftharpoons 2HF$	3.053
$F_2 + 2e \rightleftharpoons 2F^-$	2.866
$F_2O + 2H^+ + 4e \rightleftharpoons H_2O + 2F^-$	2.153
$Fe^{2+} + 2e \rightleftharpoons Fe$	−0.447
$Fe^{3+} + 3e \rightleftharpoons Fe$	−0.037
$Fe^{3+} + e \rightleftharpoons Fe^{2+}$	0.771
$2HFeO_4^- + 8H^+ + 6e \rightleftharpoons Fe_2O_3 + 5H_2O$	2.09
$HFeO_4^- + 4H^+ + 3e \rightleftharpoons FeOOH + 2H_2O$	2.08

(Continued)

Reaction	$E°$ (V)
$HFeO_4^- + 7H^+ + 3e \rightleftharpoons Fe^{3+} + 4H_2O$	2.07
$Fe_2O_3 + 4H^+ + 2e \rightleftharpoons 2FeOH^+ + H_2O$	0.16
$[Fe(CN)_6]^{3-} + e \rightleftharpoons [Fe(CN)_6]^{4-}$	0.358
$FeO_4^{2-} + 8H^+ + 3e \ Fe^{3+} + 4H_2O$	2.20
$[Fe(bipy)_2]^{3+} + e \rightleftharpoons Fe(bipy)_2]^{2+}$	0.78
$[Fe(bipy)_3]^{3+} + e \rightleftharpoons Fe(bipy)_3]^{2+}$	1.03
$Fe(OH)_3 + e \rightleftharpoons Fe(OH)_2 + OH^-$	−0.56
$[Fe(phen)_3]^{3+} + e \rightleftharpoons [Fe(phen)_3]^{2+}$	1.147
$[Fe(phen)_3]^{3+} + e \rightleftharpoons [Fe(phen)_3]^{2+}$ (1 molar H_2SO_4)	1.06
$[Ferricinium]^+ + e \rightleftharpoons ferrocene$	0.400
$Fm^{3+} + e \rightleftharpoons Fm^{2+}$	−1.1
$Fm^{3+} + 3e \rightleftharpoons Fm$	−1.89
$Fm^{2+} + 2e \rightleftharpoons Fm$	−2.30
$Fr^+ + e \rightleftharpoons Fr$	−2.9
$Ga^{3+} + 3e \rightleftharpoons Ga$	−0.549
$Ga^+ + e \rightleftharpoons Ga$	−0.2
$GaOH^{2+} + H^+ + 3e \rightleftharpoons Ga + H_2O$	−0.498
$H_2GaO_3^- + H_2O + 3e \rightleftharpoons Ga + 4OH^-$	−1.219
$Gd^{3+} + 3e \rightleftharpoons Gd$	−2.279
$Ge^{2+} + 2e \rightleftharpoons Ge$	0.24
$Ge^{4+} + 4e \rightleftharpoons Ge$	0.124
$Ge^{4+} + 2e \rightleftharpoons Ge^{2+}$	0.00
$GeO_2 + 2H^+ + 2e \rightleftharpoons GeO + H_2O$	−0.118
$H_2GeO_3 + 4H^+ + 4e \rightleftharpoons Ge + 3H_2O$	−0.182
$2H^+ + 2e \rightleftharpoons H_2$	0.00000
$H_2 + 2e \rightleftharpoons 2H^-$	−2.23
$HO_2 + H^+ + e \rightleftharpoons H_2O_2$	1.495
$2H_2O + 2e \rightleftharpoons H_2 + 2OH^-$	−0.8277
$H_2O_2 + 2H^+ + 2e \rightleftharpoons 2H_2O$	1.776
$Hf^{4+} + 4e \rightleftharpoons Hf$	−1.55
$HfO^{2+} + 2H^+ + 4e \rightleftharpoons Hf + H_2O$	−1.724
$HfO_2 + 4H^+ + 4e \rightleftharpoons Hf + 2H_2O$	−1.505
$HfO(OH)_2 + H_2O + 4e \rightleftharpoons Hf + 4OH^-$	−2.50
$Hg^{2+} + 2e \rightleftharpoons Hg$	0.851
$2Hg^{2+} + 2e \rightleftharpoons Hg_2^{2+}$	0.920
$Hg_2^{2+} + 2e \rightleftharpoons 2Hg$	0.7973
$Hg_2(ac)_2 + 2e \rightleftharpoons 2Hg + 2(ac)^-$	0.51163
$Hg_2Br_2 + 2e \rightleftharpoons 2Hg + 2Br^-$	0.13923
$Hg_2Cl_2 + 2e \rightleftharpoons 2Hg + 2Cl^-$	0.26808
$Hg_2HPO_4 + 2e \rightleftharpoons 2Hg + HPO_4^{2-}$	0.6359
$Hg_2I_2 + 2e \rightleftharpoons 2Hg + 2I^-$	−0.0405
$Hg_2O + H_2O + 2e \rightleftharpoons 2Hg + 2OH^-$	0.123
$HgO + H_2O + 2e \rightleftharpoons Hg + 2OH^-$	0.0977

(Continued)

Reaction	$E°$ (V)
$Hg(OH)_2 + 2H^+ + 2e \rightleftharpoons Hg + 2H_2O$	1.034
$Hg_2SO_4 + 2e \rightleftharpoons 2Hg + SO_4^{2-}$	0.6125
$Ho^{2+} + 2e \rightleftharpoons Ho$	−2.1
$Ho^{3+} + 3e \rightleftharpoons Ho$	−2.33
$Ho^{3+} + e \rightleftharpoons Ho^{2+}$	−2.8
$I_2 + 2e \rightleftharpoons 2 I^-$	0.5355
$I_3^- + 2e \rightleftharpoons 3 I^-$	0.536
$H_3IO_6^{2-} + 2e \rightleftharpoons IO_3^- + 3OH^-$	0.7
$H_5IO_6 + H^+ + 2e \rightleftharpoons IO_3^- + 3H_2O$	1.601
$2HIO + 2H^+ + 2e \rightleftharpoons I_2 + 2H_2O$	1.439
$HIO + H^+ + 2e \rightleftharpoons I^- + H_2O$	0.987
$IO^- + H_2O + 2e \rightleftharpoons I^- + 2OH^-$	0.485
$2 IO_3^- + 12H^+ + 10e \rightleftharpoons I_2 + 6H_2O$	1.195
$IO_3^- + 6H^+ + 6e \rightleftharpoons I^- + 3H_2O$	1.085
$IO_3^- + 2H_2O + 4e \rightleftharpoons IO^- + 4OH^-$	0.15
$IO_3^- + 3H_2O + 6e \rightleftharpoons IO^- + 6OH^-$	0.26
$In^+ + e \rightleftharpoons In$	−0.14
$In^{2+} + e \rightleftharpoons In^+$	−0.40
$In^{3+} + e \rightleftharpoons In^{2+}$	−0.49
$In^{3+} + 2e \rightleftharpoons In^+$	−0.443
$In^{3+} + 3e \rightleftharpoons In$	−0.3382
$In(OH)_3 + 3e \rightleftharpoons In + 3OH^-$	−0.99
$In(OH)_4^- + 3e \rightleftharpoons In + 4OH^-$	−1.007
$In_2O_3 + 3H_2O + 6e \rightleftharpoons 2In + 6OH^-$	−1.034
$Ir^{3+} + 3e \rightleftharpoons Ir$	1.156
$[IrCl_6]^{2-} + e \rightleftharpoons [IrCl_6]^{3-}$	0.8665
$[IrCl_6]^{3-} + 3e \rightleftharpoons Ir + 6Cl^-$	0.77
$Ir_2O_3 + 3H_2O + 6e \rightleftharpoons 2Ir + 6OH^-$	0.098
$K^+ + e \rightleftharpoons K$	−2.931
$La^{3+} + 3e \rightleftharpoons La$	−2.379
$La(OH)_3 + 3e \rightleftharpoons La + 3OH^-$	−2.90
$Li^+ + e \rightleftharpoons Li$	−3.0401
$Lr^{3+} + 3e \rightleftharpoons Lr$	−1.96
$Lu^{3+} + 3e \rightleftharpoons Lu$	−2.28
$Md^{3+} + e \rightleftharpoons Md^{2+}$	−0.1
$Md^{3+} + 3e \rightleftharpoons Md$	−1.65
$Md^{2+} + 2e \rightleftharpoons Md$	−2.40
$Mg^+ + e \rightleftharpoons Mg$	−2.70
$Mg^{2+} + 2e \rightleftharpoons Mg$	−2.372
$Mg(OH)_2 + 2e \rightleftharpoons Mg + 2OH^-$	−2.690
$Mn^{2+} + 2e \rightleftharpoons Mn$	−1.185
$Mn^{3+} + e \rightleftharpoons Mn^{2+}$	1.5415
$MnO_2 + 4H^+ + 2e \rightleftharpoons Mn^{2+} + 2H_2O$	1.224

(Continued)

Reaction	$E°$ (V)
$MnO_4^- + e \rightleftharpoons MnO_4^{2-}$	0.558
$MnO_4^- + 4H^+ + 3e \rightleftharpoons MnO_2 + 2H_2O$	1.679
$MnO_4^- + 8H^+ + 5e \rightleftharpoons Mn^{2+} + 4H_2O$	1.507
$MnO_4^- + 2H_2O + 3e \rightleftharpoons MnO_2 + 4OH^-$	0.595
$MnO_4^- + 4H_2O + 5e^- \rightleftharpoons Mn(OH)_2 + 6OH^-$	0.34
$MnO_4^{2-} + 2H_2O + 2e \rightleftharpoons MnO_2 + 4OH^-$	0.60
$Mn(OH)_2 + 2e \rightleftharpoons Mn + 2OH^-$	−1.56
$Mn(OH)_3 + e \rightleftharpoons Mn(OH)_2 + OH^-$	0.15
$Mn_2O_3 + 6H^+ + e \rightleftharpoons 2\,Mn^{2+} + 3H_2O$	1.485
$Mo^{3+} + 3e \rightleftharpoons Mo$	−0.200
$MoO_2 + 4H^+ + 4e \rightleftharpoons Mo + 4H_2O$	−0.152
$H_3Mo_7O_{24}^{3-} + 45H^+ + 42e \rightleftharpoons 7\,Mo + 24H_2O$	0.082
$MoO_3 + 6H^+ + 6e \rightleftharpoons Mo + 3H_2O$	0.075
$N_2 + 2H_2O + 6H^+ + 6e \rightleftharpoons 2\,NH_4OH$	0.092
$3\,N_2 + 2H^+ + 2e \rightleftharpoons 2HN_3$	−3.09
$N_5^+ + 3H^+ + 2e \rightleftharpoons 2\,NH_4^+$	1.275
$N_2O + 2H^+ + 2e \rightleftharpoons N_2 + H_2O$	1.766
$H_2N_2O_2 + 2H^+ + 2e \rightleftharpoons N_2 + 2H_2O$	2.65
$N_2O_4 + 2e \rightleftharpoons 2\,NO_2^-$	0.867
$N_2O_4 + 2H^+ + 2e \rightleftharpoons 2\,NHO_2$	1.065
$N_2O_4 + 4H^+ + 4e \rightleftharpoons 2\,NO + 2H_2O$	1.035
$2\,NH_3OH^+ + H^+ + 2e \rightleftharpoons N_2H_5^+ + 2H_2O$	1.42
$2\,NO + 2H^+ + 2e \rightleftharpoons N_2O + H_2O$	1.591
$2\,NO + H_2O + 2e \rightleftharpoons N_2O + 2OH^-$	0.76
$HNO_2 + H^+ + e \rightleftharpoons NO + H_2O$	0.983
$2HNO_2 + 4H^+ + 4e \rightleftharpoons H_2N_2O_2 + 2H_2O$	0.86
$2HNO_2 + 4H^+ + 4e \rightleftharpoons N_2O + 3H_2O$	1.297
$NO_2^- + H_2O + e \rightleftharpoons NO + 2OH^-$	−0.46
$2\,NO_2^- + 2H_2O + 4e \rightleftharpoons N_2O_2^{2-} + 4OH^-$	−0.18
$2\,NO_2^- + 3H_2O + 4e \rightleftharpoons N_2O + 6OH^-$	0.15
$NO_3^- + 3H^+ + 2e \rightleftharpoons HNO_2 + H_2O$	0.934
$NO_3^- + 4H^+ + 3e \rightleftharpoons NO + 2H_2O$	0.957
$2\,NO_3^- + 4H^+ + 2e \rightleftharpoons N_2O_4 + 2H_2O$	0.803
$NO_3^- + H_2O + 2e \rightleftharpoons NO_2^- + 2OH^-$	0.01
$2\,NO_3^- + 2H_2O + 2e \rightleftharpoons N_2O_4 + 4OH^-$	−0.85
$Na^+ + e \rightleftharpoons Na$	−2.71
$Nb^{3+} + 3e \rightleftharpoons Nb$	−1.099
$NbO_2 + 2H^+ + 2e \rightleftharpoons NbO + H_2O$	−0.646
$NbO_2 + 4H^+ + 4e \rightleftharpoons Nb + 2H_2O$	−0.690
$NbO + 2H^+ + 2e \rightleftharpoons Nb + H_2O$	−0.733
$Nb_2O_5 + 10H^+ + 10e \rightleftharpoons 2\,Nb + 5H_2O$	−0.644
$Nd^{3+} + 3e \rightleftharpoons Nd$	−2.323
$Nd^{2+} + 2e \rightleftharpoons Nd$	−2.1

(*Continued*)

Reaction	$E°$ (V)
$Nd^{3+} + e \rightleftharpoons Nd^{2+}$	−2.7
$Ni^{2+} + 2e \rightleftharpoons Ni$	−0.257
$Ni(OH)_2 + 2e \rightleftharpoons Ni + 2OH^-$	−0.72
$NiO_2 + 4H^+ + 2e \rightleftharpoons Ni^{2+} + 2H_2O$	1.678
$NiO_2 + 2H_2O + 2e \rightleftharpoons Ni(OH)_2 + 2OH^-$	−0.490
$No^{3+} + e \rightleftharpoons No^{2+}$	1.4
$No^{3+} + 3e \rightleftharpoons No$	−1.20
$No^{2+} + 2e \rightleftharpoons No$	−2.50
$Np^{3+} + 3e \rightleftharpoons Np$	−1.856
$Np^{4+} + e \rightleftharpoons Np^{3+}$	0.147
$NpO_2 + H_2O + H^+ + e \rightleftharpoons Np(OH)_3$	−0.962
$O_2 + 2H^+ + 2e \rightleftharpoons H_2O_2$	0.695
$O_2 + 4H^+ + 4e \rightleftharpoons 2H_2O$	1.229
$O_2 + H_2O + 2e \rightleftharpoons HO_2^- + OH^-$	−0.076
$O_2 + 2H_2O + 2e \rightleftharpoons H_2O_2 + 2OH^-$	−0.146
$O_2 + 2H_2O + 4e \rightleftharpoons 4OH^-$	0.401
$O_3 + 2H^+ + 2e \rightleftharpoons O_2 + H_2O$	2.076
$O_3 + H_2O + 2e \rightleftharpoons O_2 + 2OH^-$	1.24
$O(g) + 2H^+ + 2e \rightleftharpoons H_2O$	2.421
$OH + e \rightleftharpoons OH^-$	2.02
$HO_2^- + H_2O + 2e \rightleftharpoons 3OH^-$	0.878
$OsO_4 + 8H^+ + 8e \rightleftharpoons Os + 4H_2O$	0.838
$OsO_4 + 4H^+ + 4e \rightleftharpoons OsO_2 + 2H_2O$	1.02
$[Os(bipy)_2]^{3+} + e \rightleftharpoons [Os(bipy)_2]^{2+}$	0.81
$[Os(bipy)_3]^{3+} + e \rightleftharpoons [Os(bipy)_3]^{2+}$	0.80
$P(red) + 3H^+ + 3e \rightleftharpoons PH_3(g)$	−0.111
$P(white) + 3H^+ + 3e \rightleftharpoons PH_3(g)$	−0.063
$P + 3H_2O + 3e \rightleftharpoons PH_3(g) + 3OH^-$	−0.87
$H_2P_2^- + e \rightleftharpoons P + 2OH^-$	−1.82
$H_3PO_2 + H^+ + e \rightleftharpoons P + 2H_2O$	−0.508
$H_3PO_3 + 2H^+ + 2e \rightleftharpoons H_3PO_2 + H_2O$	−0.499
$H_3PO_3 + 3H^+ + 3e \rightleftharpoons P + 3H_2O$	−0.454
$HPO_3^{2-} + 2H_2O + 2e \rightleftharpoons H_2PO_2^- + 3OH^-$	−1.65
$HPO_3^{2-} + 2H_2O + 3e \rightleftharpoons P + 5OH^-$	−1.71
$H_3PO_4 + 2H^+ + 2e \rightleftharpoons H_3PO_3 + H_2O$	−0.276
$PO_4^{3-} + 2H_2O + 2e \rightleftharpoons HPO_3^{2-} + 3OH^-$	−1.05
$Pa^{3+} + 3e \rightleftharpoons Pa$	−1.34
$Pa^{4+} + 4e \rightleftharpoons Pa$	−1.49
$Pa^{4+} + e \rightleftharpoons Pa^{3+}$	−1.9
$Pb^{2+} + 2e \rightleftharpoons Pb$	−0.1262
$Pb^{2+} + 2e \rightleftharpoons Pb(Hg)$	−0.1205
$PbBr_2 + 2e \rightleftharpoons Pb + 2Br^-$	−0.284
$PbCl_2 + 2e \rightleftharpoons Pb + 2Cl^-$	−0.2675

(Continued)

Reaction	$E°$ (V)
$PbF_2 + 2e \rightleftharpoons Pb + 2F^-$	−0.3444
$PbHPO_4 + 2e \rightleftharpoons Pb + HPO_4^{2-}$	−0.465
$PbI_2 + 2e \rightleftharpoons Pb + 2 I^-$	−0.365
$PbO + H_2O + 2e \rightleftharpoons Pb + 2OH^-$	−0.580
$PbO_2 + 4H^+ + 2e \rightleftharpoons Pb^{2+} + 2H_2O$	1.455
$HPbO_2^- + H_2O + 2e \rightleftharpoons Pb + 3OH^-$	−0.537
$PbO_2 + H_2O + 2e \rightleftharpoons PbO + 2OH^-$	0.247
$PbO_2 + SO_4^{2-} + 4H^+ + 2e \rightleftharpoons PbSO_4 + 2H_2O$	1.6913
$PbSO_4 + 2e \rightleftharpoons Pb + SO_4^{2-}$	−0.3588
$PbSO_4 + 2e \rightleftharpoons Pb(Hg) + SO_4^{2-}$	−0.3505
$Pd^{2+} + 2e \rightleftharpoons Pd$	0.951
$[PdCl_4]^{2-} + 2e \rightleftharpoons Pd + 4 Cl^-$	0.591
$[PdCl_6]^{2-} + 2e \rightleftharpoons [PdCl_4]^{2-} + 2 Cl^-$	1.288
$Pd(OH)_2 + 2e \rightleftharpoons Pd + 2OH^-$	0.07
$Pm^{2+} + 2e \rightleftharpoons Pm$	−2.2
$Pm^{3+} + 3e \rightleftharpoons Pm$	−2.30
$Pm^{3+} + e \rightleftharpoons Pm^{2+}$	−2.6
$Po^{2+} + 2e^- \rightleftharpoons Po$	0.368
$PoO_2 + 4H^+ + 2e^- \rightleftharpoons Po^{2+} + 2H_2O$	1.095
$Pr^{4+} + e \rightleftharpoons Pr^{3+}$	3.2
$Pr^{2+} + 2e \rightleftharpoons Pr$	−2.0
$Pr^{3+} + 3e \rightleftharpoons Pr$	−2.353
$Pr^{3+} + e \rightleftharpoons Pr^{2+}$	−3.1
$Pt^{2+} + 2e \rightleftharpoons Pt$	1.18
$[PtCl_4]^{2-} + 2e \rightleftharpoons Pt + 4 Cl^-$	0.755
$[PtCl_6]^{2-} + 2e \rightleftharpoons [PtCl_4]^{2-} + 2 Cl^-$	0.68
$Pt(OH)_2 + 2e \rightleftharpoons Pt + 2OH^-$	0.14
$PtO_3 + 2H^+ + 2e \rightleftharpoons PtO_2 + H_2O$	1.7
$PtO_3 + 4H^+ + 2e \rightleftharpoons Pt(OH)_2^{2+} + H_2O$	1.5
$PtOH^+ + H^+ + 2e \rightleftharpoons Pt + H_2O$	1.2
$PtO_2 + 2H^+ + 2e \rightleftharpoons PtO + H_2O$	1.01
$PtO_2 + 4H^+ + 4e \rightleftharpoons Pt + 2H_2O$	1.00
$Pu^{3+} + 3e \rightleftharpoons Pu$	−2.031
$Pu^{4+} + e \rightleftharpoons Pu^{3+}$	1.006
$Pu^{5+} + e \rightleftharpoons Pu^{4+}$	1.099
$PuO_2(OH)_2 + 2H^+ + 2e \rightleftharpoons Pu(OH)_4$	1.325
$PuO_2(OH)_2 + H^+ + e \rightleftharpoons PuO_2OH + H_2O$	1.062
$Ra^{2+} + 2e \rightleftharpoons Ra$	−2.8
$Rb^+ + e \rightleftharpoons Rb$	−2.98
$Re^{3+} + 3e \rightleftharpoons Re$	0.300
$ReO_4^- + 4H^+ + 3e \rightleftharpoons ReO_2 + 2H_2O$	0.510
$ReO_2 + 4H^+ + 4e \rightleftharpoons Re + 2H_2O$	0.2513
$ReO_4^- + 2H^+ + e \rightleftharpoons ReO_3 + H_2O$	0.768

(*Continued*)

Reaction	$E°$ (V)
$ReO_4^- + 4H_2O + 7e \rightleftharpoons Re + 8OH^-$	-0.604
$ReO_4^- + 8H^+ + 7e \rightleftharpoons Re + 4H_2O$	0.34
$Rh^+ + e \rightleftharpoons Rh$	0.600
$Rh^{3+} + 3e \rightleftharpoons Rh$	0.758
$[RhCl_6]^{3-} + 3e \rightleftharpoons Rh + 6\ Cl^-$	0.431
$RhOH^{2+} + H^+ + 3e \rightleftharpoons Rh + H_2O$	0.83
$Ru^{2+} + 2e \rightleftharpoons Ru$	0.455
$Ru^{3+} + e \rightleftharpoons Ru^{2+}$	0.2487
$RuO_2 + 4H^+ + 2e \rightleftharpoons Ru^{2+} + 2H_2O$	1.120
$RuO_4^- + e \rightleftharpoons RuO_4^{2-}$	0.59
$RuO_4 + e \rightleftharpoons RuO_4^-$	1.00
$RuO_4 + 6H^+ + 4e \rightleftharpoons Ru(OH)_2^{2+} + 2H_2O$	1.40
$RuO_4 + 8H^+ + 8e \rightleftharpoons Ru + 4H_2O$	1.038
$[Ru(bipy)_3]^{3+} + e^- \rightleftharpoons [Ru(bipy)_3]^{2+}$	1.24
$[Ru(H_2O)_6]^{3+} + e^- \rightleftharpoons [Ru(H_2O)_6]^{2+}$	0.23
$[Ru(NH_3)_6]^{3+} + e^- \rightleftharpoons [Ru(NH_3)_6]^{2+}$	0.10
$[Ru(en)_3]^{3+} + e^- \rightleftharpoons [Ru(en)_3]^{2+}$	0.210
$[Ru(CN)_6]^{3-} + e^- \rightleftharpoons [Ru(CN)_6]^{4-}$	0.86
$S + 2e \rightleftharpoons S^{2-}$	-0.47627
$S + 2H^+ + 2e \rightleftharpoons H_2S(aq)$	0.142
$S + H_2O + 2e \rightleftharpoons SH^- + OH^-$	-0.478
$2\ S + 2e \rightleftharpoons S_2^{2-}$	-0.42836
$S_2O_6^{2-} + 4H^+ + 2e \rightleftharpoons 2H_2SO_3$	0.564
$S_2O_8^{2-} + 2e \rightleftharpoons 2\ SO_4^{2-}$	2.010
$S_2O_8^{2-} + 2H^+ + 2e \rightleftharpoons 2HSO_4^-$	2.123
$S_4O_6^{2-} + 2e \rightleftharpoons 2\ S_2O_3^{2-}$	0.08
$2H_2SO_3 + H^+ + 2e \rightleftharpoons HS_2O_4^- + 2H_2O$	-0.056
$H_2SO_3 + 4H^+ + 4e \rightleftharpoons S + 3H_2O$	0.449
$2\ SO_3^{2-} + 2H_2O + 2e \rightleftharpoons S_2O_4^{2-} + 4OH^-$	-1.12
$2\ SO_3^{2-} + 3H_2O + 4e \rightleftharpoons S_2O_3^{2-} + 6OH^-$	-0.571
$SO_4^{2-} + 4H^+ + 2e \rightleftharpoons H_2SO_3 + H_2O$	0.172
$2\ SO_4^{2-} + 4H^+ + 2e \rightleftharpoons S_2O_6^{2-} + H_2O$	-0.22
$SO_4^{2-} + H_2O + 2e \rightleftharpoons SO_3^{2-} + 2OH^-$	-0.93
$Sb + 3H^+ + 3e \rightleftharpoons SbH_3$	-0.510
$Sb_2O_3 + 6H^+ + 6e \rightleftharpoons 2\ Sb + 3H_2O$	0.152
Sb_2O_5 (senarmontite) $+ 4H^+ + 4e \rightleftharpoons Sb_2O_3 + 2H_2O$	0.671
Sb_2O_5 (valentinite) $+ 4H^+ + 4e \rightleftharpoons Sb_2O_3 + 2H_2O$	0.649
$Sb_2O_5 + 6H^+ + 4e \rightleftharpoons 2\ SbO^+ + 3H_2O$	0.581
$SbO^+ + 2H^+ + 3e \rightleftharpoons Sb + 2H_2O$	0.212
$SbO_2^- + 2H_2O + 3e \rightleftharpoons Sb + 4OH^-$	-0.66
$SbO_3^- + H_2O + 2e \rightleftharpoons SbO_2^- + 2OH^-$	-0.59
$Sc^{3+} + 3e \rightleftharpoons Sc$	-2.077
$Se + 2e \rightleftharpoons Se^{2-}$	-0.670
$Se + 2H^+ + 2e \rightleftharpoons H_2Se(aq)$	-0.399

(Continued)

Reaction	$E°$ (V)
$H_2SeO_3 + 4H^+ + 4e \rightleftharpoons Se + 3H_2O$	0.74
$Se + 2H^+ + 2e \rightleftharpoons H_2Se$	−0.082
$SeO_3^{2-} + 3H_2O + 4e \rightleftharpoons Se + 6OH^-$	−0.366
$SeO_4^{2-} + 4H^+ + 2e \rightleftharpoons H_2SeO_3 + H_2O$	1.151
$SeO_4^{2-} + H_2O + 2e \rightleftharpoons SeO_3^{2-} + 2OH^-$	0.05
$SiF_6^{2-} + 4e \rightleftharpoons Si + 6F^-$	−1.24
$SiO + 2H^+ + 2e \rightleftharpoons Si + H_2O$	−0.8
$SiO_2 \text{ (quartz)} + 4H^+ + 4e \rightleftharpoons Si + 2H_2O$	0.857
$SiO_3^{2-} + 3H_2O + 4e \rightleftharpoons Si + 6OH^-$	−1.697
$Sm^{3+} + e \rightleftharpoons Sm^{2+}$	−1.55
$Sm^{3+} + 3e \rightleftharpoons Sm$	−2.304
$Sm^{2+} + 2e \rightleftharpoons Sm$	−2.68
$Sn^{2+} + 2e \rightleftharpoons Sn$	−0.1375
$Sn^{4+} + 2e \rightleftharpoons Sn^{2+}$	0.151
$Sn(OH)_3^+ + 3H^+ + 2e \rightleftharpoons Sn^{2+} + 3H_2O$	0.142
$SnO_2 + 4H^+ + 2e^- \rightleftharpoons Sn^{2+} + 2H_2O$	−0.094
$SnO_2 + 4H^+ + 4e \rightleftharpoons Sn + 2H_2O$	−0.117
$SnO_2 + 3H^+ + 2e \rightleftharpoons SnOH^+ + H_2O$	−0.194
$SnO_2 + 2H_2O + 4e \rightleftharpoons Sn + 4OH^-$	−0.945
$HSnO_2^- + H_2O + 2e \rightleftharpoons Sn + 3OH^-$	−0.909
$Sn(OH)_6^{2-} + 2e \rightleftharpoons HSnO_2^- + 3OH^- + H_2O$	−0.93
$Sr^+ + e \rightleftharpoons Sr$	−4.10
$Sr^{2+} + 2e \rightleftharpoons Sr$	−2.899
$Sr^{2+} + 2e \rightleftharpoons Sr(Hg)$	−1.793
$Sr(OH)_2 + 2e \rightleftharpoons Sr + 2OH^-$	−2.88
$Ta_2O_5 + 10H^+ + 10e \rightleftharpoons 2\,Ta + 5H_2O$	−0.750
$Ta^{3+} + 3e \rightleftharpoons Ta$	−0.6
$Tc^{2+} + 2e \rightleftharpoons Tc$	0.400
$TcO_4^- + 4H^+ + 3e \rightleftharpoons TcO_2 + 2H_2O$	0.728
$Tc^{3+} + e \rightleftharpoons Tc^{2+}$	0.3
$TcO_4^- + 8H^+ + 7e \rightleftharpoons Tc + 4H_2O$	0.472
$Tb^{4+} + e \rightleftharpoons Tb^{3+}$	3.1
$Tb^{3+} + 3e \rightleftharpoons Tb$	−2.28
$Te + 2e \rightleftharpoons Te^{2-}$	−1.143
$Te + 2H^+ + 2e \rightleftharpoons H_2Te$	−0.793
$Te^{4+} + 4e \rightleftharpoons Te$	0.568
$TeO_2 + 4H^+ + 4e \rightleftharpoons Te + 2H_2O$	0.593
$TeO_3^{2-} + 3H_2O + 4e \rightleftharpoons Te + 6OH^-$	−0.57
$TeO_4^- + 8H^+ + 7e \rightleftharpoons Te + 4H_2O$	0.472
$H_6TeO_6 + 2H^+ + 2e \rightleftharpoons TeO_2 + 4H_2O$	1.02
$Th^{4+} + 4e \rightleftharpoons Th$	−1.899
$ThO_2 + 4H^+ + 4e \rightleftharpoons Th + 2H_2O$	−1.789
$Th(OH)_4 + 4e \rightleftharpoons Th + 4OH^-$	−2.48
$Ti^{2+} + 2e \rightleftharpoons Ti$	−1.628

(Continued)

Reaction	$E°$ (V)
$Ti^{3+} + e \rightleftharpoons Ti^{2+}$	-0.369
$TiO_2 + 4H^+ + 2e \rightleftharpoons Ti^{2+} + 2H_2O$	-0.502
$Ti^{3+} + 3e \rightleftharpoons Ti$	-1.209
$TiOH^{3+} + H^+ + e \rightleftharpoons Ti^{3+} + H_2O$	-0.055
$Tl^+ + e \rightleftharpoons Tl$	-0.336
$Tl^+ + e \rightleftharpoons Tl(Hg)$	-0.3338
$Tl^{3+} + 2e \rightleftharpoons Tl^+$	1.252
$Tl^{3+} + 3e \rightleftharpoons Tl$	0.741
$TlBr + e \rightleftharpoons Tl + Br^-$	-0.658
$TlCl + e \rightleftharpoons Tl + Cl^-$	-0.5568
$TlI + e \rightleftharpoons Tl + I^-$	-0.752
$Tl_2O_3 + 3H_2O + 4e \rightleftharpoons 2\,Tl^+ + 6OH^-$	0.02
$TlOH + e \rightleftharpoons Tl + OH^-$	-0.34
$Tl(OH)_3 + 2e \rightleftharpoons TlOH + 2OH^-$	-0.05
$Tl_2SO_4 + 2e \rightleftharpoons Tl + SO_4^{2-}$	-0.4360
$Tm^{3+} + e \rightleftharpoons Tm^{2+}$	-2.2
$Tm^{3+} + 3e \rightleftharpoons Tm$	-2.319
$Tm^{2+} + 2e \rightleftharpoons Tm$	-2.4
$U^{3+} + 3e \rightleftharpoons U$	-1.66
$U^{4+} + e \rightleftharpoons U^{3+}$	-0.52
$UO_2^+ + 4H^+ + e \rightleftharpoons U^{4+} + 2H_2O$	0.612
$UO_2^{2+} + e \rightleftharpoons UO^+_2$	0.16
$UO_2^{2+} + 4H^+ + 2e \rightleftharpoons U^{4+} + 2H_2O$	0.327
$UO_2^{2+} + 4H^+ + 6e \rightleftharpoons U + 2H_2O$	-1.444
$V^{2+} + 2e \rightleftharpoons V$	-1.175
$V^{3+} + e \rightleftharpoons V^{2+}$	-0.255
$VO^{2+} + 2H^+ + e \rightleftharpoons V^{3+} + H_2O$	0.337
$VO_2^+ + 2H^+ + e \rightleftharpoons VO^{2+} + H_2O$	0.991
$V_2O_5 + 6H^+ + 2e \rightleftharpoons 2VO^{2+} + 3H_2O$	0.957
$V_2O_5 + 10H^+ + 10e \rightleftharpoons 2V + 5H_2O$	-0.242
$V(OH)_4^+ + 2H^+ + e \rightleftharpoons VO^{2+} + 3H_2O$	1.00
$V(OH)_4^+ + 4H^+ + 5e \rightleftharpoons V + 4H_2O$	-0.254
$[V(phen)_3]^{3+} + e \rightleftharpoons [V(phen)_3]^{2+}$	0.14
$W^{3+} + 3e \rightleftharpoons W$	0.1
$W_2O_5 + 2H^+ + 2e \rightleftharpoons 2\,WO_2 + H_2O$	-0.031
$WO_2 + 4H^+ + 4e \rightleftharpoons W + 2H_2O$	-0.119
$WO_3 + 6H^+ + 6e \rightleftharpoons W + 3H_2O$	-0.090
$WO_3 + 2H^+ + 2e \rightleftharpoons WO_2 + H_2O$	0.036
$2\,WO_3 + 2H^+ + 2e \rightleftharpoons W_2O_5 + H_2O$	-0.029
$H_4XeO_6 + 2H^+ + 2e \rightleftharpoons XeO_3 + 3H_2O$	2.42
$XeO_3 + 6H^+ + 6e \rightleftharpoons Xe + 3H_2O$	2.10
$XeF + e \rightleftharpoons Xe + F^-$	3.4
$Y^{3+} + 3e \rightleftharpoons Y$	-2.372

(Continued)

Reaction	$E°$ (V)
$Yb^{3+} + e \rightleftharpoons Yb^{2+}$	−1.05
$Yb^{3+} + 3e \rightleftharpoons Yb$	−2.19
$Yb^{2+} + 2e \rightleftharpoons Yb$	−2.76
$Zn^{2+} + 2e \rightleftharpoons Zn$	−0.7618
$Zn^{2+} + 2e \rightleftharpoons Zn(Hg)$	−0.7628
$ZnO_2^{2-} + 2H_2O + 2e \rightleftharpoons Zn + 4OH^-$	−1.215
$ZnSO_4 \cdot 7H_2O + 2e = Zn(Hg) + SO_4^{2-} + 7H_2O$ (Saturated $ZnSO_4$)	−0.7993
$ZnOH^+ + H^+ + 2e \rightleftharpoons Zn + H_2O$	−0.497
$Zn(OH)_4^{2-} + 2e \rightleftharpoons Zn + 4OH^-$	−1.199
$Zn(OH)_2 + 2e \rightleftharpoons Zn + 2OH^-$	−1.249
$ZnO + H_2O + 2e \rightleftharpoons Zn + 2OH^-$	−1.260
$ZrO_2 + 4H^+ + 4e \rightleftharpoons Zr + 2H_2O$	−1.553
$ZrO(OH)_2 + H_2O + 4e \rightleftharpoons Zr + 4OH^-$	−2.36
$Zr^{4+} + 4e \rightleftharpoons Zr$	−1.45

10.14   DISSOCIATION CONSTANTS OF INORGANIC ACIDS AND BASES[15]

The data in this table are presented as values of pK_a, defined as the negative decimal logarithm of the acid dissociation constant K_a for the reaction

$$BH(aq) \rightleftharpoons B^-(aq) + H^+(aq)$$

Thus $pK_a = -\log_{10} K_a$, and the hydrogen ion concentration [H⁺] can be calculated from

$$K_a = \frac{H^+ \; B^-}{[BH]}$$

In the case of bases, the entry in the table is for the conjugate acid; e.g., ammonium ion for ammonia. The OH⁻(aq) concentration in the system

$$NH_3(aq) + H_2O(aq) \rightleftharpoons NH_4^+(aq) + OH^-(aq)$$

can be calculated from the equation

$$K_b = K_{water}/K_a = \frac{OH^- \; NH_4^+}{[NH_3]}$$

where $K_{water} = 1.01 \times 10^{-14}$ at 298.15 K (Table 10.6). Note that $pK_a + pK_b = pK_{water}$.

All values refer to standard aqueous solutions at the temperature indicated. The standard state is the hypothetical ideal solution with molality equals 1 mol kg⁻¹.

[15] Adapted from W.M. Haynes (Editor-in-Chief), *CRC Handbook of Chemistry and Physics*, 95th edition, CRC Press, Boca Raton, 2014, pp. 5–92.

Name	Formula	Step	T (K)	pK_a
Ammonia	NH_3		298.15	9.25
Boric acid	H_3BO_3	1	293.15	9.27
		2	293.15	>14
Carbonic acid	H_2CO_3	1	298.15	6.35
		2	298.15	10.33
Hydrogen sulfide	H_2S	1	298.15	7.05
		2	298.15	19
		3	298.15	12.32
Phosphorous acid	H_3PO_3	1	293.15	1.3
		2	293.15	6.70
		2	298.15	8.32
Silicic acid	H_4SiO_4	1	303.15	9.9
		2	303.15	11.8
		3	303.15	12
		4	303.15	12
Sulfuric acid	H_2SO_4	2	298.15	1.99
Water	H_2O		298.15	13.995

10.15 DISSOCIATION CONSTANTS OF ORGANIC ACIDS AND BASES[16]

This table lists the dissociation (ionization) constants of a number of organic compounds. All data apply to dilute aqueous solutions and are presented as values of pK_a, which is defined as the negative of the decimal logarithm of the equilibrium constant K_a for the reaction

$$HA(aq) = H^+(aq) + A^-(aq)$$

i.e.,

$$K_a = H^+ \ A^- /[HA]$$

where [H^+], etc. represent the concentrations of the respective species in mol kg^{-1}. It follows that $pK_a = pH + \log[HA] - \log[A^-]$, so that a solution with 50% dissociation has pH equal to the pK_a of the acid.

Data for bases are presented as pK_a values for the conjugate acid, i.e., for the reaction

$$BH^+(aq) = H^+(aq) + B(aq)$$

In older literature, an ionization constant K_b was used for the reaction B(aq) + H_2O (aq) = BH$^+$(aq) + OH$^-$(aq). This is related to K_a by

$$pK_a + pK_b = pK_{water} = 14.00 (at 298.15 K) \quad (Table 10.6)$$

[16] Adapted from W.M. Haynes (Editor-in-Chief), *CRC Handbook of Chemistry and Physics*, 95th edition, CRC Press, Boca Raton, 2014, pp. 5–94.

Mol. form.	Name	Step	T (K)	pK_a
CH_2O_2	Formic acid		298.15	3.75
CH_4O	Methanol		298.15	15.5
$C_2H_2O_4$	Oxalic acid	1	298.15	1.25
		2	298.15	3.81
$C_2H_4O_2$	Acetic acid		298.15	4.756
C_2H_6O	Ethanol		298.15	15.5

10.16   ACTIVITY COEFFICIENTS OF AQUEOUS ACIDS, BASES, AND SALTS[17]

This table gives mean activity coefficients at 298.15 K for molalities in the range of 0.1–1.0 mol kg^{-1}.

	0.1	0.2	0.3	0.4	0.5	0.6	0.7	0.8	0.9	1.0
$AgNO_3$	0.734	0.657	0.606	0.567	0.536	0.509	0.485	0.464	0.446	0.429
$AlCl_3$	0.337	0.305	0.302	0.313	0.331	0.356	0.388	0.429	0.479	0.539
$BaCl_2$	0.500	0.444	0.419	0.405	0.397	0.391	0.391	0.391	0.392	0.395
$CaCl_2$	0.518	0.472	0.455	0.448	0.448	0.453	0.460	0.470	0.484	0.500
$CuCl_2$	0.508	0.455	0.429	0.417	0.411	0.409	0.409	0.410	0.413	0.417
$CuSO_4$	0.150	0.104	0.0829	0.0704	0.0620	0.0559	0.0512	0.0475	0.0446	0.0423
$FeCl_2$	0.5185	0.473	0.454	0.448	0.450	0.454	0.463	0.473	0.488	0.506
HCl	0.796	0.767	0.756	0.755	0.757	0.763	0.772	0.783	0.795	0.809
H_2SO_4	0.2655	0.2090	0.1826	—	0.1557	—	0.1417	—	—	0.1316
KCl	0.770	0.718	0.688	0.666	0.649	0.637	0.626	0.618	0.610	0.604
$K_3Fe(CN)_6$	0.268	0.212	0.184	0.167	0.155	0.146	0.140	0.135	0.131	0.128
$K_4Fe(CN)_6$	0.139	0.0993	0.0808	0.0693	0.0614	0.0556	0.0512	0.0479	0.0454	—
KOH	0.798	0.760	0.742	0.734	0.732	0.733	0.736	0.742	0.749	0.756
K_2SO_4	0.441	0.360	0.316	0.286	0.264	0.246	0.232	—	—	—
LiCl	0.790	0.757	0.744	0.740	0.739	0.743	0.748	0.755	0.764	0.774
NH_4Cl	0.770	0.718	0.687	0.665	0.649	0.636	0.625	0.617	0.609	0.603
NaCl	0.778	0.735	0.710	0.693	0.681	0.673	0.667	0.662	0.659	0.657
NaOH	0.766	0.727	0.708	0.697	0.690	0.685	0.681	0.679	0.678	0.678
Na_2SO_4	0.445	0.365	0.320	0.289	0.266	0.248	0.233	0.221	0.210	0.201
$ZnCl_2$	0.515	0.462	0.432	0.411	0.394	0.380	0.369	0.357	0.348	0.339
$ZnSO_4$	0.150	0.10	0.0835	0.0714	0.0630	0.0569	0.0523	0.0487	0.0458	0.0435

[17] Adapted from W.M. Haynes (Editor-in-Chief), *CRC Handbook of Chemistry and Physics*, 95th edition, CRC Press, Boca Raton, 2014, pp. 5–104.

10.17 MEAN ACTIVITY COEFFICIENTS OF AQUEOUS ELECTROLYTES[18]

This table gives the mean activity coefficients of a number of electrolytes in aqueous solution as a function of concentration, expressed in molality (mol kg⁻¹). All values refer to a temperature of 298.15 K.

b (mol kg⁻¹)	BaCl₂	CaCl₂	CuCl₂	HCl	H₂SO₄	KCl
0.001	0.887	0.888	0.887	0.965	0.804	0.965
0.002	0.849	0.851	0.849	0.952	0.740	0.951
0.005	0.782	0.787	0.783	0.929	0.634	0.927
0.010	0.721	0.727	0.722	0.905	0.542	0.901
0.020	0.653	0.664	0.654	0.876	0.445	0.869
0.050	0.559	0.577	0.561	0.832	0.325	0.816
0.100	0.492	0.517	0.495	0.797	0.251	0.768
0.200	0.436	0.469	0.441	0.768	0.195	0.717
0.500	0.391	0.444	0.401	0.759	0.146	0.649
1.000	0.393	0.495	0.405	0.811	0.125	0.604
2.000		0.784	0.453	1.009	0.119	0.573
5.000		5.907	0.601	2.380	0.197	0.593
10.000		43.1		10.4	0.527	
15.000					1.077	
20.000					1.701	

b (mol kg⁻¹)	KOH	LiCl	NaCl	NaOH	Na₂SO₄	ZnCl₂
0.001	0.965	0.965	0.965	0.965	0.886	0.887
0.002	0.952	0.952	0.952	0.952	0.846	0.847
0.005	0.927	0.928	0.928	0.927	0.777	0.781
0.010	0.902	0.904	0.903	0.902	0.712	0.719
0.020	0.871	0.874	0.872	0.870	0.637	0.652
0.050	0.821	0.827	0.822	0.819	0.529	0.561
0.100	0.779	0.789	0.779	0.775	0.446	0.499
0.200	0.740	0.756	0.734	0.731	0.366	0.447
0.500	0.710	0.739	0.681	0.685	0.268	0.384
1.000	0.733	0.775	0.657	0.674	0.204	0.330
2.000	0.860	0.924	0.668	0.714	0.155	0.283
5.000	1.697	2.000	0.874	1.076		0.342
10.000	6.110	9.600		3.258		0.876
15.000	19.9	30.9		9.796		1.914
20.000	46.4			19.410		2.968

[18] Adapted from W.M. Haynes (Editor-in-Chief), *CRC Handbook of Chemistry and Physics,* 95th edition, CRC Press, Boca Raton, 2014, pp. 5–106.

10.18   CONCENTRATIVE PROPERTIES OF AQUEOUS SOLUTIONS[19]

This table gives properties of aqueous solutions as a function of concentration. All data refer to a temperature of 293.15 K. The properties are:

Mass %:	Mass of solute divided by total mass of solution, expressed as percent (pph).
b	Molality (moles of solute per kg of water).
c	Molarity (moles of solute per liter of solution).
ρ	Density of solution in g cm^{-3}.
n	Index of refraction, relative to air, at a wavelength of 589 nm (sodium D line); the index of pure water at 293.15 K is 1.3330.
Δ	Freezing point depression in K relative to pure water.
η	Absolute (dynamic) viscosity in mPa s (equal to centipoise, cP); the viscosity of pure water at 293.15 is 1.002 mPa s.

Solute	Mass %	b (mol kg^{-1})	c (mol L^{-1})	ρ (g cm^{-3})	n	Δ (K)	η (mPa s)
Acetic acid CH$_3$COOH	0.5	0.084	0.083	0.9989	1.3334	0.16	1.012
	1.0	0.168	0.166	0.9996	1.3337	0.32	1.022
	2.0	0.340	0.333	1.0011	1.3345	0.63	1.042
	3.0	0.515	0.501	1.0025	1.3352	0.94	1.063
	4.0	0.694	0.669	1.0038	1.3359	1.26	1.084
	5.0	0.876	0.837	1.0052	1.3366	1.58	1.105
	6.0	1.063	1.006	1.0066	1.3373	1.90	1.125
	7.0	1.253	1.175	1.0080	1.3381	2.23	1.143
	8.0	1.448	1.345	1.0093	1.3388	2.56	1.162
	9.0	1.647	1.515	1.0107	1.3395	2.89	1.186
	10.0	1.850	1.685	1.0121	1.3402	3.23	1.210
	12.0	2.271	2.028	1.0147	1.3416	3.91	1.253
	14.0	2.711	2.372	1.0174	1.3430	4.61	1.298
	16.0	3.172	2.718	1.0200	1.3444	5.33	1.341
	18.0	3.655	3.065	1.0225	1.3458	6.06	1.380
	20.0	4.163	3.414	1.0250	1.3472	6.81	1.431
	22.0	4.697	3.764	1.0275	1.3485	7.57	1.478
	24.0	5.259	4.116	1.0299	1.3498	8.36	1.525
	26.0	5.851	4.470	1.0323	1.3512	9.17	1.572
	28.0	6.476	4.824	1.0346	1.3525	10.00	1.613
	30.0	7.137	5.180	1.0369	1.3537	10.84	1.669
	32.0	7.837	5.537	1.0391	1.3550	11.70	1.715

(*Continued*)

[19] Adapted from W.M. Haynes (Editor-in-Chief), *CRC Handbook of Chemistry and Physics,* 95th edition, CRC Press, Boca Raton, 2014, pp. 5–123.

Solute	Mass %	b (mol kg^{-1})	c (mol L^{-1})	ρ (g cm^{-3})	n	Δ (K)	η (mPa s)
	34.0	8.579	5.896	1.0413	1.3562	12.55	1.762
	36.0	9.367	6.255	1.0434	1.3574	13.38	1.812
	38.0	10.207	6.615	1.0454	1.3586		1.852
	40.0	11.102	6.977	1.0474	1.3598		1.912
	50.0	16.653	8.794	1.0562	1.3653		2.158
	60.0	24.979	10.620	1.0629	1.3700		2.409
	70.0	38.857	12.441	1.0673	1.3738		2.629
	80.0	66.611	14.228	1.0680	1.3767		2.720
	90.0	149.875	15.953	1.0644	1.3771		2.386
	92.0	191.507	16.284	1.0629	1.3766		2.240
	94.0	260.894	16.602	1.0606	1.3759		2.036
	96.0	399.667	16.911	1.0578	1.3748		1.813
	98.0	815.987	17.198	1.0538	1.3734		1.535
	100.0		17.447	1.0477	1.3716		1.223
Calcium chloride CaCl$_2$	0.5	0.045	0.045	1.0024	1.3342	0.22	1.015
	1.0	0.091	0.091	1.0065	1.3354	0.44	1.028
	2.0	0.184	0.183	1.0148	1.3378	0.88	1.050
	3.0	0.279	0.277	1.0232	1.3402	1.33	1.078
	4.0	0.375	0.372	1.0316	1.3426	1.82	1.110
	5.0	0.474	0.469	1.0401	1.3451	2.35	1.143
	6.0	0.575	0.567	1.0486	1.3475	2.93	1.175
	7.0	0.678	0.667	1.0572	1.3500	3.57	1.208
	8.0	0.784	0.768	1.0659	1.3525	4.28	1.242
	9.0	0.891	0.872	1.0747	1.3549	5.04	1.279
	10.0	1.001	0.976	1.0835	1.3575	5.86	1.319
	12.0	1.229	1.191	1.1014	1.3625	7.70	1.408
	14.0	1.467	1.413	1.1198	1.3677	9.83	1.508
	16.0	1.716	1.641	1.1386	1.3730	12.28	1.625
	18.0	1.978	1.878	1.1579	1.3784	15.11	1.764
	20.0	2.253	2.122	1.1775	1.3839	18.30	1.930
	22.0	2.541	2.374	1.1976	1.3895	21.70	2.127
	24.0	2.845	2.634	1.2180	1.3951	25.30	2.356
	26.0	3.166	2.902	1.2388	1.4008	29.70	2.645
	28.0	3.504	3.179	1.2600	1.4066	34.70	3.000
	30.0	3.862	3.464	1.2816	1.4124	41.00	3.467
	32.0	4.240	3.759	1.3036	1.4183	49.70	4.035
	34.0	4.642	4.062	1.3260	1.4242		4.820
	36.0	5.068	4.375	1.3488	1.4301		5.807
	38.0	5.522	4.698	1.3720	1.4361		7.321
	40.0	6.007	5.030	1.3957	1.4420		8.997
Copper sulfate CuSO$_4$	0.5	0.031	0.031	1.0033	1.3339	0.08	1.017
	1.0	0.063	0.063	1.0085	1.3348	0.14	1.036
	2.0	0.128	0.128	1.0190	1.3367	0.26	1.084
	3.0	0.194	0.194	1.0296	1.3386	0.37	1.129
	4.0	0.261	0.261	1.0403	1.3405	0.48	1.173

(Continued)

Solute	Mass %	b (mol kg^{-1})	c (mol L^{-1})	ρ (g cm^{-3})	n	Δ (K)	η (mPa s)
	5.0	0.330	0.329	1.0511	1.3424	0.59	1.221
	6.0	0.400	0.399	1.0620	1.3443	0.70	1.276
	7.0	0.472	0.471	1.0730	1.3462	0.82	1.336
	8.0	0.545	0.543	1.0842	1.3481	0.93	1.400
	9.0	0.620	0.618	1.0955	1.3501	1.05	1.469
	10.0	0.696	0.694	1.1070	1.3520	1.18	1.543
	12.0	0.854	0.850	1.1304	1.3560	1.45	1.701
	14.0	1.020	1.013	1.1545	1.3601	1.75	1.889
	16.0	1.193	1.182	1.1796	1.3644		2.136
	18.0	1.375	1.360	1.2059	1.3689		2.449
Ethanol CH$_3$CH$_2$OH	0.5	0.109	0.108	0.9973	1.3333	0.20	1.023
	1.0	0.219	0.216	0.9963	1.3336	0.40	1.046
	2.0	0.443	0.432	0.9945	1.3342	0.81	1.095
	3.0	0.671	0.646	0.9927	1.3348	1.23	1.140
	4.0	0.904	0.860	0.9910	1.3354	1.65	1.183
	5.0	1.142	1.074	0.9893	1.3360	2.09	1.228
	6.0	1.385	1.286	0.9878	1.3367	2.54	1.279
	7.0	1.634	1.498	0.9862	1.3374	2.99	1.331
	8.0	1.887	1.710	0.9847	1.3381	3.47	1.385
	9.0	2.147	1.921	0.9833	1.3388	3.96	1.442
	10.0	2.412	2.131	0.9819	1.3395	4.47	1.501
	12.0	2.960	2.551	0.9792	1.3410	5.56	1.627
	14.0	3.534	2.967	0.9765	1.3425	6.73	1.761
	16.0	4.134	3.382	0.9739	1.3440	8.01	1.890
	18.0	4.765	3.795	0.9713	1.3455	9.40	2.019
	20.0	5.427	4.205	0.9687	1.3469	10.92	2.142
	22.0	6.122	4.613	0.9660	1.3484	12.60	2.259
	24.0	6.855	5.018	0.9632	1.3498	14.47	2.370
	26.0	7.626	5.419	0.9602	1.3511	16.41	2.476
	28.0	8.441	5.817	0.9571	1.3524	18.43	2.581
	30.0	9.303	6.212	0.9539	1.3535	20.47	2.667
	32.0	10.215	6.601	0.9504	1.3546	22.44	2.726
	34.0	11.182	6.987	0.9468	1.3557	24.27	2.768
	36.0	12.210	7.370	0.9431	1.3566	25.98	2.803
	38.0	13.304	7.747	0.9392	1.3575	27.62	2.829
	40.0	14.471	8.120	0.9352	1.3583	29.26	2.846
	42.0	15.718	8.488	0.9311	1.3590	30.98	2.852
	44.0	17.055	8.853	0.9269	1.3598	32.68	2.850
	46.0	18.490	9.213	0.9227	1.3604	34.36	2.843
	48.0	20.036	9.568	0.9183	1.3610	36.04	2.832
	50.0	21.706	9.919	0.9139	1.3616	37.67	2.813
	60.0	32.559	11.605	0.8911	1.3638	44.93	2.547
	70.0	50.648	13.183	0.8676	1.3652		2.214
	80.0	86.824	14.649	0.8436	1.3658		1.881
	90.0	195.355	15.980	0.8180	1.3650		1.542

(Continued)

Solute	Mass %	b (mol kg^{-1})	c (mol L^{-1})	ρ (g cm^{-3})	n	Δ (K)	η (mPa s)
	92.0	249.620	16.225	0.8125	1.3646		1.475
	94.0	340.062	16.466	0.8070	1.3642		1.407
	96.0	520.946	16.697	0.8013	1.3636		1.342
	98.0		16.920	0.7954	1.3630		1.273
	100.0		17.133	0.7893	1.3614		1.203
Ferric chloride FeCl$_3$	0.5	0.031	0.031	1.0025	1.3344	0.21	1.024
	1.0	0.062	0.062	1.0068	1.3358	0.39	1.047
	2.0	0.126	0.125	1.0153	1.3386	0.75	1.093
	3.0	0.191	0.189	1.0238	1.3413	1.15	1.139
	4.0	0.257	0.255	1.0323	1.3441	1.56	1.187
	5.0	0.324	0.321	1.0408	1.3468	2.00	1.238
	6.0	0.394	0.388	1.0493	1.3496	2.48	1.292
	7.0	0.464	0.457	1.0580	1.3524	2.99	1.350
	8.0	0.536	0.526	1.0668	1.3552	3.57	1.412
	9.0	0.610	0.597	1.0760	1.3581	4.19	1.480
	10.0	0.685	0.669	1.0853	1.3611	4.85	1.553
	12.0	0.841	0.817	1.1040	1.3670	6.38	1.707
	14.0	1.004	0.969	1.1228	1.3730	8.22	1.879
	16.0	1.174	1.126	1.1420		10.45	2.080
	18.0	1.353	1.289	1.1615		13.08	2.311
	20.0	1.541	1.457	1.1816		16.14	2.570
	24.0	1.947	1.810	1.2234		23.79	3.178
	28.0	2.398	2.189	1.2679		33.61	4.038
	32.0	2.901	2.595	1.3153		49.16	5.274
	36.0	3.468	3.030	1.3654			7.130
	40.0	4.110	3.496	1.4176			9.674
Hydrochloric acid HCl	0.5	0.138	0.137	1.0007	1.3341	0.49	1.008
	1.0	0.277	0.275	1.0031	1.3353	0.99	1.015
	2.0	0.560	0.553	1.0081	1.3376	2.08	1.029
	3.0	0.848	0.833	1.0130	1.3399	3.28	1.044
	4.0	1.143	1.117	1.0179	1.3422	4.58	1.059
	5.0	1.444	1.403	1.0228	1.3445	5.98	1.075
	6.0	1.751	1.691	1.0278	1.3468	7.52	1.091
	7.0	2.064	1.983	1.0327	1.3491	9.22	1.108
	8.0	2.385	2.277	1.0377	1.3515	11.10	1.125
	9.0	2.713	2.574	1.0426	1.3538	13.15	1.143
	10.0	3.047	2.873	1.0476	1.3561	15.40	1.161
	12.0	3.740	3.481	1.0576	1.3607	20.51	1.199
	14.0	4.465	4.099	1.0676	1.3653		1.239
	16.0	5.224	4.729	1.0777	1.3700		1.282
	18.0	6.020	5.370	1.0878	1.3746		1.326
	20.0	6.857	6.023	1.0980	1.3792		1.374
	22.0	7.736	6.687	1.1083	1.3838		1.426
	24.0	8.661	7.362	1.1185	1.3884		1.483
	26.0	9.636	8.049	1.1288	1.3930		1.547

(Continued)

Solute	Mass %	b (mol kg^{-1})	c (mol L^{-1})	ρ (g cm^{-3})	n	Δ (K)	η (mPa s)
	28.0	10.666	8.748	1.1391	1.3976		1.620
	30.0	11.754	9.456	1.1492	1.4020		1.705
	32.0	12.907	10.175	1.1594	1.4066		1.799
	34.0	14.129	10.904	1.1693	1.4112		1.900
	36.0	15.427	11.642	1.1791	1.4158		2.002
	38.0	16.810	12.388	1.1886	1.4204		2.105
	40.0	18.284	13.140	1.1977	1.4250		
Lithium chloride LiCl	0.5	0.119	0.118	1.0012	1.3341	0.42	1.019
	1.0	0.238	0.237	1.0041	1.3351	0.84	1.037
	2.0	0.481	0.476	1.0099	1.3373	1.72	1.072
	3.0	0.730	0.719	1.0157	1.3394	2.68	1.108
	4.0	0.983	0.964	1.0215	1.3415	3.73	1.146
	5.0	1.241	1.211	1.0272	1.3436	4.86	1.185
	6.0	1.506	1.462	1.0330	1.3457	6.14	1.226
	7.0	1.775	1.715	1.0387	1.3478	7.56	1.269
	8.0	2.051	1.971	1.0444	1.3499	9.11	1.313
	9.0	2.333	2.230	1.0502	1.3520	10.79	1.360
	10.0	2.621	2.491	1.0560	1.3541	12.61	1.411
	12.0	3.217	3.022	1.0675	1.3583	16.59	1.522
	14.0	3.840	3.564	1.0792	1.3625	21.04	1.647
	16.0	4.493	4.118	1.0910	1.3668		1.787
	18.0	5.178	4.683	1.1029	1.3711		1.942
	20.0	5.897	5.260	1.1150	1.3755		2.128
	22.0	6.653	5.851	1.1274	1.3799		2.341
	24.0	7.449	6.453	1.1399	1.3844		2.600
	26.0	8.288	7.069	1.1527	1.3890		2.925
	28.0	9.173	7.700	1.1658	1.3936		3.318
	30.0	10.109	8.344	1.1791	1.3983		3.785
Methanol CH$_3$OH	0.5	0.157	0.156	0.9973	1.3331	0.28	1.022
	1.0	0.315	0.311	0.9964	1.3332	0.56	1.040
	2.0	0.637	0.621	0.9947	1.3334	1.14	1.070
	3.0	0.965	0.930	0.9930	1.3336	1.75	1.100
	4.0	1.300	1.238	0.9913	1.3339	2.37	1.131
	5.0	1.643	1.544	0.9896	1.3341	3.02	1.163
	6.0	1.992	1.850	0.9880	1.3343	3.71	1.196
	7.0	2.349	2.155	0.9864	1.3346	4.41	1.229
	8.0	2.714	2.459	0.9848	1.3348	5.13	1.264
	9.0	3.087	2.762	0.9832	1.3351	5.85	1.297
	10.0	3.468	3.064	0.9816	1.3354	6.60	1.329
	12.0	4.256	3.665	0.9785	1.3359	8.14	1.389
	14.0	5.081	4.262	0.9755	1.3365	9.72	1.446
	16.0	5.945	4.856	0.9725	1.3370	11.36	1.501
	18.0	6.851	5.447	0.9695	1.3376	13.13	1.554
	20.0	7.803	6.034	0.9666	1.3381	15.02	1.604
	22.0	8.803	6.616	0.9636	1.3387	16.98	1.652

(*Continued*)

Solute	Mass %	b (mol kg^{-1})	c (mol L^{-1})	ρ (g cm^{-3})	n	Δ (K)	η (mPa s)
	24.0	9.856	7.196	0.9606	1.3392	19.04	1.697
	26.0	10.966	7.771	0.9576	1.3397	21.23	1.735
	28.0	12.138	8.341	0.9545	1.3402	23.59	1.769
	30.0	13.376	8.908	0.9514	1.3407	25.91	1.795
	32.0	14.688	9.470	0.9482	1.3411	28.15	1.814
	34.0	16.078	10.028	0.9450	1.3415	30.48	1.827
	36.0	17.556	10.580	0.9416	1.3419	32.97	1.835
	38.0	19.129	11.127	0.9382	1.3422	35.60	1.839
	40.0	20.807	11.669	0.9347	1.3425	38.60	1.837
	50.0	31.211	14.288	0.9156	1.3431	54.50	1.761
	60.0	46.816	16.749	0.8944	1.3426	74.50	1.600
	70.0	72.826	19.040	0.8715	1.3411		1.368
	80.0	124.844	21.144	0.8468	1.3385		1.128
	90.0	280.899	23.045	0.8204	1.3348		0.861
	100.0		24.710	0.7917	1.3290		0.586
Phosphoric acid H$_3$PO$_4$	0.5	0.051	0.051	1.0010	1.3335	0.12	1.010
	1.0	0.103	0.102	1.0038	1.3340	0.24	1.020
	2.0	0.208	0.206	1.0092	1.3349	0.46	1.050
	3.0	0.316	0.311	1.0146	1.3358	0.69	1.079
	4.0	0.425	0.416	1.0200	1.3367	0.93	1.108
	5.0	0.537	0.523	1.0254	1.3376	1.16	1.138
	6.0	0.651	0.631	1.0309	1.3385	1.38	1.169
	7.0	0.768	0.740	1.0363	1.3394	1.62	1.200
	8.0	0.887	0.850	1.0418	1.3403	1.88	1.232
	9.0	1.009	0.962	1.0474	1.3413	2.16	1.267
	10.0	1.134	1.075	1.0531	1.3422	2.45	1.303
	12.0	1.392	1.304	1.0647	1.3441	3.01	1.382
	14.0	1.661	1.538	1.0765	1.3460	3.76	1.469
	16.0	1.944	1.777	1.0885	1.3480	4.45	1.565
	18.0	2.240	2.022	1.1009	1.3500	5.25	1.671
	20.0	2.551	2.273	1.1135	1.3520	6.23	1.788
	22.0	2.878	2.529	1.1263	1.3540	7.38	1.914
	24.0	3.223	2.791	1.1395	1.3561	8.69	2.049
	26.0	3.585	3.059	1.1528	1.3582	10.12	2.198
	28.0	3.968	3.333	1.1665	1.3604	11.64	2.365
	30.0	4.373	3.614	1.1804	1.3625	13.23	2.553
	32.0	4.802	3.901	1.1945	1.3647	14.94	2.766
	34.0	5.257	4.194	1.2089	1.3669	16.81	3.001
	36.0	5.740	4.495	1.2236	1.3691	18.85	3.260
	38.0	6.254	4.803	1.2385	1.3713	21.09	3.544
	40.0	6.803	5.117	1.2536	1.3735	23.58	3.856
Potassium chloride KCl	0.5	0.067	0.067	1.0014	1.3337	0.23	1.000
	1.0	0.135	0.135	1.0046	1.3343	0.46	0.999
	2.0	0.274	0.271	1.0110	1.3357	0.92	0.999
	3.0	0.415	0.409	1.0174	1.3371	1.38	0.998

(Continued)

Solute	Mass %	b (mol kg^{-1})	c (mol L^{-1})	ρ (g cm^{-3})	n	Δ (K)	η (mPa s)
	4.0	0.559	0.549	1.0239	1.3384	1.85	0.997
	5.0	0.706	0.691	1.0304	1.3398	2.32	0.996
	6.0	0.856	0.835	1.0369	1.3411	2.80	0.994
	7.0	1.010	0.980	1.0434	1.3425	3.29	0.992
	8.0	1.166	1.127	1.0500	1.3438	3.80	0.990
	9.0	1.327	1.276	1.0566	1.3452	4.30	0.989
	10.0	1.490	1.426	1.0633	1.3466	4.81	0.988
	12.0	1.829	1.733	1.0768	1.3493	5.88	0.990
	14.0	2.184	2.048	1.0905	1.3521		0.994
	16.0	2.555	2.370	1.1043	1.3549		0.999
	18.0	2.944	2.701	1.1185	1.3577		1.004
	20.0	3.353	3.039	1.1328	1.3606		1.012
	22.0	3.783	3.386	1.1474	1.3635		1.024
	24.0	4.236	3.742	1.1623	1.3665		1.040
Potassium hydroxide KOH	0.5	0.090	0.089	1.0025	1.3340	0.30	1.010
	1.0	0.180	0.179	1.0068	1.3350	0.61	1.019
	2.0	0.364	0.362	1.0155	1.3369	1.24	1.038
	3.0	0.551	0.548	1.0242	1.3388	1.89	1.058
	4.0	0.743	0.736	1.0330	1.3408	2.57	1.079
	5.0	0.938	0.929	1.0419	1.3427	3.36	1.102
	6.0	1.138	1.124	1.0509	1.3445	4.14	1.126
	7.0	1.342	1.322	1.0599	1.3464	4.92	1.151
	8.0	1.550	1.524	1.0690	1.3483		1.177
	9.0	1.763	1.729	1.0781	1.3502		1.205
	10.0	1.980	1.938	1.0873	1.3520		1.233
	12.0	2.431	2.365	1.1059	1.3558		1.294
	14.0	2.902	2.806	1.1246	1.3595		1.361
	16.0	3.395	3.261	1.1435	1.3632		1.436
	18.0	3.913	3.730	1.1626	1.3670		1.521
	20.0	4.456	4.213	1.1818	1.3707		1.619
	22.0	5.027	4.711	1.2014	1.3744		1.732
	24.0	5.629	5.223	1.2210	1.3781		1.861
	26.0	6.262	5.750	1.2408	1.3818		2.006
	28.0	6.931	6.293	1.2609	1.3854		2.170
	30.0	7.639	6.851	1.2813	1.3889		2.357
	40.0	11.882	9.896	1.3881	1.4068		3.879
	50.0	17.824	13.389	1.5024	1.4247		7.892
Sodium acetate CH$_3$COONa	0.5	0.061	0.061	1.0008	1.3337	0.22	1.021
	1.0	0.123	0.122	1.0034	1.3344	0.43	1.040
	2.0	0.249	0.246	1.0085	1.3358	0.88	1.080
	3.0	0.377	0.371	1.0135	1.3372	1.34	1.124
	4.0	0.508	0.497	1.0184	1.3386	1.82	1.171
	5.0	0.642	0.624	1.0234	1.3400	2.32	1.222
	6.0	0.778	0.752	1.0283	1.3414	2.85	1.278
	7.0	0.918	0.882	1.0334	1.3428	3.40	1.337

(Continued)

Solute	Mass %	b (mol kg^{-1})	c (mol L^{-1})	ρ (g cm^{-3})	n	Δ (K)	η (mPa s)
	8.0	1.060	1.013	1.0386	1.3442	3.98	1.401
	9.0	1.206	1.145	1.0440	1.3456	4.57	1.468
	10.0	1.354	1.279	1.0495	1.3470		1.539
	12.0	1.662	1.552	1.0607	1.3498		1.688
	14.0	1.984	1.829	1.0718	1.3526		1.855
	16.0	2.322	2.112	1.0830	1.3554		2.054
	18.0	2.676	2.400	1.0940	1.3583		2.284
	20.0	3.047	2.694	1.1050	1.3611		2.567
	22.0	3.438	2.993	1.1159	1.3639		2.948
	24.0	3.849	3.297	1.1268	1.3666		3.400
	26.0	4.283	3.606	1.1377	1.3693		3.877
	28.0	4.741	3.921	1.1488	1.3720		4.388
	30.0	5.224	4.243	1.1602	1.3748		4.940
Sodium chloride NaCl	0.5	0.086	0.086	1.0018	1.3339	0.30	1.011
	1.0	0.173	0.172	1.0053	1.3347	0.59	1.020
	2.0	0.349	0.346	1.0125	1.3365	1.19	1.036
	3.0	0.529	0.523	1.0196	1.3383	1.79	1.052
	4.0	0.713	0.703	1.0268	1.3400	2.41	1.068
	5.0	0.901	0.885	1.0340	1.3418	3.05	1.085
	6.0	1.092	1.069	1.0413	1.3435	3.70	1.104
	7.0	1.288	1.256	1.0486	1.3453	4.38	1.124
	8.0	1.488	1.445	1.0559	1.3470	5.08	1.145
	9.0	1.692	1.637	1.0633	1.3488	5.81	1.168
	10.0	1.901	1.832	1.0707	1.3505	6.56	1.193
	12.0	2.333	2.229	1.0857	1.3541	8.18	1.250
	14.0	2.785	2.637	1.1008	1.3576	9.94	1.317
	16.0	3.259	3.056	1.1162	1.3612	11.89	1.388
	18.0	3.756	3.486	1.1319	1.3648	14.04	1.463
	20.0	4.278	3.928	1.1478	1.3684	16.46	1.557
	22.0	4.826	4.382	1.1640	1.3721	19.18	1.676
	24.0	5.403	4.847	1.1804	1.3757		1.821
	26.0	6.012	5.326	1.1972	1.3795		1.990
Sodium hydroxide NaOH	0.5	0.126	0.125	1.0039	1.3344	0.43	1.027
	1.0	0.253	0.252	1.0095	1.3358	0.86	1.054
	2.0	0.510	0.510	1.0207	1.3386	1.74	1.112
	3.0	0.773	0.774	1.0318	1.3414	2.64	1.176
	4.0	1.042	1.043	1.0428	1.3441	3.59	1.248
	5.0	1.316	1.317	1.0538	1.3467	4.57	1.329
	6.0	1.596	1.597	1.0648	1.3494	5.60	1.416
	7.0	1.882	1.883	1.0758	1.3520	6.69	1.510
	8.0	2.174	2.174	1.0869	1.3546	7.87	1.616
	9.0	2.473	2.470	1.0979	1.3572	9.12	1.737
	10.0	2.778	2.772	1.1089	1.3597	10.47	1.882
	12.0	3.409	3.393	1.1309	1.3648	13.42	2.201
	14.0	4.070	4.036	1.1530	1.3697	16.76	2.568

(Continued)

Solute	Mass %	b (mol kg^{-1})	c (mol L^{-1})	ρ (g cm^{-3})	n	Δ (K)	η (mPa s)
	15.0	4.412	4.365	1.1640	1.3722		2.789
	16.0	4.762	4.701	1.1751	1.3746		3.043
	18.0	5.488	5.387	1.1971	1.3793		3.698
	20.0	6.250	6.096	1.2192	1.3840		4.619
	22.0	7.052	6.827	1.2412	1.3885		5.765
	24.0	7.895	7.579	1.2631	1.3929		7.100
	26.0	8.784	8.352	1.2848	1.3971		8.744
	28.0	9.723	9.145	1.3064	1.4012		10.832
	30.0	10.715	9.958	1.3277	1.4051		13.517
	32.0	11.766	10.791	1.3488	1.4088		16.844
	34.0	12.880	11.643	1.3697	1.4123		20.751
	36.0	14.064	12.512	1.3901	1.4156		25.290
	38.0	15.324	13.398	1.4102	1.4186		30.461
	40.0	16.668	14.300	1.4299	1.4215		36.312
Sulfuric acid H_2SO_4	0.5	0.051	0.051	1.0016	1.3336	0.21	1.010
	1.0	0.103	0.102	1.0049	1.3342	0.42	1.019
	2.0	0.208	0.206	1.0116	1.3355	0.80	1.036
	3.0	0.315	0.311	1.0183	1.3367	1.17	1.059
	4.0	0.425	0.418	1.0250	1.3379	1.60	1.085
	5.0	0.537	0.526	1.0318	1.3391	2.05	1.112
	6.0	0.651	0.635	1.0385	1.3403	2.50	1.136
	7.0	0.767	0.746	1.0453	1.3415	2.95	1.159
	8.0	0.887	0.858	1.0522	1.3427	3.49	1.182
	9.0	1.008	0.972	1.0591	1.3439	4.08	1.206
	10.0	1.133	1.087	1.0661	1.3451	4.64	1.230
	12.0	1.390	1.322	1.0802	1.3475	5.93	1.282
	14.0	1.660	1.563	1.0947	1.3500	7.49	1.337
	16.0	1.942	1.810	1.1094	1.3525	9.26	1.399
	18.0	2.238	2.064	1.1245	1.3551	11.29	1.470
	20.0	2.549	2.324	1.1398	1.3576	13.64	1.546
	22.0	2.876	2.592	1.1554	1.3602	16.48	1.624
	24.0	3.220	2.866	1.1714	1.3628	19.85	1.706
	26.0	3.582	3.147	1.1872	1.3653	24.29	1.797
	28.0	3.965	3.435	1.2031	1.3677	29.65	1.894
	30.0	4.370	3.729	1.2191	1.3701	36.21	2.001
	32.0	4.798	4.030	1.2353	1.3725	44.76	2.122
	34.0	5.252	4.339	1.2518	1.3749	55.28	2.255
	36.0	5.735	4.656	1.2685	1.3773		2.392
	38.0	6.249	4.981	1.2855	1.3797		2.533
	40.0	6.797	5.313	1.3028	1.3821		2.690
	42.0	7.383	5.655	1.3205	1.3846		2.872
	44.0	8.011	6.005	1.3386	1.3870		3.073
	46.0	8.685	6.364	1.3570	1.3895		3.299
	48.0	9.411	6.734	1.3759	1.3920		3.546
	50.0	10.196	7.113	1.3952	1.3945		3.826

(Continued)

Solute	Mass %	b (mol kg^{-1})	c (mol L^{-1})	ρ (g cm^{-3})	n	Δ (K)	η (mPa s)
	52.0	11.045	7.502	1.4149	1.3971		4.142
	54.0	11.969	7.901	1.4351	1.3997		4.499
	56.0	12.976	8.312	1.4558	1.4024		4.906
	58.0	14.080	8.734	1.4770	1.4050		5.354
	60.0	15.294	9.168	1.4987	1.4077		5.917
	70.0	23.790	11.494	1.6105			
	80.0	40.783	14.088	1.7272			
	90.0	91.762	16.649	1.8144			
	92.0	117.251	17.109	1.8240			
	94.0	159.734	17.550	1.8312			
	96.0	244.698	17.966	1.8355			
	98.0	499.592	18.346	1.8361			
	100.0		18.663	1.8305			
Zinc sulfate ZnSO$_4$	0.5	0.031	0.031	1.0034	1.3339	0.08	1.021
	1.0	0.063	0.062	1.0085	1.3348	0.15	1.040
	2.0	0.126	0.126	1.0190	1.3366	0.28	1.081
	3.0	0.192	0.191	1.0296	1.3384	0.41	1.126
	4.0	0.258	0.258	1.0403	1.3403	0.53	1.175
	5.0	0.326	0.326	1.0511	1.3421	0.65	1.227
	6.0	0.395	0.395	1.0620	1.3439	0.77	1.283
	7.0	0.466	0.465	1.0730	1.3457	0.89	1.341
	8.0	0.539	0.537	1.0842	1.3475	1.01	1.403
	9.0	0.613	0.611	1.0956	1.3494	1.14	1.470
	10.0	0.688	0.686	1.1071	1.3513	1.27	1.545
	12.0	0.845	0.840	1.1308	1.3551	1.55	1.716
	14.0	1.008	1.002	1.1553	1.3590	1.89	1.918
	16.0	1.180	1.170	1.1806	1.3630	2.31	2.152

10.19 AQUEOUS SOLUBILITY OF INORGANIC COMPOUNDS AT VARIOUS TEMPERATURES[20]

The solubility of some compounds in water is tabulated here as a function of temperature. Solubility is defined as the concentration of the compound in a solution that is in equilibrium with a solid phase at the specified temperature. In this table the solid phase is generally the most stable crystalline phase at the temperature in question.

The solubility values are expressed as mass percent (pph) of solute.

---

[20] Adapted from W.M. Haynes (Editor-in-Chief), *CRC Handbook of Chemistry and Physics*, 95th edition, CRC Press, Boca Raton, 2014, pp. 5–190.

Compound	273.15 K	283.15 K	293.15 K	298.15 K	303.15 K	313.15 K	323.15 K	333.15 K	343.15 K	353.15 K	363.15 K	373.15 K
Ag$_2$SO$_4$	0.56	0.67	0.78	0.83	0.88	0.97	1.05	1.13	1.20	1.26	1.32	1.39
AlCl$_3$	30.84	30.91	31.03	31.10	31.18	31.37	31.60	31.87	32.17	32.51	32.90	33.32
Ba(OH)$_2$	1.67			4.68	8.4	19	33	52	74	100		
CaCl$_2$	36.70	39.19	42.13	44.83*	49.12*	52.85*	56.05*	56.73	57.44	58.21	59.04	59.94
CaSO$_4$	0.174	0.191	0.202	0.205	0.208	0.210	0.207	0.201	0.193	0.184	0.173	0.163
CuCl$_2$	40.8	41.7	42.6	43.1	43.7	44.8	46.0	47.2	48.5	49.9	51.3	52.7
CuSO$_4$	12.4	14.4	16.7	18.0	19.3	22.2	25.4	28.8	32.4	36.3	40.3	43.5
FeCl$_2$	33.2*			39.4*								48.7*
FeCl$_3$	42.7	44.9	47.9	47.7	51.6	74.8	76.7	84.6	84.3	84.3	84.4	84.7
Hg$_2$Cl$_2$				0.0004								
KCl	21.74	23.61	25.39	26.22	27.04	28.59	30.04	31.40	32.66	33.86	34.99	36.05
KOH	48.7	50.8	53.2	54.7	56.1	57.9	58.6	59.5	60.6	61.8	63.1	64.6
K$_3$Fe(CN)$_6$	23.9	27.6	31.1	32.8	34.3	37.2	39.6	41.7	43.5	45.0	46.1	47.0
K$_4$Fe(CN)$_6$	12.5	17.3	22.0	23.9	25.6	29.2	32.5	35.5	38.2	40.6	41.4	43.1
LaCl$_3$	49.0	48.5	48.6	48.9	49.3	50.5	52.1	54.0	56.3	58.9	61.7	
LiCl	40.45	42.46*	45.29*	45.81	46.25	47.30	48.47	49.78	51.27	52.98	54.98*	56.34*
NH$_4$Cl	22.92	25.12	27.27	28.34	29.39	31.46	33.50	35.49	37.46	39.40	41.33	43.24
NaCl	26.28	26.32	26.41	26.45	26.52	26.67	26.84	27.03	27.25	27.50	27.78	28.05
NaOH	30	39	46	50	53	58	63	67	71	74	76	79
Na$_2$CO$_3$	6.44	10.8	17.9	23.5	28.7	32.8	32.2	31.7	31.3	31.1	30.9	30.9
Na$_2$SO$_4$			16.13	21.94	29.22*	32.35*	31.55	30.90	30.39	30.02	29.79	29.67
ZnCl$_2$		76.6	79.0	80.3	81.4	81.8	82.4	83.0	83.7	84.4	85.2	86.0
ZnSO$_4$	29.1	32.0	35.0	36.6	38.2	41.3	43.0	42.1	41.0	39.9	38.8	37.6

10.20   SOLUBILITY PRODUCT CONSTANTS[21]

The solubility product constant K_{sp} is a useful parameter for calculating the aqueous solubility of sparingly soluble compounds under various conditions. It may be determined by direct measurement or calculated from the standard Gibbs energies of formation $\Delta_f G°$ of the species involved at their standard states. Thus if $K_{sp} = [M^+]^m \times [A^-]^n$ is the equilibrium constant for the reaction

$$M_m A_n (s) \rightleftharpoons mM^+ (aq) + nA^- (aq),$$

where $M_m A_n$ is the slightly soluble substance and M^+ and A^- are the ions produced in solution by the dissociation of $M_m A_n$, then the Gibbs energy change is

$$\Delta_r G° = m\Delta_f G°\left(M^+, aq\right) + n\Delta_f G°\left(A^-, aq\right) - \Delta_f G°\left(M_m A_n, s\right)$$

The solubility product constant is calculated from the equation

$$\ln K_{sp} = -\Delta_r G° / (RT)$$

The table below gives selected values of K_{sp} at 298.15 K.

Compound	Formula	K_{sp}
Barium carbonate	$BaCO_3$	2.58×10^{-9}
Barium sulfate	$BaSO_4$	1.08×10^{-10}
Cadmium hydroxide	$Cd(OH)_2$	7.2×10^{-15}
Calcium carbonate (calcite)	$CaCO_3$	3.36×10^{-9}
Calcium hydroxide	$Ca(OH)_2$	5.02×10^{-6}
Calcium phosphate	$Ca_3(PO_4)_2$	2.07×10^{-33}
Calcium sulfate	$CaSO_4$	4.93×10^{-5}
Copper(I) chloride	$CuCl$	1.72×10^{-7}
Iron(II) carbonate	$FeCO_3$	3.13×10^{-11}
Iron(II) hydroxide	$Fe(OH)_2$	4.87×10^{-17}
Iron(III) hydroxide	$Fe(OH)_3$	2.79×10^{-39}
Mercury(I) chloride	Hg_2Cl_2	1.43×10^{-18}
Silver(I) chloride	$AgCl$	1.77×10^{-10}
Zinc hydroxide	$Zn(OH)_2$	3×10^{-17}

[21] Adapted from W.M. Haynes (Editor-in-Chief), *CRC Handbook of Chemistry and Physics*, 95th edition, CRC Press, Boca Raton, 2014, pp. 5–196.

10.21 SOLUBILITY OF COMMON SALTS AT AMBIENT TEMPERATURES[22]

This table gives the aqueous solubility of selected salts at temperatures from 283.15 to 313.15 K. Values are given in molality (mol kg^{-1}).

Salt	283.15 K	288.15 K	293.15 K	298.15 K	303.15 K	308.15 K	313.15 K
BaCl$_2$	1.603	1.659	1.716	1.774	1.834	1.895	1.958
CuSO$_4$	1.055	1.153	1.260	1.376	1.502	1.639	
NaCl	6.110	6.121	6.136	6.153	6.174	6.197	6.222
ZnSO$_4$	2.911	3.116	3.336	3.573	3.827	4.099	4.194

10.22 THERMOPHYSICAL PROPERTIES OF WATER AND STEAM AT TEMPERATURES UP TO 100°C[23]

These table summarize the thermophysical properties of water and steam at temperatures up to 100°C (373.15 K) and pressure of 1 bar in liquid state. The tabulated properties are pressure (P), density (ρ), enthalpy (H), entropy (S), isochoric heat capacity (C_v), isobaric heat capacity (C_p), speed of sound (u), viscosity (η), thermal conductivity (λ), and static dielectric constant (D).

t (°C)	P (MPa)	ρ (kg m^{-3})	H (kJ kg^{-1})	S (kJ kg^{-1} K^{-1})	C_v (kJ kg^{-1} K^{-1})	C_p (kJ kg^{-1} K^{-1})	u (m s^{-1})	D	η (µPa s)	λ (mW m^{-1} K^{-1})
0.01	0.1	999.84	0.10186	0.000007	4.2170	4.2194	1402.4	87.899	1791.1	561.09
10	0.1	999.70	42.118	0.15108	4.1906	4.1952	1447.3	83.974	1305.9	580.05
20	0.1	998.21	84.006	0.29646	4.1567	4.1841	1482.3	80.223	1001.6	598.46
25	0.1	997.05	104.92	0.36720	4.1376	4.1813	1496.7	78.408	890.02	607.19
30	0.1	995.65	125.82	0.43673	4.1172	4.1798	1509.2	76.634	797.22	615.50
40	0.1	992.22	167.62	0.57237	4.0734	4.1794	1528.9	73.201	652.73	630.63
50	0.1	988.03	209.42	0.70377	4.0262	4.1813	1542.6	69.916	546.52	643.59
60	0.1	983.20	251.25	0.83125	3.9765	4.1850	1551.0	66.774	466.03	654.39
70	0.1	977.76	293.12	0.95509	3.9251	4.1901	1554.7	63.770	403.55	663.13
80	0.1	971.79	335.05	1.0755	3.8728	4.1968	1554.4	60.898	354.05	670.01
90	0.1	965.31	377.06	1.1928	3.8204	4.2052	1550.4	58.152	314.17	675.27
99.606	0.1	958.63	417.50	1.3028	3.7702	4.2152	1543.5	55.628	282.75	678.97
99.606	0.1	0.59034	2674.9	7.3588	1.5548	2.0784	471.99	1.0058	12.218	25.053
100	0.1	0.58967	2675.8	7.3610	1.5535	2.0766	472.28	1.0058	12.234	25.079

[22] Adapted from W.M. Haynes (Editor-in-Chief), *CRC Handbook of Chemistry and Physics,* 95th edition, CRC Press, Boca Raton, 2014, pp. 5–199.

[23] Adapted from W.M. Haynes (Editor-in-Chief), *CRC Handbook of Chemistry and Physics,* 95th edition, CRC Press, Boca Raton, 2014, pp. 6–1.

10.23   VAPOR PRESSURE AND OTHER SATURATION PROPERTIES OF WATER AT TEMPERATURES UP TO 100°C (373.15 K)[24]

This table summarizes the vapor pressure and enthalpy of vaporization of water as a function of temperature.

t (°C)	P (kPa)	$\Delta_{vap}H$ (kJ kg^{-1})
0.01	0.61165	2500.9
2	0.70599	2496.2
4	0.81355	2491.4
6	0.93536	2486.7
8	1.0730	2481.9
10	1.2282	2477.2
12	1.4028	2472.5
14	1.5990	2467.7
16	1.8188	2463.0
18	2.0647	2458.3
20	2.3393	2453.5
22	2.6453	2448.8
24	2.9858	2444.0
25	3.1699	2441.7
26	3.3639	2439.3
28	3.7831	2434.6
30	4.2470	2429.8
32	4.7596	2425.1
34	5.3251	2420.3
36	5.9479	2415.5
38	6.6328	2410.8
40	7.3849	2406.0
42	8.2096	2401.2
44	9.1124	2396.4
46	10.099	2391.6
48	11.177	2386.8
50	12.352	2381.9
52	13.631	2377.1
54	15.022	2372.3
56	16.533	2367.4
58	18.171	2362.5
60	19.946	2357.7
62	21.867	2352.8
64	23.943	2347.8
66	26.183	2342.9

(Continued)

[24] Adapted from W.M. Haynes (Editor-in-Chief), *CRC Handbook of Chemistry and Physics*, 95th edition, CRC Press, Boca Raton, 2014, pp. 6–5.

t (°C)	P (kPa)	$\Delta_{vap}H$ (kJ kg^{-1})
68	28.599	2338.0
70	31.201	2333.0
72	34.000	2328.1
74	37.009	2323.1
76	40.239	2318.1
78	43.703	2313.0
80	47.414	2308.0
82	51.387	2302.9
84	55.635	2297.9
86	60.173	2292.8
88	65.017	2287.6
90	70.182	2282.5
92	75.684	2277.3
94	81.541	2272.1
96	87.771	2266.9
98	94.390	2261.7
100	101.42	2256.4

10.24 VAN DER WAALS CONSTANTS FOR GASES[25]

The van der Waals equation of state for a real gas is

$$\left(P + n^2 a/V^2\right)(V - nb) = nRT$$

where P is the pressure, V the volume, T the thermodynamic temperature, n the amount of substance (in moles), and R the molar gas constant. The van der Waals constants a and b are characteristic of the substance and are independent of temperature. They are related to the critical temperature and pressure, T_c and P_c, by

$$a = 27R^2 T_c^2/(64P_c) \quad b = RT_c/(8P_c)$$

This table gives values of a and b for some common gases.

Substance	a (bar L^{2}mol^{-2})	b (L mol^{-1})
Acetic acid	17.71	0.1065
Ammonia	4.225	0.0371
Argon	1.355	0.0320
Carbon dioxide	3.658	0.0429
Carbon monoxide	1.472	0.0395

(*Continued*)

[25] Adapted from W.M. Haynes (Editor-in-Chief), *CRC Handbook of Chemistry and Physics,* 95th edition, CRC Press, Boca Raton, 2014, pp. 6–56.

Substance	a (bar L^{2}mol^{-2})	b (L mol^{-1})
Chlorine	6.343	0.0542
Ethanol	12.56	0.0871
Helium	0.0346	0.0238
Hydrogen	0.2452	0.0265
Hydrogen chloride	3.700	0.0406
Hydrogen sulfide	4.544	0.0434
Methane	2.303	0.0431
Methanol	9.476	0.0659
Nitrogen	1.370	0.0387
Oxygen	1.382	0.0319
Propane	9.39	0.0905
Water	5.537	0.0305

10.25   VAPOR PRESSURE OF SATURATED SALT SOLUTIONS[26]

This table gives the vapor pressure in kPa of water above saturated solutions of some common salts at ambient temperatures. Data on pure water are given on the last line for comparison.

Salt	283.15 K	288.15 K	293.15 K	298.15 K	303.15 K	308.15 K	313.15 K
BaCl$_2$	0.971	1.443	2.073	2.887	3.903	5.133	6.576
CuSO$_4$	1.113	1.574	2.189	2.996	4.037	5.363	
LiCl	0.128	0.193	0.279	0.384			
NaCl	0.921	1.285	1.768	2.401	3.218	4.262	5.581
ZnSO$_4$	0.945	1.401	1.986	2.698	3.523	4.431	5.382
Water	1.228	1.706	2.339	3.169	4.246	5.627	7.381

10.26   ELECTRIC RESISTIVITY OF PURE METALS[27]

The table gives the electric resistivity, in unit of 10^{-8} Ω m, as a function of temperature. The data refer to polycrystalline samples.

[26] Adapted from W.M. Haynes (Editor-in-Chief), *CRC Handbook of Chemistry and Physics,* 95th edition, CRC Press, Boca Raton, 2014, pp. 6–126.
[27] Adapted from W.M. Haynes (Editor-in-Chief), *CRC Handbook of Chemistry and Physics,* 95th edition, CRC Press, Boca Raton, 2014, pp. 12–41.

			Electric Resistivity in 10^{-8} Ω m				
T (K)	Aluminum	Copper	Gold	Iron	Molybdenum	Nickel	Palladium
273	2.417	1.543	2.051	8.57	4.85	6.16	9.78
293	2.650	1.678	2.214	9.61	5.34	6.93	10.54
298	2.709	1.712	2.255	9.87	5.47	7.12	10.73
300	2.733	1.725	2.271	9.98	5.52	7.20	10.80
400	3.87	2.402	3.107	16.1	8.02	11.8	14.48
500	4.99	3.090	3.97	23.7	10.6	17.7	17.94
600	6.13	3.792	4.87	32.9	13.1	25.5	21.2
700	7.35	4.514	5.82	44.0	15.8	32.1	24.2
800	8.70	5.262	6.81	57.1	18.4	35.5	27.1
900	10.18	6.041	7.86		21.2	38.6	29.4
T (K)	Platinum	Silver	Tungsten	Vanadium	Zinc	Zirconium	
273	9.6	1.467	4.82	18.1	5.46	38.8	
293	10.5	1.587	5.28	19.7	5.90	42.1	
298	10.7	1.617	5.39	20.1	6.01	42.9	
300	10.8	1.629	5.44	20.2	6.06	43.3	
400	14.6	2.241	7.83	28.0	8.37	60.3	
500	18.3	2.87	10.3	34.8	10.82	76.5	
600	21.9	3.53	13.0	41.1	13.49	91.5	
700	25.4	4.21	15.7	47.2		104.2	
800	28.7	4.91	18.6	53.1		114.9	
900	32.0	5.64	21.5	58.7		123.1	

10.27 COMPOSITION OF SEAWATER AND IONIC STRENGTH ON MOLALITY SCALE AT VARIOUS SALINITIES S IN PARTS PER THOUSAND[28]

Constituent	S = 30	S = 35	S = 40
Cl^-	0.482	0.562	0.650
Br^-	0.00074	0.00087	0.00100
F^-		0.00007	
SO_4^{2-}	0.0104	0.0114	0.0122
HCO_3^-	0.00131	0.00143	0.00100
$NaSO_4^-$	0.0085	0.0108	0.0139
KSO_4^-	0.00010	0.00012	0.00015
Na^+	0.405	0.472	0.544
K^+	0.00892	0.01039	0.01200

(Continued)

[28] Adapted from W.M. Haynes (Editor-in-Chief), *CRC Handbook of Chemistry and Physics,* 95th edition, CRC Press, Boca Raton, 2014, pp. 14–16.

Constituent	$S = 30$	$S = 35$	$S = 40$
Mg^{2+}	0.0413	0.0483	0.0561
Ca^{2+}	0.00131	0.00143	0.00154
Sr^{2+}	0.00008	0.00009	0.00011
$MgHCO_3^+$	0.00028	0.00036	0.00045
$MgSO_4$	0.00498	0.00561	0.00614
$CaSO_4$	0.00102	0.00115	0.00126
$NaHCO_3$	0.00015	0.00020	0.00024
H_3BO_3	0.00032	0.00037	0.00042
Ionic strength	0.5736	0.6675	0.7701

10.28   ERROR FUNCTION[29]

Definition: $\mathrm{erf}(x) = \dfrac{2}{\sqrt{\pi}} \displaystyle\int_0^x e^{-t^2}\, dt$

Series: $\mathrm{erf}(x) = \dfrac{2}{\sqrt{\pi}} \left(x - \dfrac{x^3}{3} + \dfrac{1}{2!}\dfrac{x^5}{5} - \dfrac{1}{3!}\dfrac{x^7}{7} + \cdots \right)$

Property: $\mathrm{erf}(x) = -\mathrm{erf}(-x)$

Relationship with Normal Probability Function $f(t)$: $\displaystyle\int_0^x f(t)\, dt = \dfrac{1}{2}\mathrm{erf}\left(\dfrac{x}{\sqrt{2}}\right)$ To

evaluate erf(2,3), one proceeds as follows: For $\dfrac{x}{\sqrt{2}} = 2.3$, one finds $x = (2.3)\left(\sqrt{2} = 3.25\right)$. In the normal probability function table, one finds the entry 0.4994 opposite the value 3.25. Thus erf $(2.3) = 2(0.4994) = 0.9988$.

$$\mathrm{erfc}(z) = 1 - \mathrm{erf}(z) = \frac{2}{\sqrt{\pi}} \int_0^x e^{-t^2}\, dt$$

is known as the complementary error function.

10.29   PERIODIC TABLE OF ELEMENTS

---

[29] Adapted from W.M. Haynes (Editor-in-Chief), *CRC Handbook of Chemistry and Physics*, 95th edition, CRC Press, Boca Raton, 2014, pp. A–69.

Periodic Table of the Elements

Key to Chart

Label	Example
Atomic Number	50
Symbol	Sn
2016 Atomic Weight	118.71
Electron Configuration	-18-18-4
Oxidation States	+2 +4

Legend: Gases · Liquids · Non-metallic solids · Metallic solids

Main Table

Each cell: Atomic Number, Symbol, Oxidation States, Atomic Weight, Electron Configuration (Shell).

Group 1 (Shell K, K-L, K-L-M, ...)

Z	Symbol	Oxidation States	Atomic Weight	Config
1	H	+1, −1	[1.0078, 1.0082]	1
3	Li	+1	[6.938, 6.997]	2-1
11	Na	+1	22.990	2-8-1
19	K	+1	39.098	-8-8-1
37	Rb	+1	85.468	-18-8-1
55	Cs	+1	132.91	-18-8-1
87	Fr	+1		-18-8-1

Group 2

Z	Symbol	Oxidation States	Atomic Weight	Config
4	Be	+2	9.0122	2-2
12	Mg	+2	[24.304, 24.307]	2-8-2
20	Ca	+2	40.078	-8-8-2
38	Sr	+2	87.62	-18-8-2
56	Ba	+2	137.33	-18-8-2
88	Ra	+2		-18-8-2

Groups 3–12 (transition metals)

Z	Symbol	Oxidation States	Atomic Weight	Config
21	Sc	+3	44.956	-8-9-2
39	Y	+3	88.906	-18-9-2
57-71	La	(Lanthanoids)		
89-103	Ac	(Actinoids)		
22	Ti	+2 +3 +4	47.867	-8-10-2
40	Zr	+4	91.224	-18-10-2
72	Hf	+4	178.49	-32-10-2
104	Rf	+4		-32-10-2
23	V	+2 +3 +4 +5	50.942	-8-11-2
41	Nb	+3 +5	92.906	-18-12-1
73	Ta	+5	180.95	-32-11-2
105	Db			-32-11-2
24	Cr	+2 +3 +6	51.996	-8-13-1
42	Mo	+6	95.95	-18-13-1
74	W	+6	183.84	-32-12-2
106	Sg			-32-12-2
25	Mn	+2 +3 +4 +6 +7	54.938	-8-13-2
43	Tc	+4 +6 +7		-18-13-2
75	Re	+4 +6 +7	186.21	-32-13-2
107	Bh			-32-13-2
26	Fe	+2 +3	55.845	-8-14-2
44	Ru	+3 +4 +6 +8	101.07	-18-15-1
76	Os	+3 +4 +6 +8	190.23	-32-14-2
108	Hs			-32-14-2
27	Co	+2 +3	58.933	-8-15-2
45	Rh	+3	102.91	-18-16-1
77	Ir	+3 +4	192.22	-32-15-2
109	Mt			-32-15-2
28	Ni	+2 +3	58.693	-8-16-2
46	Pd	+2 +4	106.42	-18-18-0
78	Pt	+2 +4	195.08	-32-17-1
110	Ds			-32-16-2
29	Cu	+1 +2	63.546	-8-18-1
47	Ag	+1	107.87	-18-18-1
79	Au	+1 +3	196.97	-32-18-1
111	Rg			-32-17-2
30	Zn	+2	65.38	-8-18-2
48	Cd	+2	112.41	-18-18-2
80	Hg	+1 +2	200.59	-32-18-2
112	Cn			-32-18-2

Groups 13–18

Z	Symbol	Oxidation States	Atomic Weight	Config
5	B	+3	[10.806, 10.821]	2-3
13	Al	+3	26.982	2-8-3
31	Ga	+3	69.723	-8-18-3
49	In	+3	114.82	-18-18-3
81	Tl	+1 +3	[204.38, 204.39]	-32-18-3
113	Nh			-32-18-3
6	C	+2 +4 −4	[12.009, 12.012]	2-4
14	Si	+2 +4 −4	[28.084, 28.086]	2-8-4
32	Ge	+4	72.630	-8-18-4
50	Sn	+2 +4	118.71	-18-18-4
82	Pb	+2 +4	207.2	-32-18-4
114	Fl			-32-18-4
7	N	+1 +2 +3 +4 +5 −1 −2 −3	[14.006, 14.008]	2-5
15	P	+3 +5 −3	30.974	2-8-5
33	As	+3 +5 −3	74.922	-8-18-5
51	Sb	+3 +5	121.76	-18-18-5
83	Bi	+3 +5	208.98	-32-18-5
115	Mc			-32-18-5
8	O	+1 +2 −1 −2	[15.999, 16.000]	2-6
16	S	+4 +6 −2	[32.059, 32.076]	2-8-6
34	Se	+4 +6 −2	78.971	-8-18-6
52	Te	+4 +6 −2	127.60	-18-18-6
84	Po	+2 +4		-32-18-6
116	Lv	+3		-32-18-6
9	F	−1	18.998	2-7
17	Cl	+1 +5 +7 −1	[35.446, 35.457]	2-8-7
35	Br	+1 +5 −1	[79.901, 79.907]	-8-18-7
53	I	+1 +5 +7 −1	126.90	-18-18-7
85	At			-32-18-7
117	Ts			-32-18-7
2	He	0	4.0026	2
10	Ne	0	20.180	2-8
18	Ar	0	39.948	2-8-8
36	Kr	0	83.798	-8-18-8
54	Xe	0	131.29	-18-18-8
86	Rn	0		-32-18-8
118	Og			-32-18-8

Lanthanoids

Z	Symbol	Oxidation States	Atomic Weight	Config
57	La	+3	138.91	-18-9-2
58	Ce	+3 +4	140.12	-19-9-2
59	Pr	+3 +4	140.91	-21-8-2
60	Nd	+3	144.24	-22-8-2
61	Pm	+3		-23-8-2
62	Sm	+2 +3	150.36	-24-8-2
63	Eu	+2 +3	151.96	-25-8-2
64	Gd	+3	157.25	-25-9-2
65	Tb	+3	158.93	-27-8-2
66	Dy	+3	162.50	-28-8-2
67	Ho	+3	164.93	-29-8-2
68	Er	+3	167.26	-30-8-2
69	Tm	+3	168.93	-31-8-2
70	Yb	+2 +3	173.05	-32-8-2
71	Lu	+3	174.97	-32-9-2

Actinoids

Z	Symbol	Oxidation States	Atomic Weight	Config
89	Ac	+3		-18-9-2
90	Th	+4	232.04	-18-10-2
91	Pa	+4 +5	231.04	-20-9-2
92	U	+3 +4 +5 +6	238.03	-21-9-2
93	Np	+3 +4 +5 +6		-22-9-2
94	Pu	+3 +4 +5 +6		-24-8-2
95	Am	+3 +4 +5 +6		-25-8-2
96	Cm	+3		-25-9-2
97	Bk	+3 +4		-27-8-2
98	Cf	+3 +4		-28-8-2
99	Es	+3		-29-8-2
100	Fm	+3		-30-8-2
101	Md	+2 +3		-31-8-2
102	No	+2 +3		-32-8-2
103	Lr	+3		-32-8-3

Atomic weights are IUPAC 2016 values abridged to five significant digits. See "Standard Atomic Weights" in Sec. 1 for an explanation of the IUPAC notation for atomic weight ranges. Physical state and chemical behavior of Ts and Og are uncertain.

Appendix A: Quizzes

All questions in Appendix A are *multiple choice.*

A.1 QUIZ 1: ELECTROLYTE SOLUTIONS

1. The absolute value of the chemical potential
   a. Is known
   b. Is approximately known
   c. Is not known
2. In practical calculations, instead of the standard chemical potential, we should use
   a. Standard entropy
   b. Standard Gibbs energy of formation
   c. Standard enthalpy of formation
3. The chemical potential
   a. Depends on the concentration scale
   b. Does not depend on the concentration scale
   c. Slightly depends on the concentration scale
4. The activity coefficient
   a. Depends on the concentration scale
   b. Does not depend on the concentration scale
   c. None of these
5. The standard chemical potential
   a. Depends on the concentration scale
   b. Does not depend on the concentration scale
   c. None of the these
6. The chemical potential depends on
   a. Concentration only
   b. Temperature and pressure only
   c. Concentration, temperature and pressure
7. In aqueous electrolyte solutions, the standard chemical potential of water is defined when
   a. Mole fraction of water $\to 0$
   b. Mole fraction of electrolyte $\to 1$
   c. Mole fraction of water $\to 1$
8. In aqueous solutions, the standard chemical potential of electrolyte is defined when
   a. Molality of the electrolyte $\to 0$
   b. Molality of the electrolyte $\to 1$
   c. Mole fraction of water $\to 0$

9. Activity of a component in aqueous solution
    a. Is dimensionless
    b. Has dimensions of molality
    c. Has dimensions of molarity
10. If molality of $CuSO_4(aq)$ is 0.005 mol kg^{-1}, the dimensionless ionic strength of the solution, I, is
    a. 0.005
    b. 0.01
    c. 0.02
11. The ionic strength, I_b, of an aqueous solution that consists of 0.01 mol kg^{-1} of HCl and 0.02 mol kg^{-1} of $CaCl_2$ is
    a. 0.06 mol kg^{-1}
    b. 0.07 mol kg^{-1}
    c. 0.08 mol kg^{-1}
12. The dissociation constant of the acetic acid can be found in Chapter 10. Calculate pH of 0.02 mol kg^{-1} $CH_3COOH(aq)$ solution assuming the activity coefficients of ions equal 1.
    a. pH = 1.23
    b. pH = 2.23
    c. pH = 3.23
13. For a $CaCl_2(aq)$ solution, if the individual ion activity coefficients are equal, then
    a. $\gamma_+ = \gamma_\pm$
    b. $\gamma_- = \gamma_\pm$
    c. Both of these
14. The difference between the experimental (Chapter 10) and calculated (using Debye–Hückel limiting law) mean activity coefficient of 0.01 mol kg^{-1} NaCl(aq) is about
    a. 0.136
    b. 0.0136
    c. 0.00136
15. For 0.005 mol kg^{-1} $CuSO_4(aq)$ solution, the difference between $\gamma_\pm$ calculated by the first and second approximations of the Debye–Hückel theory is
    a. −0.0625
    b. −0.1625
    c. +0.1625
16. The electrolyte activity coefficient can be
    a. <1
    b. >1
    c. Both of these
17. If concentrations of acid and salt in the CH_3COOH/CH_3OONa buffer solution are equal, pH of the buffer solution is about
    a. 2.78
    b. 4.78
    c. 6.78

18. Calculate the dissociation constant of $NH_4OH(aq)$ if the degree of dissociation of $0.006\,mol\,kg^{-1}$ solution is 0.053, and the activity coefficients of all species equal 1.
    a. 1.78×10^{-4}
    b. 1.78×10^{-5}
    c. 1.78×10^{-6}
19. If pH of a strong electrolyte aqueous solution of $NaOH(aq)$ at 25°C is 12, the concentration of $OH^-(aq)$ ions is
    a. Exactly $10^{-2}\,mol\,kg^{-1}$
    b. Slightly less than $10^{-2}\,mol\,kg^{-1}$
    c. Slightly larger than $10^{-2}\,mol\,kg^{-1}$
20. The mole fraction of water in $3\,mol\,kg^{-1}$ $KCl(aq)$ solution is
    a. 0.949
    b. 0.0949
    c. 0.00949
21. In $0.03\,mol\,kg^{-1}$ $CaCl_2(aq)$ solution, the concentration of $Cl^-(aq)$ ions is
    a. $0.03\,mol\,kg^{-1}$
    b. $0.06\,mol\,kg^{-1}$
    c. $0.09\,mol\,kg^{-1}$

A.2   QUIZ 2: ELECTROCHEMICAL CELLS

1. Electric current density is
    a. A core SI unit
    b. A derived SI unit
    c. Not an SI unit
2. The amount of electric charge is
    a. Current × time
    b. Potential difference × time
    c. Resistance × time
3. The electrochemical system that produces electricity by consuming chemicals is
    a. An electrolytic (electrolysis) cell
    b. A galvanic cell
    c. A chemical cell
4. The electrochemical system that produces chemicals by consuming electricity is
    a. An electrolytic (electrolysis) cell
    b. A galvanic cell
    c. A chemical cell
5. At the anode, a reaction of
    a. Reduction takes place
    b. Oxidation takes place
    c. Neutralization takes place

6. In an electrolytic cell, the cathode is
   a. Positively charged
   b. Negatively charged
   c. Either positively or negatively charged
7. During water electrolysis, if 2 L of hydrogen is produced, then the volume of oxygen produced is
   a. 1 L
   b. 2 L
   c. 3 L
8. What is the internal resistance of a galvanic Daniell cell if the cell potential difference is 0.771 V, the equilibrium potential difference is 1.101 V, and the circuit current is 1.5 mA?
   a. 220 Ω
   b. 22 Ω
   c. 2.2 Ω
9. $Cu^{2+}(aq) + 2e^- = Cu(s)$ is
   a. An oxidation reaction
   b. A reduction reaction
   c. A neutralization reaction
10. $Zn(s) = Zn^{2+}(aq) + 2e^-$ is
    a. An oxidation reaction
    b. A reduction reaction
    c. A neutralization reaction
11. In a galvanic cell, the magnitude of the cell potential is always
    a. Larger than the magnitude of the equilibrium potential
    b. Smaller than the magnitude of the equilibrium potential
    c. The same as the magnitude of the equilibrium potential
12. In an electrolytic cell, the magnitude of the applied potential should always be
    a. Larger than the magnitude of the decomposition potential
    b. Smaller than the magnitude of the decomposition potential
    c. The same as the magnitude of the decomposition potential
13. What is the applied potential of the Daniell cell in an electrolytic mode if the decomposition potential is -1.101 V, the circuit current is -3 mA, and the internal resistance is 100 Ω?
    a. -1.401 V
    b. -0.801 V
    c. -0.401 V
14. It is good practice if on both ends of an electrochemical system the metal terminals are
    a. Made by Al(s)
    b. Different
    c. The same
15. The salt bridge in the Daniell cell consists of
    a. LiCl(aq)
    b. NaCl(aq)
    c. KCl(aq)

16. The power (P) in an electrochemical cell is defined as
    a.   Current divided by potential difference
    b.   Current multiplied by potential difference
    c.   Current plus potential difference

17. Using Chapter 10, calculate the resistance of an aluminum wire (at 20°C) with the length of 1 m and cross section of 1 mm^2.
    a. 2.65×10^{-1} Ω
    b. 2.65×10^{-2} Ω
    c. 2.65×10^{-3} Ω

18. If 50 mg of Zn is electrochemically deposited in the Daniell cell, how much charge should be passed through the cell assuming that there are not any parasitic reactions in the electrolysis?
    a.   1.48 C
    b.   14.8 C
    c.   148 C

19. What is molal concentration of 20% (by mass) KCl(aq) solution in a salt bridge?
    a.   1.353 mol kg^{-1}
    b.   2.353 mol kg^{-1}
    c.   3.353 mol kg^{-1}

20. How much H_2O(l) would you need to electrolytically produce 2 m^3 of H_2(g) at 25°C and 1 bar?
    a.   14.53 g
    b.   145.3 g
    c.   1453 g

A.3   QUIZ 3: ELECTRICAL CONDUCTIVITY

1. To correctly measure the conductivity of an electrolyte solution
   a.   ac should be used
   b.   dc should be used
   c.   Either of this can be used

2. If the length of a wire increases, the wire resistance
   a.   Increases
   b.   Decreases
   c.   Does not change

3. If the cross section of a wire increases, the wire resistivity
   a.   Increases
   b.   Decreases
   c.   Does not change

4. The dimension of conductance is
   a. Ω
   b.   S
   c.   V A^{-1}

5. Conductance of distilled water is
   a.   Higher than conductance of tap water

    b. Lower than conductance of tap water

    c. About the same as conductance of tap water

6. The conductivity of Millipore water is around

    a. 5.5×10^{-6} S m^{-1}

    b. $5.5 \times$ S m^{-1}

    c. 5.5×10^{6} S m^{-1}

7. Measuring the impedance of a conductivity cell with an aqueous solution, the solution resistance can be obtained by extrapolating the frequency to

    a. ∞

    b. 0

    c. Both of these

8. Using Chapter 10, find the limiting molar conductivity of $CuSO_4$(aq).

    a. 267.2×10^{-4} S m^{2}mol^{-1}

    b. 133.6×10^{-4} S m^{2}mol^{-1}

    c. 534.4×10^{-4} S m^{2}mol^{-1}

9. For NH_4OH(aq) solution, the $\Lambda(c^{1/2})$ dependence is

    a. Linear

    b. Nonlinear

    c. None of these

10. The molar limiting conductivity, Λ^o, is

    a. An additive physical quantity

    b. Not an additive physical quantity

    c. A multiplicative physical quantity

11. Conductivity of an individual ion (λ_+ or λ_-)

    a. Can be experimentally obtained

    b. Cannot be experimentally obtained

    c. Can be calculated only theoretically

12. The hopping (Grotthuss) mechanism can be used to explain the conductivity of

    a. H^+(aq)

    b. OH^-(aq)

    c. Both of these

13. Walden's rule can be formulated as

    a. The product of the electrolyte limiting mobility and the solvent viscosity is constant

    b. The product of the electrolyte limiting conductivity and the solvent viscosity is constant

    c. Both of these

14. If the conductivity of KOH(aq) solution in an alkaline electrolyzer increases, the applied electrical power to keep the same hydrogen production

    a. Increases

    b. Decreases

    c. Does not change

15. Using the molar conductivity of LiCl(aq) to be found in Chapter 10, find the limiting ionic conductivities of Li$^+$(aq) and Cl$^-$(aq) if the Li$^+$(aq) transport number is 0.33.

a. 38×10^{-4} and 77×10^{-4} S m^{2}mol^{-1}
b. 77×10^{-4} and 38×10^{-4} S m^{2}mol^{-1}
c. 57.5×10^{-4} and 57.5×10^{-4} S m^{2}mol^{-1}

16. The limiting ionic conductivities of Li$^+$(aq) and Rb$^+$(aq) are, respectively, 38.7×10^{-4} and 77.8×10^{-4} S m^{2}mol^{-1}, so we can conclude that
    a. The hydration shell of Li$^+$(aq) is larger than the hydration shell of Rb$^+$(aq)
    b. The hydration shell of Rb$^+$(aq) is larger than the hydration shell of Li$^+$(aq)
    c. The hydration shells of Rb$^+$(aq) and Li$^+$(aq) are about the same

17. Using the limiting ionic conductivities of H$^+$(aq) and SO$_4^{2-}$(aq) from Chapter 10, the limiting molar conductivity ($\Lambda°$) of H$_2$SO$_4$(aq) can be calculated as
    a. 511.4×10^{-4} S m^{2}mol^{-1}
    b. 859.3×10^{-4} S m^{2}mol^{-1}
    c. 1207.2×10^{-4} S m^{2}mol^{-1}

18. What is conductance of 0.1 mol kg^{-1} KCl(aq) solution at 25°C (use Chapter 10) in a conductivity cell that consists of two parallel 1 cm^2 electrodes that are 2 cm apart?
    a. 0.641×10^{-1} S
    b. 0.641×10^{-2} S
    c. 0.641×10^{-3} S

19. Someone put 0.0001 g of NaCl(s) into 1 L of Millipore water, so the conductivity of water contaminated by NaCl became
    a. 5.46×10^{-6} S m^{-1}
    b. 21.6×10^{-6} S m^{-1}
    c. 27.1×10^{-6} S m^{-1}

20. The electric conductivity of Cu(s) is provided by
    a. Electrons
    b. Ions
    c. Both of these

21. The electric conductivity in KCl(aq) solution is provided by
    a. Ions
    b. Electrons
    c. Both of these

A.4   QUIZ 4: EQUILIBRIUM ELECTROCHEMISTRY

1. The standard Gibbs energy of formation of any chemical element in its reference state is
   a. $= 0$
   b. $\neq 0$
   c. None of these

2. Absolute value of the Galvani potential difference (electrical potential difference between two phases)
   a. Is known

b. Is unknown

c. Can be thermodynamically calculated

3. When temperature increases, the standard OCP of an H_2/O_2 fuel cell
   a. Increases
   b. Decreases
   c. Does not change

4. If one uses data of Chapter10, it can be found that when temperature increases, the standard Gibbs energy of a $C(s)/O_2(g)$ fuel cell reaction
   a. Increases
   b. Decreases
   c. Does not change

5. If a fuel cell reaction is $(1/2)CH_4(g) + O_2(g) = (1/2)CO_2(g) + H_2O(g)$, the standard Gibbs energy of the reaction is defined as
   a. $\Delta_f G^\circ\left(H_2O,g\right)+\Delta_f G^\circ\left(CO_2,g\right)-\Delta_f G^\circ\left(CH_4,g\right)-\Delta_f G^\circ\left(O_2,g\right)$
   b. $\Delta_f G^\circ\left(H_2O,g\right)+0.5\Delta_f G^\circ\left(CO_2,g\right)-0.5\Delta_f G^\circ\left(CH_4,g\right)-\Delta_f G^\circ\left(O_2,g\right)$
   c. $\Delta_f G^\circ\left(H_2O,g\right)+0.5\Delta_f G^\circ\left(CO_2,g\right)+0.5\Delta_f G^\circ\left(CH_4,g\right)+\Delta_f G^\circ\left(O_2,g\right)$

6. The precise relationship between Gibbs energy ($\Delta_r G$), enthalpy ($\Delta_r H$), and entropy ($\Delta_r S$) of a chemical reaction at temperature T is
   a. $\Delta_r G_T = \Delta_r H_T - T\Delta_r S_T$
   b. $\Delta_r G_T = \Delta_r H_T + T\Delta_r S_T$
   c. $\Delta_r G_T = \Delta_r H_T - T_o\Delta_r S_T$, where $T_o = 298.15$ K

7. If a fuel cell reaction is $H_2(g) + (1/2)O_2(g) = H_2O(g)$, the Nernst equation (at low pressure) can be written as
   a. $E = E^0 + \left[RT/(2F)\right]\ln\left[(pH_2)(pO_2)^{0.5}/pH_2O\right]$
   b. $E = E^0 - \left[RT/(2F)\right]\ln\left[(pH_2)(pO_2)^{0.5}/pH_2O\right]$
   c. $E = E^0 \pm \left[RT/(2F)\right]\ln\left[(pH_2)(pO_2)^{0.5}/pH_2O\right]$

8. The absolute entropy of any chemical is
   a. >0
   b. <0
   c. $=0$

9. The following half-reaction takes place at the oxygen gas electrode: $O_2(g) + 2H^+(aq) + 2e^- = H_2O(l)$. If one uses the data from Chapter 10, the standard electrode potential corresponding to this half-reaction at 100°C and 1 bar is
   a. 1.067 V
   b. 1.167 V
   c. 1.267 V

10. In a carbon (graphite)/oxygen fuel cell, the cathodic reaction is $O_2(g) + 4e^- = 2O^{2-}$ and the anodic one is $CO_2(g) + 4e^- = C(s) + 2O^{2-}$. If one uses data from Chapter10, the standard electrode potential difference of the fuel cell at 1000 K and 1 bar is
    a. 1.026 V
    b. 1.126 V
    c. 1.226 V

11. What is the OCP of the H_2/Cl_2 electrochemical cell [Cu|Pt|H_2(g)|HCl(aq)|C l_2(g)|Pt|Cu] at 25°C, if partial pressure of hydrogen and chlorine is 0.5 bar and concentration of HCl(aq) is 5 mol kg^{-1}?
    a. 1.358 V
    b. 1.213 V
    c. 1.426 V
12. Using the ionic activity coefficients calculated from the Debye–Hückel limiting law, one can calculate the OCP of a Daniell cell (at 25°C and 1 bar) with $ZnSO_4$(aq, 0.01 mol kg^{-1}) on the left-hand side and $CuSO_4$(aq, 0.002 mol kg^{-1}) on the right-hand side as
    a. 1.102 V
    b. 1.088 V
    c. 1.117 V
13. A person body can approximately be considered as
    a. In equilibrium state
    b. In steady state
    c. None of these
14. Any thermodynamic function can be calculated if
    a. G is known as a function of T and P
    b. G is known as a function of T and V
    c. G is known as a function of P and V
15. The OCP of the Daniell cell measured at ambient conditions
    a. Can be used to immediately calculate the Gibbs energy of the corresponding electrochemical reaction
    b. Can be used to immediately calculate the enthalpy of the corresponding electrochemical reaction
    c. Can be used to immediately calculate the entropy of the corresponding electrochemical reaction
16. A two-phase electrochemical system is in equilibrium if
    a. Chemical potentials of the same species in both phases are equal
    b. Electrochemical potentials of the same species in both phases are equal
    c. Electric potentials of the same species in both phases are equal
17. The selected electrode to conventionally define the Galvani potential difference is
    a. Silver–silver chloride
    b. Oxygen
    c. Hydrogen
18. Gibbs energy for the following electrochemical half-reaction is zero.
    a. H^+(aq) + e$^-$ = 0.5H_2(g)
    b. $0.5O_2$(g) + 2H^+(aq) + 2e$^-$ = H_2O(l)
    c. AgCl + e$^-$ = Ag(s) + Cl$^-$(aq)
19. If the activity of HCl(aq) in the Harned cell, Cu(s)|Pt|H_2(g)|HCl(aq)|AgCl(s) |Ag(s)|Cu, increases, the cell potential difference, $E = E_R - E_L$,
    a. Increases
    b. Decreases
    c. Does not change

20. The standard electrode potential of $0.5O_2(g) + 2H^+(aq) + 2e^- = H_2O(l)$ and $O_2(g) + 4H^+(aq) + 4e^- = 2H_2O(l)$ is
    a. The same
    b. Slightly different
    c. Different

A.5 QUIZ 5: ELECTROCHEMICAL TECHNIQUES I

1. The hydrogen electrode potential depends on
   a. pH
   b. Hydrogen partial pressure
   c. Both of these
2. The RH electrode is Zn(s) in $0.5\,\text{mol kg}^{-1}$ $ZnSO_4(aq)$, and the LH electrode is silver/silver chloride reference electrode in $1\,\text{mol kg}^{-1}$ KCl(aq). Using data from Chapter 10, calculate the potential difference between the RH and LH electrodes at 25°C.
   a. −0.984 V
   b. −1.041 V
   c. +1.041 V
3. In this course, potentiometry is used to experimentally determine
   a. A standard electrode potential
   b. pH
   c. Both of these
4. pH of DI water, which is in contact with pure hydrogen, should be
   a. =7
   b. >7
   c. <7
5. If one uses $2\,\text{mol kg}^{-1}$ KCl(aq) for preparing a silver/silver chloride reference electrode, the expected electrode potential at 25°C is
   a. 0.222 V
   b. 0.234 V
   c. 0.219 V
6. Using data from Chapter 10, one can calculate the electrode potential of the calomel electrode in $0.1\,\text{mol kg}^{-1}$ KCl(aq) at 25°C as
   a. 0.334 V
   b. 0.281 V
   c. 0.268 V
7. Using data from Chapter 10, one can calculate the standard electrode potential of Ag/AgCl electrode at a temperature of 50°C as
   a. 0.204 V
   b. 0.252 V
   c. 0.157 V
8. Using a glass electrode at 25°C, pH and potential of the standard solution are $pH_S = 8$ and $E_S = 400\,\text{mV}$, respectively. What is the pH of a test solution if its potential (E_X) is 800 mV?
   a. 1.24

b. 1.38

c. 1.01

9. Using a glass electrode at 15°C, the pH_S and potential (E_S) of the standard solution are 8 and 400 mV, respectively. What is the pH of a test solution if its potential (E_X) is 800 mV?

   a. 1.24

   b. 1.38

   c. 1.01

10. In order to avoid error from a diffusion potential,

   a. The Harned cell can be used

   b. A cell without transfer can be used

   c. Both of these

11. If a small amount of water is added to a buffer solution, its pH will

   a. Not practically be changed

   b. Be changed

   c. Significantly be changed

12. In the SHE, platinum is used because the metal is highly

   a. Chemically stable

   b. Catalytically active

   c. Both of these

13. A disadvantage of the SHE is

   a. Complicated and expensive design

   b. Contamination of the electrode by impurities

   c. Both of these

14. The common commercial Ag/AgCl reference electrode can properly work up to

   a. 30°C

   b. 80°C

   c. 110°C

15. Precision of a commercial Ag/AgCl reference electrode should be about

   a. ±0.01 mV

   b. ±0.1 mV

   c. ±10 mV

16. The silver–silver chloride reference electrode is a typical

   a. Redox electrode

   b. Metal electrode

   c. Metal/salt electrode

17. The standard electrode potentials in Electrochemical Series are given at

   a. 25°C and 1.013 bar

   b. 25°C and 1 atm

   c. Either of these

18. Using data from Chapter 10, calculate the diffusion potential of HCl (0.1 mol L⁻¹, Sol. I)|KCl (0.1 mol L⁻¹, Sol. II) liquid junction at 25°C.

   a. +28.5 mV

   b. −28.5 mv

   c. 0 mV

19. Using data from Chapter 10 and the Debye–Hückel limiting law, calculate the diffusion potential of the HCl (0.001 mol kg^{-1}, Sol. I)|HCl (0.01 mol kg^{-1}, Sol. II) liquid junction at 25°C.
    a. −36.6 mV
    b. −38.6 mv
    c. −41.6 mV

20. Using data from Chapter 10, find the mean activity coefficient of 0.1 mol kg^{-1} HCl(aq), and calculate the potential difference of the Harned cell (Ag/AgCl electrode on the right) at a temperature of 25°C and at hydrogen partial pressure of 1 bar.
    a. 0.352 V
    b. 0.252 V
    c. 0.152 V

A.6   QUIZ 6: ELECTROCHEMICAL KINETICS

1. Overpotential of a cell is
   a. A deviation of the electrode potential from its equilibrium value
   b. A deviation of the electrode potential from its equilibrium value excluding the potential drop due to the cell internal resistance
   c. A deviation of the electrode potential from the SHE potential

2. To measure the overpotential of a single electrode,
   a. A two-electrode cell should be used
   b. A three-electrode cell should be used
   c. A one-electrode cell should be used

3. The three-electrode cell consists of
   a. Two working electrodes and a reference electrode
   b. A working electrode, a counter electrode, and a reference electrode
   c. A working electrode and two reference electrodes

4. Positive deviation from the zero overpotential is called
   a. Cathodic polarization
   b. Anodic polarization
   c. Electrode polarization

5. The counter electrode should
   a. Have a large surface
   b. Not degrade
   c. Both of these

6. The reference electrode should have
   a. Nernstian behavior
   b. A stable electrode potential
   c. Both of these

7. It is usually assumed that the main contribution to the total electrode overpotential is coming from
   a. An electron transfer process
   b. A mass transfer process
   c. Both of these

8. IUPAC recommends the cathodic current to be
   a. Positive
   b. Negative
   c. Either positive or negative
9. If the exchange current density is 10^{-1} A cm^{-2}, the electrochemical reaction is rather
   a. Fast
   b. Slow
   c. Very slow
10. If the exchange current density is smaller, the electrochemical reaction should
   a. Be faster
   b. Be slower
   c. Has the same rate
11. The symmetry coefficient ranges from
   a. -1 to 1
   b. -1 to 0
   c. 0 to 1
12. Based on the Tafel equation, when the current goes to zero, the half-reaction overpotential goes to
   a. Zero
   b. One
   c. Either plus or minus infinity
13. Based on the Butler–Volmer equation, when current goes to zero, the half-reaction overpotential goes to
   a. Zero
   b. One
   c. Either plus or minus infinity
14. The standard value of the exchange current density is the exchange current density when
   a. Concentration of Ox equals 1 mol L^{-1}
   b. Concentration of Red equals 1 mol L^{-1}
   c. Both of these
15. The current density approaches its limiting value (limiting current density) when surface concentration is
   a. The same as in the bulk
   b. Zero
   c. Between the bulk concentration and zero
16. Using available experimental data on the exchange current densities, calculate the area-specific charge transfer resistance of dissolution of Cd(s) in $[10^{-2}$ mol L^{-1} $CdSO_4$(aq) $+ 0.4$ mol L^{-1} K_2SO_4(aq)] solution at 25°C.
   a. 8.564 Ω cm^2
   b. 7.564 Ω cm^2
   c. 6.564 Ω cm^2
17. Using available experimental data on the exchange current densities and symmetry coefficients, calculate coefficients A_T and B_T' of Tafel's equation

for anodic polarization of $Fe^{3+}(aq) + e^- = Fe^{2+}(aq)$ reaction on Pt in 1 mol
L^{-1} $H_2SO_4(aq)$ solution at 25°C.
a. $A_T = 0.275$ V and $B_T' = 0.102$ V dec^{-1}
b. $B_T' = 0.275$ V and $A_T = 0.102$ V dec^{-1}
c. $A_T = 0.275$ V dec^{-1} and $B_T' = 0.102$ V

18. What is the diffusion overpotential in deposition reaction, $Cu^{2+}(aq) + 2e^-$
= $Cu(s)$, at a current density of 0.5 A cm^{-2} and a temperature of 25°C, if the
limiting current density is 0.6 A cm^{-2}?
a. -13 mV
b. -23 mV
c. -33 mV

19. Using available experimental data on the exchange current densities and
symmetry coefficients and employing the Butler–Volmer equation, calculate
the current density of the hydrogen half-reaction, $2H^+(aq) + 2e^- = H_2(g)$,
on Pt(s) electrode in 1 mol L^{-1} H_2SO_4 (aq) at an anodic polarization of 0.1 V.
a. 0.049 A cm^{-2}
b. 0.49 A cm^{-2}
c. 4.9 A cm^{-2}

20. Based on the generalized Butler–Volmer equation, when current goes to its
anodic limiting value, the potential goes to
a. Plus infinity
b. Minus infinity
c. Either plus or minus infinity

A.7   QUIZ 7: ELECTROCHEMICAL TECHNIQUES II

1. Temperature of a body is
a. A scalar
b. A vector
c. A curl

2. From the potentiostatic current–time measurements
a. The diffusion layer thickness can be obtained
b. The diffusion coefficient can be obtained
c. Either of these

3. From the galvanostatic current–time measurements
a. The diffusion coefficient can be obtained if bulk concentration and the
transport number are known
b. The bulk concentration can be obtained if the current density is measured and the diffusion coefficient and the transport number are known
c. Either of these

4. When stirring in an electrochemical system is more intensive, the Nernst
diffusion layer is
a. Smaller
b. Larger
c. Not changing

5. In an aqueous electrochemical cell at ambient conditions, the Nernst diffusion layer is about
   a. Ten times smaller than the Prandtl layer
   b. Ten times larger than the Prandtl layer
   c. The same as the Prandtl layer
6. In a simple electrochemical cell with stirring, the Nernst diffusion layer thickness can
   a. Be precisely estimated
   b. Be approximately estimated
   c. Not be estimated
7. The Nernst diffusion layer thickness can be calculated
   a. In a simple three-electrode cell
   b. In an RDE cell
   c. In a simple two-electrode cell
8. The main feature of an RDE cell is
   a. Precise current measurement
   b. Precise potential measurement
   c. Equally accessible electrode surface
9. The limiting current density vs. square root of the rotation rate in an RDE cell is
   a. A parabola
   b. A hyperbola
   c. A straight line
10. When the rotation rate of an RDE goes to infinity, the diffusion layer thickness approaches
    a. Infinity
    b. Zero
    c. A constant value
11. Calculate the Nernst diffusion layer thickness on an RDE if an electrochemical reaction takes place in a dilute aqueous solution at 25°C and 1 bar ($v = 10^{-6} m^2 s^{-1}$), the angular velocity is $100 s^{-1}$, and diffusion coefficient of the electrochemically active species is $10^{-9} m^2 s^{-1}$.
    a. 0.161 micron
    b. 1.61 micron
    c. 16.1 micron
12. Calculate the limiting current for the hydrogen half-reaction, $H^+(aq) + e^- = 0.5 H_2(g)$, which takes place at an RDE in HCl(aq) solution with pH = 4 at 25°C and 1 bar ($v = 10^{-6} m^2 s^{-1}$), at an angular velocity of $100 rad s^{-1}$. The diffusion coefficient of proton can be found in Chapter 10.
    a. 2.65×10^{-2} A cm^{-2}
    b. 2.65×10^{-3} A cm^{-2}
    c. 2.65×10^{-4} A cm^{-2}
13. Based on the vector analysis, the variation of molar concentration with time is
    a. Negative divergence of the flux vector
    b. Negative gradient of the flux vector
    c. Divergence of the flux vector

14. Using an RDE to take polarization curves at a steady state, when the rotation rate increases, the limiting current
    a. Increases
    b. Decreases
    c. Does not change
15. The $[Fe(CN)_6]^{3-} + e^- = [Fe(CN)_6]^{4-}$ reaction on Pt(s) is
    a. A relatively fast reaction
    b. A quite slow reaction
    c. A very slow reaction
16. To employ the cyclic voltammetry method
    a. A three-electrode cell should be used
    b. A two-electrode cell should be used
    c. Both of these
17. A maximum or a minimum on CV polarization curves can appear due to coupling of
    a. Diffusion of the electrochemically active species and electron transfer reaction
    b. Convection of the electrochemically active solution and electron transfer reaction
    c. Convection of the electrochemically active solution and diffusion of the electrochemically active species
18. When the CV peak potential separation depends on the scan rate, the electrochemical reaction is
    a. Fast
    b. Slow
    c. Very fast
19. When the CV peak current ratio is the same regardless of the scan rate, the electrochemical reaction is
    a. Irreversible
    b. Slow
    c. Reversible
20. From CV curves for the $[Fe(CN)_6]^{3-}/[Fe(CN)_6]^{4-}$ couple on Pt(s), one can see that the electrode potential minimum and maximum
    a. Are significantly shifting with changing the scan rate
    b. Are slightly shifting with changing the scan rate
    c. Are not shifting with changing the scan rate
21. A useful application of the EIS is a possibility to extract the solution
    a. resistance
    b. chemical composition
    c. viscosity

A.8 QUIZ 8: ELECTROCHEMICAL ENERGY CONVERSION

1. The fuel cell was invented by
   a. M. Faraday
   b. W. Nernst
   c. C.F. Schönbein

2. The fuel cell was discovered in
    a. 18th century
    b. 19th century

    c. 20th century
3. Aqueous KOH solution is used as electrolyte in
    a. PEMFC
    b. SOFC
    c. AFC
4. DMFC is a kind of
    a. PEMFC
    b. SOFC
    c. AFC
5. An anion conductive membrane is used in
    a. PEMFC
    b. DMFC
    c. SOFC
6. In DMFC water is
    a. Consumed at the anode
    b. Produced at the cathode
    c. Both of these
7. Carbon monoxide can be used as a fuel in
    a. PEMFC
    b. DMFC
    c. SOFC
8. At ambient temperature and pressure, for the PEMFC cathode, the main challenge to be addressed is the
    a. Two-phase boundary problem
    b. Three-phase boundary problem
    c. Four-phase boundary problem
9. At ambient pressure, and temperature above 100°C, for the PEMFC cathode, the main challenge to be addressed is the
    a. Two-phase boundary problem
    b. Three-phase boundary problem
    c. Four-phase boundary problem
10. In PEMFC, the electrolyte is
    a. An ion conductive aqueous solution
    b. An ion conductive ceramic membrane
    c. An ion conductive polymeric membrane
11. Using data from Chapter 10, estimate the standard value of the Gibbs energy of an H_2/O_2 PEMFC [$H_2(g) + 0.5O_2(g) = H_2O(g)$] at 1000 K.
    a. $-228.6 \, kJ \, mol^{-1}$
    b. $-192.6 \, kJ \, mol^{-1}$
    c. $-164.4 \, kJ \, mol^{-1}$
12. Using data from Chapter 10, calculate the standard value of the decomposition potential (DP) of the H_2O electrolysis [$H_2O(l) = H_2(g) + (1/2)O_2(g)$] at 100°C.

    a. −1.229 V

    b. −1.167 V

    c. +1.167 V

13. Using data from Chapter 10, the standard thermodynamic efficiency of a C/O_2 fuel cell [$C(s) + O_2(g) = CO_2(g)$] at 900 K is

    a. <100%

    b. >100%

    c. =100%

14. Using data from Chapter 10, calculate the voltage efficiency of a C/O_2 fuel cell [$C(s) + O_2(g) = CO_2(g)$] at 1000 K when the fuel cell potential is 0.5 V and all activities equal 1.

    a. 41%

    b. 49%

    c. 55%

15. If the thermodynamic, voltage, and current efficiencies are, respectively, 93, 51, and 95%, the total efficiency of an electrochemical energy conversion system (without heat exchange) is

    a. 35%

    b. 45%

    c. 55%

16. Using the following parameters of the potential–current density equation ($A_{FC} = 0.08$ V, $B_{FC} = 0.05$ V, $j_o = 0.001$ A cm^{-2}, $j_p = 0.01$ A cm^{-2}, $j_{lim} = 1.0$ A cm^{-2}, $r = 0.2$ Ω cm^2) for an H_2/O_2 PEMFC, [$H_2(g) + (1/2)O_2(g) = H_2O(l)$, 25°C], calculate the cell potential, E_{FC}, at (I) 0 A cm^{-2} and (II) 0.7 A cm^{-2} with all species at unit activity.

    a. (I) 1.229 V and (II) 0.505 V

    b. (I) 1.045 V and (II) 0.705 V

    c. (I) 1.045 V and (II) 0.504 V

17. The fuel cell was developed by

    a. M. Faraday

    b. W. Nernst

    c. W. Grove

18. In a water electrolyzer at ambient conditions and at low current, the heat is

    a. Released

    b. Consumed

    c. Neither released nor consumed

19. In an H_2/O_2 fuel cell at ambient conditions and at low current, the amount of heat released due to electrochemical reactions is about

    a. 38.7 kJ mol^{-1}

    b. 48.7 kJ mol^{-1}

    c. 58.7 kJ mol^{-1}

20. Using data from Chapter 10, calculate the standard thermodynamic efficiency of a water electrolyzer at ambient conditions.

    a. 120%

    b. 83%

    c. 100%

A.9   QUIZ 9: ELECTROCHEMICAL CORROSION

1. The Flade potential is due to metal
   a.   Activation
   b.   Transpassivation
   c.   Passivation
2. The electrochemical corrosion involves
   a.   A cathodic reaction
   b.   An anodic reaction
   c.   Both of these
3. In an electrochemical corrosion process, the current of the anodic and cathodic reactions should be
   a.   Zero
   b.   Equal
   c.   Different
4. In an electrochemical corrosion measurement
   a.   The corrosion current can be measured
   b.   The corrosion potential can be measured
   c.   Both of these
5. To correctly study metal corrosion rates, the electrochemical cell should have
   a.   Two electrodes
   b.   Three electrodes
   c.   Either of these
6. The current density in the passive region of a corrosion polarization curve is
   a.   Zero
   b.   Small
   c.   Large
7. If the corrosion current is $0.1 \, mA \, cm^{-2}$, the corrosion rate of iron due to $Fe(s) \rightarrow Fe^{2+}(aq) + 2e^-$ reaction is
   a. $1.16 \, mm \, year^{-1}$
   b. $0.116 \, mm \, year^{-1}$
   c. $0.0116 \, mm \, year^{-1}$
8. The potential–pH (Pourbaix) diagram can
   a.   Be used to estimate the corrosion rate
   b.   Not be used to estimate the corrosion rate
   c.   Be used to approximately estimate the corrosion rate
9. Immunity of Al(s) at ambient conditions in pure water can be achieved at Eh
   a. $> -1.9 \, V$
   b. $< -1.9 \, V$
   c. $< 0 \, V$
10. In the Cu Pourbaix diagram, the line corresponding to the $Cu^{2+}(aq) + 2H_2O$ (l) $= Cu(OH)_2(s) + 2H^+(aq)$ reaction is
   a.   Horizontal
   b.   Vertical
   c.   Inclined

11. In the Cu Pourbaix diagram, the line corresponding to the $Cu^{2+}(aq) + 2e^-$ = Cu(s) half-reaction is
    a. Horizontal
    b. Vertical
    c. Inclined
12. In the Cu Pourbaix diagram, the line corresponding to the $Cu_2O(s) + 2H^+(aq) + 2e^- = 2Cu(s) + H_2O(l)$ half-reaction is
    a. Horizontal
    b. Vertical
    c. Inclined
13. When pH = 0 and temperature is 25°C, the oxygen reduction line $[2H^+(aq) + (1/2)O_2(g) + 2e^- = H_2O(l)]$ in the Pourbaix diagram intersects the Eh axis at
    a. 0 V
    b. 1.23 V
    c. −1.23 V
14. When pH = 7 and temperature is 25°C, the hydrogen evolution line $[2H^+(aq) + 2e^- = H_2(g)]$ in the Pourbaix diagram intersects the Eh axis at
    a. About 0.0 V
    b. About −0.4 V
    c. About −0.8 V
15. Mitigation of corrosion can be provided by
    a. Forming a protective film
    b. Electrochemical cathodic protection
    c. Both of these
16. If Zn(s) and Fe(s) are galvanically coupled, the metal to be corroded is
    a. Fe(s)
    b. Zn(s)
    c. Both of these
17. The cathodic protection of Cu(s) can be provided if Cu(s) is galvanically connected to
    a. Ag(s)
    b. Au(s)
    c. Zn(s)
18. In any Pourbaix diagram, a line represents
    a. A mass action law equation
    b. A Nernst's equation
    c. Either of these
19. To calculate the corrosion current,
    a. The corrosion polarization resistance should be known
    b. Tafel constants should be known
    c. Both of these should be known
20. Transpassivation occurs due to
    a. Another anodic electrochemical reaction
    b. Breakdown of a passive film
    c. Either of these

Appendix B: Video-Based Assignments

B.1 ASSIGNMENT 1: ELECTROLYTE SOLUTIONS

Video URL: http://youtu.be/fEtMwU32odo

1. Solutions made on the molar concentration scale should be prepared in
   a. Graduated cylinders
   b. Erlenmeyer flasks
   c. Volumetric flasks
   d. Any of the above
2. The mass of water added must be known when making a solution on the molal scale.
   a. True
   b. False
3. How many acceptable methods of mixing in volumetric flasks are mentioned in the video?
   a. 1
   b. 2
   c. 3
   d. 4
4. When preparing a solution on the molal scale, the mass of water to be added can be adjusted precisely with
   a. A pipette
   b. A beaker
   c. A graduated cylinder
   d. A volumetric flask
5. In this video, when adding components together to be mixed, the salt is added first.
   a. True
   b. False
6. It is important to measure volume when making a solution on the molal scale.
   a. True
   b. False
7. The mass of water added must be known when making a solution on the molar scale.
   a. True
   b. False
8. When preparing a molar solution, the solute was added first.
   a. True
   b. False

9. When making a solution on the molar scale, water must be added until the solution reaches
   a. The base of the flask neck
   b. The etched line on the flask neck
   c. The top of the flask neck
   d. The desired mass
10. When making a solution on the molal scale, water must be added until the solvent reaches
   a. The base of the flask neck
   b. The etched line on the flask neck
   c. The top of the flask neck
   d. The desired mass
11. When the volumetric flask is filled to the etched line, the acceptable method of mixing is
   a. Vortexing
   b. Inversion
   c. Either inversion or vortexing
   d. Neither inversion or vortexing
12. When using a gas regulator, the first valve to be opened is
   a. The pressure dial
   b. The regulator spring
   c. The cylinder valve
   d. The needle valve
13. When the cylinder valve is opened, the gauge reading the cylinder pressure should be
   a. The low pressure gauge
   b. The high pressure gauge
   c. Either of these
14. The item for fine-tuning the outlet pressure from a gas cylinder is
   a. The spring dial
   b. The cylinder valve
   c. The needle valve
15. When fine-tuning the regulator attached to a gas cylinder, the outlet tubing should be vented to a fume hood.
   d. True
   e. False
16. When bubbling solution into an experimental system, which item should be turned to start the flow of gas into the experimental system?
   a. The spring dial
   b. The cylinder valve
   c. The needle valve
17. This lab video indicates that a _____ is used to de-aerate a solution after it is used for several minutes.
   a. Vacuum pump
   b. Stirring

    c. Plunger

    d. Sparger

18. When de-pressurizing a gas cylinder, the _____ should be closed first.

    a. The spring dial

    b. The cylinder valve

    c. The needle valve

    d. None of the above

19. After the needle valve is opened during the de-pressurization procedure, the low pressure gauge drops first and then the high pressure gauge drops.

    a. True

    b. False

20. According to the video, bubbling excess gas from a test vessel through a separate beaker is one way to test for

    a. A de-aerated solution

    b. A negative pressure

    c. A leakless vessel

    d. A positive pressure

B.2   ASSIGNMENT 2: ELECTROCHEMICAL CELLS

Video URL: http://youtu.be/XSRFUZ9txdA

1. The mode of cell operation shown in the lab video is

    a. Equilibrium

    b. Galvanic

    c. Electrolytic

    d. All of these

2. In the Daniell cell, the electrodes are placed in a mixture of $ZnSO_4$(aq) and $CuSO_4$(aq).

    a. True

    b. False

3. In the Daniell cell, the copper metal bar is always a positive electrode.

    a. True

    b. False

4. The salt bridge in the Daniell cell is for

    a. Conducting ions

    b. Separating the electrode chambers

    c. Both of these

5. The instrument used in the Daniell Cell lab for accurate measurement of the cell potential is

    a. High-impedance electrometer

    b. Ammeter

    c. Power supply

6. Choose a schematic of the experimental setup used as a galvanic cell.

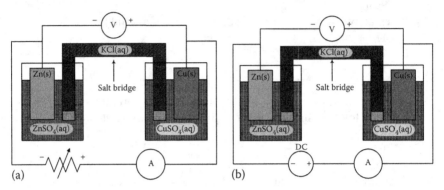

(a)   (b)

7. What is the purpose of using a dc power supply in the Daniell Cell lab?
   a.   To measure current
   b.   To measure the cell potential
   c.   To provide electrical power
8. Select three reasons for polishing the metal electrodes with sand paper before performing electrochemical measurements are to
   a.   Remove the metal oxide layer
   b.   Provide safety
   c.   Avoid an undesirable reaction
   d.   Remove the impurities
   e.   Avoid additional overpotentials
9. In the Daniell cell, power is produced when the cell is in
   a.   The equilibrium mode
   b.   The galvanic mode
   c.   The electrolytic mode
10. Select the right mathematical relation among power (P), current (I), and cell potential (E).
   a. $P = IE$
   b. $P = I/E$
   c. $E = PI$
11. The correct expression for Faraday's law of electrolysis is
   a. $m_i = M_i Q/F$
   b. $m_i = nM_i Q/F$
   c. $m_i = v_i M_i It/(Fn)$
12. Calculate the total charge passed in the Daniell cell during 20 minutes using the data given in the following table.

Time, min	Current, mA
0	21
5	22
10	23
15	24
20	25

    a.   25.2 C
    b.   27.6 C
    c.   30.0 C

13. Calculate the mass of copper dissolved during electrolysis in the Daniell cell if the charge passed through the electrode is 30.62 C.
    a.   0.0101 g
    b.   0.101 g
    c.   1.010 g

14. The reaction $H_2O(l) \rightarrow (1/2) O_2(g) + H_2(g)$ takes place in
    a.   An equilibrium cell
    b.   An electrolytic cell
    c.   A galvanic cell

15. Arrange in an order the experimental procedures for the open circuit potential measurement.
    a.   Assembling the cell
    b.   Polishing electrodes
    c.   Connecting electrochemical devices
    d.   Weighing electrodes

16. What is the polarity of the zinc electrode during the electrolysis process in the Daniell Cell?
    a.   Negative
    b.   Positive
    c.   Either negative or positive

17. How should waste solutions from the Daniell Cell lab be dealt with?
    a.   Should be diluted and drained
    b.   Should be collected in special containers
    c.   Should be left in the lab as they are

18. Using the electrolytic polarization plot shown in the following figure, estimate the decomposition potential and the internal resistance of a Daniell cell.

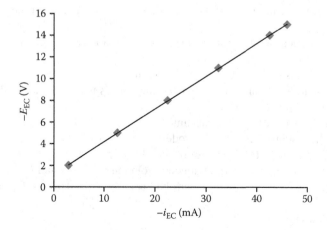

    a.   −1.2 V and 300 Ω
    b.   +1.2 V and 300 Ω
    c.   −1.2 V and 600 Ω

19. What is the power density generated by a fuel cell producing 0.5 A cm^{-2} at 0.7 V?
    a.   0.35 W cm^{-2}
    b.   0.35 J cm^{-2}s^{-1}
    c.   Both of these
20. The reaction $Zn(s) + CuSO_4(aq) \rightarrow Cu(s) + ZnSO_4(aq)$ takes place in
    a.   An equilibrium cell
    b.   An electrolytic cell
    c.   A galvanic cell

B.3   ASSIGNMENT 3: ELECTRICAL CONDUCTIVITY

Video URL: http://youtu.bc/GpdErVzWaQU

1. Resistance is related to conductance by
   a. $1/S = 1/\Omega$
   b. $R = 1/L$
   c. $R = G$
   d.   All of these
2. Electrolyte conductivity is determined by (select all applicable)
   a.   Mobility of the ions
   b.   Charge of the ions
   c.   Concentration of the ions
   d.   Hydration shell of the ions
   e.   All of them
3. The relation between resistance and conductivity is
   a. $R = A/(l\ k)$
   b. $R = l/(A\ k)$
   c. $R = k/(A\ l)$
4. For best practices with conductivity measurements
   a.   Start measurements at the lowest concentration and then proceed to the highest
   b.   Start measurements at the highest concentration and then proceed to the lowest.
   c.   No measurement order is recommended
5. A standard solution with known conductivity is used prior to conductivity measurements to determine
   a.   The conductivity cell constant
   b.   The distance between electrodes in the conductivity cell
   c.   The cross-sectional area of the conductivity cell
6. At ambient conditions, if an aqueous solution with 0.01 mol L^{-1} of sodium chloride is compared to an electrolyte solution with 0.01 mol L^{-1} of acetic acid, the sodium chloride has a molar conductivity that is
   a.   The same
   b.   Lower
   c.   Higher

7. What is the mass of dry sodium chloride that should be used to prepare 500 mL of 0.01 mol L^{-1} NaCl(aq) solution?
   a. 0.585 g
   b. 0.346 g
   c. 0.293 g
   d. 0.201 g
   e. 0.393 g

8. Electrochemical impedance spectroscopy can be used to gain an estimate for an electrolyte resistance when extrapolating the cell impedance to
   a. A low frequency
   b. A high frequency
   c. None of these

9. For solutions with a molar concentration less than 4 mol L^{-1}, as the concentration of sodium chloride increases, the solution conductivity
   a. Increases
   b. Decreases
   c. Does not change

10. Molar conductivity is related to conductivity as
    a. $\Lambda = 1/\kappa$
    b. $\Lambda = \kappa c$
    c. $\Lambda = \kappa/c$

11. Kohlrausch's law, $\Lambda = \Lambda^\circ - K(c)^{\frac{1}{2}}$, is valid for NH$_4$OH(aq) solution.
    a. True
    b. False

12. Kohlrausch's law, $\Lambda = \Lambda^\circ - K(c)^{\frac{1}{2}}$, is valid for NaOH(aq) solution.
    a. True
    b. False

13. The use of soap for cleaning the conductivity cell between measurements is highly encouraged.
    a. True
    b. False

14. Electrolyte conductivity depends on temperature.
    a. True
    b. False

15. The acetic acid and sodium chloride solutions used in this lab can be disposed of down the drain.
    a. True
    b. False

16. At ambient conditions, the infinitely dilute solution of sodium chloride, when compared with the infinitely dilute solution of acetic acid, has a molar conductivity that is
    a. The same
    b. Lower
    c. Higher

17. For solutions with a molar concentration more than 6 mol L^{-1}, as the concentration of hydrochloric acid increases, the solution conductivity

  a. Increases
  b. Decreases
  c. Does not change
18. Nafion is
  a. An ion conductive polymer
  b. An ion conductive ceramic material
  c. An electron conductor
19. When measuring conductivity, temperature should be
  a. Constant
  b. Precisely defined
  c. Both of these
20. When measuring conductivity, the test solution
  a. Should be free of atmospheric CO_2
  b. Can have some atmospheric CO_2
  c. Should be saturated with atmospheric CO_2

B.4 ASSIGNMENT 4: EQUILIBRIUM ELECTROCHEMISTRY

Video URL: http://youtu.be/BvMIhH2tiiw

  1. Select all electrodes that are used in this lab?
     a. Hydrogen electrode
     b. Silver/silver chloride electrode
     c. Copper metal electrode
     d. Calomel electrode
     e. Zinc metal electrode
  2. Select the correct Nernst equations for the copper metal electrode reduction
     reaction.
     a. $E = E^0 - [RT/(F)] \ln (a_{Cu2+})$
     b. $E = E^0 + [RT/(2F)] \ln (1/a_{Cu2+})$
     c. $E = E^0 - [RT/(2F)] \ln (1/a_{Cu2+})$
     d. $E = E^0 + [RT/(F)] \ln (a_{Cu2+})$
     e. $E = E^0 + [RT/(2F)] \ln (a_{Cu2+})$
  3. Solution concentrations should be tested from
     a. Lowest to highest
     b. Highest to lowest
     c. In either direction; there is no advantage to any measurement order
  4. The equilibrium electrode potential measurements can be made using
     a. A zero resistance ammeter
     b. A digibridge
     c. A high-resistance electrometer
     d. A power supply
  5. The standard electrode potential of the Ag/AgCl reference electrode at 25°C
     and 1 bar is
     a. 0.7996 V
     b. 0.2223 V

    c. 0.1976 V

    d. 0.2368 V

6. The standard electrode potential of the $Cu^{2+}(aq)/Cu(s)$ electrode at 25°C and 1 bar is

    a. 0.521 V

    b. 0.153 V

    c. 0.342 V

    d. 0.445 V

7. The dimensional ionic strength of $0.005\,mol\,kg^{-1}$ $CuSO_4(aq)$ is

    a. $0.005\,mol\,kg^{-1}$

    b. $0.02\,mol\,kg^{-1}$

    c. $0.04\,mol\,kg^{-1}$

    d. $0.01\,mol\,kg^{-1}$

8. Calculate the mean activity coefficient of $0.005\,mol\,kg^{-1}$ $CuSO_4(aq)$ at 25°C and 1 bar.

    a. 0.577

    b. 0.453

    c. 0.738

    d. 0.982

9. Calculate the electrode potential of the $Cu^{2+}(aq)/Cu(s)$ electrode at 25°C and 1 bar, which is in contact with $0.005\,mol\,kg^{-1}$ $CuSO_4(aq)$.

    a. 270 mV

    b. 263 mV

    c. 274 mV

    d. 267 mV

10. Copper sulfate can be disposed of down the drain.

    a. True

    b. False

11. When the copper electrode is on the right side with respect to a reference electrode and the concentration of copper sulfate solution is increased, the cell potential difference between the right- and left-hand electrodes is decreased.

    a. True

    b. False

12. If the solution density of $4.0\,mol\,L^{-1}$ KCl(aq) at 25°C is $1173\,g\,L^{-1}$, what is molality of the solution?

    a. $4.00\,mol\,kg^{-1}$

    b. $3.43\,mol\,kg^{-1}$

    c. $4.27\,mol\,kg^{-1}$

    d. $4.57\,mol\,kg^{-1}$

13. Calculate the electrode potential of the $Ag/AgCl$ reference electrode with a $4\,mol\,L^{-1}$ KCl(aq) solution at 25°C and 1 bar.

    a. 0.2223 V

    b. 0.2002 V

    c. 0.1968 V

    d. 0.2444 V

14. Conventionally, the cell potential difference for this electrochemical cell should be calculated as

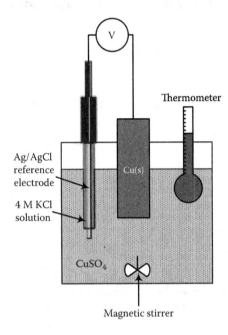

    a. E_{cell} = Cu²⁺/Cu electrode potential–Ag/AgCl electrode potential
    b. E_{cell} = Ag/AgCl electrode potential–Cu²⁺/Cu electrode potential
    c. E_{cell} = Cu²⁺/Cu electrode potential + Ag/AgCl electrode potential

15. From a thermodynamic standpoint, the relationship between OCP and DP is
    a. OCP = DP
    b. OCP = –DP
    c. OCP ≠ DP

16. Using the electrochemical series, one could find that the standard Ag⁺/Ag electrode has a potential that is
    a. More positive then Cu²⁺/Cu electrode
    b. More negative then Cu²⁺/Cu electrode
    c. The same as Cu²⁺/Cu electrode

17. Using thermodynamic data from Chapter 10, one can calculate the standard electrode potential of the $AgCl(s) + e^- = Ag(s) + Cl^-(aq)$ half-reaction at a temperature of 100°C and 1 bar as
    a. 0.157 V
    b. 0.222 V
    c. 0.284 V

18. The standard electrode potentials of $0.5O_2(g) + 2H^+(aq) + 2e^- = H_2O(l)$ and $0.5O_2(g) + 2H^+(aq) + 2e^- = H_2O(g)$ at ambient conditions are
    a. The same
    b. Different
    c. Almost the same

19. Using the Pourbaix diagram for Cu(s), one can conclude that in the presence of hydrogen, this metal is thermodynamically stable at
   a. pH > 7
   b. pH < 7
   c. Any pH
20. A Cu^{2+}(aq)/Cu^+(aq) electrode is considered a
   a. Metal electrode
   b. Redox electrode
   c. Gas electrode
   d. Metal/salt electrode
   e. Metal/oxide electrode

B.5   ASSIGNMENT 5: ELECTROCHEMICAL TECHNIQUES I

Video URL: http://youtu.be/LzjL2jkaUuQ

1. Select all concentrations of KCl(aq) are used in this lab?
   a. 1 mol L^{-1} KCl(aq)
   b. 2 mol L^{-1} KCl(aq)
   c. 3 mol L^{-1} KCl(aq)
   d. 4 mol L^{-1} KCl(aq)
   e. 5 mol L^{-1} KCl(aq)
2. Silver chloride powder degrades slowly in
   a. Air
   b. Water
   c. Light
3. What hydrocarbon is used in the torch?
   a. Acetylene
   b. Butane
   c. Propane
   d. Decane
4. Select all reasons for sanding the silver wire
   a. Remove oxides
   b. Remove contaminates
   c. Polish it
5. The HNO_3(aq) for removing impurities can be reused.
   a. True
   b. False
6. When making an Ag/AgCl electrode, silver powder is melted over the silver chloride bulb.
   a. True
   b. False
7. After the silver/silver chloride electrodes are constructed, the electrodes are set to soak in
   a. HCl(aq) for 24 h; then KCl(aq) for 24 h
   b. KCl(aq) for 24 h; then HCl(aq) for 24 h
   c. HNO_3(aq) for 24 h; then HCl(aq) for 24 h

8. What potential difference between the commercial reference electrode and the custom reference electrode, both in contact with the same solutions, will indicate an error of less than 10%?
    a. 2 mV
    b. 4 mV
    c. 6 mV

9. The custom reference electrode in the lab is in contact with _____ during measurements against the commercial reference electrodes.
    a. 1 mol L⁻¹ HCl(aq)
    b. 1 mol L⁻¹ KCl(aq)
    c. 4 mol L⁻¹ KCl(aq)
    d. 1 mol L⁻¹ HNO₃(aq)

10. Calculate the electrode potential of the Ag/AgCl reference electrode that is in contact with a 1 mol kg⁻¹ KCl(aq) solution at 25°C and 1 bar.
    a. 0.2345 V
    b. 0.2353 V
    c. 0.2368 V
    d. 0.2373 V

11. Calculate the electrode potential of the Ag/AgCl reference electrode in contact with a 1.88 mol L⁻¹ KCl(aq) solution that has a solution density of 1081.26 g L⁻¹ at 25°C and 1 bar.
    a. 0.2345 V
    b. 0.2223 V
    c. 0.2204 V
    d. 0.2188 V

12. Calculate the diffusion potential at 25°C and 1 bar when solution (I) in Eq. 5.7 is KCl(aq) with an activity of 2.965 and solution (II) is KCl(aq) with an activity of 1.146.
    a. −0.46 mV
    b. +0.46 mV
    c. −4.6 mV
    d. +4.6 mV

13. Calculate the diffusion potential at 25°C and 1 bar when solution (I) in Eq. 5.6 is 0.001 mol L⁻¹ KCl(aq) and solution (II) is 0.001 mol L⁻¹ HCl(aq).
    a. −2.7 mV
    b. +2.7 mV
    c. −27 mV
    d. +27 mV

14. For solutions with the same concentrations at 25°C and 1 bar, the calculated diffusion potential of
    a. HCl(aq) | KCl(aq) > HCl(aq) | LiCl(aq)
    b. NaCl(aq) | KCl(aq) > HCl(aq) | KCl(aq)
    c. HCl(aq) | KCl(aq) > HCl(aq) | NaCl(aq)
    d. HCl(aq) | KCl(aq) < HCl(aq) | LiCl(aq)

15. The diffusion potential can correctly be derived using
    a. Equilibrium thermodynamics
    b. Nonequilibrium thermodynamics
    c. Hydrodynamics
16. In order to minimize the diffusion (liquid junction) potential, the salt bridge should consist of
    a. LiCl(aq)
    b. NaCl(aq)
    c. KCl(aq)
17. Which reference electrode is the safest one to use in an educational electrochemical lab
    a. Silver/silver chloride
    b. Calomel
    c. Hydrogen
18. In order to minimize the diffusion potential, the salt bridge should consist of
    a. A high concentrated KCl(aq) solution
    b. A low concentrated KCl(aq) solution
    c. A very low concentrated KCl(aq) solution
19. The Ag/AgCl reference electrode can
    a. Be prepared in a university lab
    b. Not be prepared in a university lab
    c. Only be commercially prepared
20. In order to use a glass electrode,
    a. One buffer solution can be used
    b. More than one buffer solutions can be used
    c. Both of these

B.6   ASSIGNMENT 6: ELECTROCHEMICAL KINETICS

Video URL: http://youtu.be/LYUEFuAM62c

1. What does RDE stand for?
    a. Rapid diffusion electrode
    b. Reference disc electrode
    c. Rotating disc electrode
    d. Reference diffusion electrode
2. What does LSV stand for?
    a. Linear sweep voltammetry
    b. Laplace sinewave voltammetry
    c. Lateral sweep voltammetry
    d. Logarithmic sweep voltammetry
3. What is the background solution used in this lab?
    a. $6.4\,mmol\,L^{-1}$ $K_3Fe(CN)_6$(aq)
    b. $1\,mol\,L^{-1}$ KNO_3(aq)

    c. 6.4 mol L^{-1} K$_3$Fe(CN)$_6$(aq)

    d. 1 mmol L^{-1} KNO$_3$(aq)

4. What is the test solution used in this lab?

    a. 6.4 mmol L^{-1} K$_3$Fe(CN)$_6$(aq)

    b. 1 mol L^{-1} KNO$_3$(aq)

    c. 6.4 mol L^{-1} K$_3$Fe(CN)$_6$(aq)

    d. 1 mmol L^{-1} KNO$_3$(aq)

5. What is the reference electrode used in this lab?

    a. Calomel with KNO$_3$(aq)

    b. Calomel with KCl(aq)

    c. AgCl/Ag with KNO$_3$(aq)

    d. AgCl/Ag with KCl(aq)

6. What is the metal used for the counter and working electrodes in this lab?

    a. Ruthenium

    b. Palladium

    c. Platinum

    d. Silver

7. What is the maximum speed allowable for the RDE in this lab?

    a. 400 RPM

    b. 900 RPM

    c. 1600 RPM

    d. 2500 RPM

8. As the RDE speed increases, the limiting current

    a. Increases

    b. Decreases

    c. Does not change

9. Decreasing convection in the electrolyte increases the mass transport.

    a. False

    b. True

10. The counter electrode should be rotating.

    a. False

    b. True

11. Potassium ferricyanide and potassium nitrate are toxic and cannot be dumped down the drain.

    a. True

    b. False

12. A polarization curve can be obtained by performing

    a. Electrochemical impedance spectroscopy

    b. Open circuit potential measurements

    c. Linear sweep voltammetry

    d. Scanning electron microscopy

13. With an area-specific charge transfer resistance of 2.569 Ω cm^2 and an over-potential of 9 mV, the resulting current density is

    a. 23.1 A cm^{-2}

    b. 3.5 A cm^{-2}

    c. 23.1 mA cm^{-2}

    d. 3.5 mA cm^{-2}

14. At very small charge transfer overpotentials ($\eta < 10\,\text{mV}$), this equation can be used to accurately describe the relationship between overpotential and current density.
    a. $\eta = jRT/(j_o\, nF)$
    b. $\eta = [RT/(\beta nF)]\ln j_o - [RT/(\beta nF)]\ln |j|$
    c. $\eta = [RT/(nF)]\ln [1 - (j/j_{lim})]$
    d. $\eta = \{RT/[(1-\beta)nF)]\}\ln j - \{RT/[(1-\beta)nF)]\}\ln j_o$

15. For an electrode under charge transfer control with an overpotential $\eta > 30\,\text{mV}$, this equation can be used to accurately describe the relationship between overpotential and current density.
    a. $\eta = jRT/(j_o\, nF)$
    b. $\eta = [RT/(\beta nF)]\ln j_o - [RT/(\beta nF)]\ln |j|$
    c. $\eta = [RT/(nF)]\ln [1 - (j/j_{lim})]$
    d. $\eta = \{RT/[(1-\beta)nF)]\}\ln j - \{RT/[(1-\beta)nF)]\}\ln j_o$

16. For an electrode under charge transfer control, this equation can be used to accurately describe the relationship between overpotential and current density at any overpotential.
    a. $\eta = jRT/(j_o\, nF)$
    b. $\eta = [RT/(\beta nF)]\ln j_o - [RT/(\beta nF)]\ln |j|$
    c. $\eta = [RT/(nF)]\ln [1 - (j/j_{lim})]$
    d. $\eta = \{RT/[(1-\beta)nF)]\}\ln j - \{RT/[(1-\beta)nF)]\}\ln j_o$
    e. None of these

17. For an electrode with diffusion mass transfer and charge transfer control, this equation can be used to accurately describe the relationship between overpotential and current density at any overpotential.
    a. $j = j_o\,[\exp\{(1-\beta)nF\eta/(RT)\} - \exp\{-\beta nF\eta/(RT)\}$
    b. $j = j_o[(1 - (j/j_{lim,\,a}))\exp\{(1-\beta)nF\eta/(RT)\} - (1 - (j/j_{lim,\,c}))\exp\{-\beta nF\eta/(RT)\}]$
    c. $j = j_o\, nF\eta/(RT)$
    d. $j = -j_o\exp[-\beta nF\eta/(RT)]$

18. For a polarized electrode only under the influence of a charge transfer overpotential, select all parameters that can be changed in value with the electrode material?
    a. Symmetry coefficient
    b. Limiting current density
    c. Exchange current density

19. For a polarized electrode only under the influence of a mass transfer overpotential, select all parameters that will change in value with changes in the Nernst diffusion layer?
    a. Current density
    b. Limiting current density
    c. Exchange current density
    d. Symmetry coefficient

20. For a polarized electrode with both mass and charge transfer overpotentials, select all parameters that will change in value with changes in the bulk solution concentration?
    a. Current density
    b. Limiting current density

    c. Exchange current density
    d. Standard exchange current density

B.7 ASSIGNMENT 7: ELECTROCHEMICAL TECHNIQUES II

Video URL: http://youtu.be/4fjVCuZEMz0

1. What reference electrode does this lab use?
    a. Silver/silver chloride
    b. Calomel
    c. Hydrogen
    d. Platinum
2. What working electrode does this lab use?
    a. Silver/silver chloride
    b. Calomel
    c. Hydrogen
    d. Platinum
3. What counter electrode does this lab use?
    a. Silver/silver chloride
    b. Calomel
    c. Hydrogen
    d. Platinum
4. What is the electrochemical couple studied in this lab?
    a. $[Fe(CN)_6]^{3-}(aq) + e^- = [Fe(CN)_6]^{4-}(aq)$
    b. $2H^+(aq) + 2e^- = H_2(g)$
    c. $Hg_2Cl_2(s) + 2e^- = 2Hg(l) + 2Cl^-(aq)$
5. The $[Fe(CN)_6]^{3-}(aq) + e^- = [Fe(CN)_6]^{4-}(aq)$ reaction is considered an irreversible one.
    a. True
    b. False
6. If the peak potential separation between $E_{c(min)}$ and $E_{a(max)}$ on a CV curve depends on the scan rate, the electron transfer reaction is considered fast.
    a. True
    b. False
7. When performing cyclic voltammetry, the scan rate should usually be
    a. $> 5\,mV\ s^{-1}$
    b. $< 5\,mV\ s^{-1}$
    c. $= 5\,mV\ s^{-1}$
8. With cyclic voltammetry, it is always possible to determine if the charge transfer reaction is fast or slow from the ratio of the $j_{a(max)}$ and $j_{c(max)}$ current peaks.
    a. True
    b. False
9. In cyclic voltammetry, potential is the
    a. Dependent variable
    b. Independent variable
    c. A constant parameter

10. If the peak potential difference on a single cyclic voltammetry curve is 58 mV, and the peak currents are proportional to the scan rate, this confirms that
    a. The charge transfer reaction is fast
    b. The charge transfer reaction uses one electron
    c. The charge transfer reaction is irreversible
    d. All of the above

11. For the $[Fe(CN)_6]^{3-}(aq) + e^- = [Fe(CN)_6]^{4-}(aq)$ reaction, if the scan rate increases in cyclic voltammetry, the peak current magnitudes for the anodic and cathodic sweeps
    a. Increase
    b. Decrease
    c. Remain the same

12. The RDE surface
    a. Should be polished
    b. Should not be polished
    c. Can be any

13. To properly employ the RDE system,
    a. A potentiostat should be used
    b. Only a voltmeter and an amperemeter can be used
    c. Only an electrometer can be used

14. To employ the RDE system,
    a. A three-electrode cell should be used
    b. A two-electrode cell should be used
    c. Both of these

15. Convert 60 RPM to rad s^{-1}.
    a. 5.28
    b. 6.28
    c. 7.28

B.8 ASSIGNMENT 8: ELECTROCHEMICAL ENERGY CONVERSION

Video URL: http://youtu.be/EVM04HEc8u0

1. What are the electrochemical systems demonstrated in the video?
   a. Direct methanol fuel cell
   b. PEM electrolytic cell
   c. PEM fuel cell
   d. Alkaline electrolytic cell

2. Which is the mode demonstrated first in the video?
   a. Galvanic mode
   b. Electrolytic mode
   c. Equilibrium mode
   d. None of the above

3. What are the potential and current limits advised for the electrolyzer?
   a. 3 V and 1 A
   b. 2 V and 0.5 A
   c. 1.23 V and 0.25 A
   d. 0.5 V and 1 A

4. When connecting a voltmeter to the electrolytic system, the voltmeter is connected in series.
   a. True
   b. False
5. When operating the electrolyzer, it is recommended that the circuit be broken by
   a. Disconnecting a lead from the power supply
   b. Disconnecting a lead from the voltmeter
   c. Connecting a lead to the electronic load
   d. Connecting a lead to the ammeter
6. Before measuring the gases produced for current efficiency calculations, the anode and cathode chambers should be vented to the 0.0 mL level.
   a. True
   b. False
7. When measuring the amount of gases produced via electrolysis, it was suggested that the electrolyzer should be run for
   a. 6 min
   b. 4 min
   c. 2 min
8. The two fuel cells tested in this video are connected in series.
   a. True
   b. False
9. If fuel cells are connected in series select all effects on the maximum current and potential.
   a. The maximum potential output is increased
   b. The maximum current output is increased
   c. The maximum current output is the same
   d. The maximum potential output is the same
10. The open circuit option on the electronic load
   a. Results in a large current through the fuel cells
   b. Results in a very small amount of current through the fuel cells
   c. Results in absolutely no current through the fuel cells
11. As the electronic load resistance was decreased, the potential from the two fuel cells
   a. Increased
   b. Decreased
   c. Remained the same
12. As the electronic load resistance was decreased, the current from the two fuel cells
   a. Increased
   b. Decreased
   c. Remained the same
13. In the electrolytic cell, oxygen is produced at the
   a. Anode
   b. Cathode
   c. Neither of these

14. In an H_2/O_2 fuel cell, the positive electrode is the electrode that
    a. Consumes water
    b. Consumes hydrogen
    c. Consumes oxygen
15. When modeling the polarization curve for a fuel cell with Eq. 8.1 and close to OCP,
    a. The mass transport resistances dominate
    b. The charge transfer resistances dominate
    c. The ohmic resistances dominate
16. When modeling the polarization curve for a fuel cell with Eq. 8.1 and the cell potential approaches zero,
    a. The mass transport resistances dominate
    b. The charge transfer resistances dominate
    c. The ohmic resistances dominate
17. When modeling the polarization curve for a fuel cell with Eq. 8.1 and the current response vs. cell potential is linear,
    a. The mass transport resistances dominate
    b. The charge transfer resistances dominate
    c. The ohmic resistances dominate
18. A power supply is used to study
    a. Galvanic mode
    b. Electrolytic mode
    c. Equilibrium mode
19. An electronic load is used to study
    a. Galvanic mode
    b. Electrolytic mode
    c. Equilibrium mode
20. The equilibrium mode should be studied using
    a. A very high resistance of the electronic load
    b. A low resistance of the electronic load
    c. A very low resistance of the electronic load

B.9 ASSIGNMENT 9: ELECTROCHEMICAL CORROSION

Video URL: http://youtu.be/kQTdqmFx_KQ

1. In a corrosion process, metal is dissolved due to
    a. Cathodic reaction
    b. Anodic reaction
    c. Either the anodic or cathodic reaction
2. In a corrosion process, electrons are consumed in the
    a. Cathodic reaction
    b. Anodic reaction
    c. Either the anodic or cathodic reaction
3. In a corrosion process, electrons are produced in the
    a. Cathodic reaction

b. Anodic reaction
c. Either the anodic or cathodic reaction

4. Fundamentally, a corrosion process is
   a. An anodic half-reaction coupled with a cathodic half-reaction
   b. An anodic half-reaction only
   c. A cathodic half-reaction only

5. When a metal is corroding in a deaerated acidic solution, the corrosion potential
   a. Can be greater or less than the hydrogen electrode potential
   b. Is less than the hydrogen electrode potential
   c. Is greater than the hydrogen electrode potential

6. If the pH of sea water is 8, then a boat hull undergoing corrosion is most likely due to the following reaction.
   a. $2H^+(aq) + 2e^- \rightarrow H_2(g)$
   b. $2H^+(aq) + 0.5O_2(g) + 2e^- \rightarrow H_2O(l)$
   c. $2H_2O(l) + 2e^- \rightarrow H_2(g) + 2OH^-(aq)$
   d. $H_2O(l) + 0.5O_2(g) + 2e^- \rightarrow 2OH^-(aq)$

7. If the pH of a solution is 2, and the solution is under a pure argon blanket, then a metal undergoing corrosion is most likely due to the following reaction.
   a. $2H^+(aq) + 2e^- \rightarrow H_2(g)$
   b. $2H^+(aq) + 0.5O_2(g) + 2e^- \rightarrow H_2O(l)$
   c. $2H_2O(l) + 2e^- \rightarrow H_2(g) + 2OH^-(aq)$
   d. $H_2O(l) + 0.5O_2(g) + 2e^- \rightarrow 2OH^-(aq)$

8. Pourbaix diagrams are based on
   a. Chemical kinetics
   b. Chemical thermodynamics
   c. Both of these
   d. Neither of these

9. The immunity region on a Pourbaix diagram indicates
   a. Corrosion is not possible due to a protective oxide layer
   b. Corrosion is slow due to an oxide layer
   c. Corrosion is not possible due to thermodynamic limitations

10. If the current density in a corrosion process is 0.01 A cm^{-2} and the overpotential is 100 mV, the area-specific corrosion polarization resistance is
    a. $10 \, \Omega \, cm^{-2}$
    b. $0.1 \, \Omega \, cm^{-2}$
    c. $10,000 \, \Omega \, cm^{-2}$

11. In a corrosion process, the anodic and cathodic reactions should have
    a. The same symmetry parameter
    b. The same potential
    c. The same current

12. Iron will have cathodic protection when in contact with
    a. Cd
    b. Zn
    c. Either of these

13. When zinc and iron form a galvanic couple,
    a. Zinc is the cathode
    b. Iron is the negative electrode
    c. Iron is the positive electrode
14. When copper and iron form a galvanic couple,
    a. Copper is corroded
    b. Iron is corroded
    c. Copper and iron are corroded
15. For the corrosion process with two electron half-reactions at 25°C, if the overpotential from E_{corr} is 20 mV and the measured current density response is 10 mA cm^{-2}, with the anodic and cathodic symmetry parameters, respectively, 0.75 and 0.25, the corrosion current density, j_{corr}, using the Stern–Geary equation is
    a. 3.11 mA cm^{-2}
    b. 6.42 mA cm^{-2}
    c. 9.21 mA cm^{-2}
16. What is the metal studied in the corrosion lab?
    a. Copper
    b. Platinum
    c. Iron
17. What is the solution used in the corrosion lab?
    a. $CuSO_4(aq)$
    b. $KNO_3(aq)$
    c. $K_3[Fe(CN)_6](aq)$
18. In the corrosion lab, the working electrode
    a. Should be polished
    b. Should not be polished
    c. Should be used as is
19. In the corrosion lab,
    a. A three-electrode cell is used
    b. A two-electrode cell is used
    c. A four-electrode cell is used
20. In the corrosion lab, the reference electrode is
    a. Hg/Hg_2Cl_2
    b. $Ag/AgCl$
    c. Pt

Appendix C: Laboratory Instructions

SAFETY GUIDELINES

- Learn safety rules before working in the lab.
- Wear protective goggles, gloves, and lab coat when working in the lab.
- Follow guidelines for chemical waste disposal.

LABORATORY: DANIELL CELL IN EQUILIBRIUM GALVANIC AND ELECTROLYTIC MODES

Video URL: http://youtu.be/XSRFUZ9txdA
Video-Based Assignment 2 (Appendix B)

OBJECTIVES

- Learn about electrolytic and galvanic cells.
- Learn how to make a classic Daniell cell.
- Learn how to measure the open-circuit potential.
- Learn how to find charge passed through the cell.
- Learn how to apply Faraday's law of electrolysis.
- Understand the function of a salt bridge.
- Follow the Safety Guidelines in carrying out this lab.

MATERIALS AND TOOLS

• Zn electrode	• Analytical balance (with precision to 0.0001 g)
• 1 M (mol L^{-1}) $ZnSO_4$(aq) solution	• High-impedance electrometer
• Cu electrode	• Ammeter (current meter)
• 1 M $CuSO_4$(aq) solution + H_2SO_4(aq) to make pH = 2	• Variable electronic load
	• DC power supply
• Salt bridge containing 1 M KCl(aq)	• Distilled water for rinsing
• Two 150 mL glass beakers	• Stopwatch
• Sandpaper	• Cables and connection clips

EXPERIMENTAL SETUP

1. Electrode Preparation and Weighing
   - Thoroughly polish the surface of zinc (Zn) and copper (Cu) electrodes with sandpaper; rinse them with DI water and dry.

351

- Take extra care to ensure that no residual moisture/sand is anywhere on the electrode or on the wire connected to the electrode. Failure to remove all contaminates will adversely affect the weighing.
- Weigh each of the electrodes using an analytical balance with an accuracy of at least 1 mg. Take five repeated readings after each 30 seconds, and calculate the average values and standard deviation of the data.

2. Assemble the Daniell Cell
   - Add approximately 80 mL of 1 M $CuSO_4$(aq) solution to one 150 mL beaker and 80 mL of 1 M $ZnSO_4$(aq) solution to another 150 mL beaker.
   - Place the Zn electrode in the $ZnSO_4$ solution and the Cu electrode in the $CuSO_4$ solution.
   - Make sure a rubber stopper is in the top of the KCl bridge. Insert the KCl salt bridge into both beakers.

Experimental Procedure

A. Open-circuit potential (OCP) of the Daniell cell
   1. Connect the high-impedance electrometer (Figure C.1.1).
      - Connect the positive (red) lead to the copper electrode and the negative (black) lead to the zinc electrode.
      - Power on the electrometer; press the zero check button.
      - The electrometer is now measuring the open-circuit potential of the cell. Take three repeated readings with 1 minute intervals and calculate the average value and standard deviation.

B. Faraday's Law of Electrolysis
   1. Assemble the Daniell cell for current controlled electrolysis, as shown in Figure C.1.2.
      - Attach the negative terminal of the DC power supply to the Zn electrode, and the positive terminal (through ammeter) to the Cu electrode.

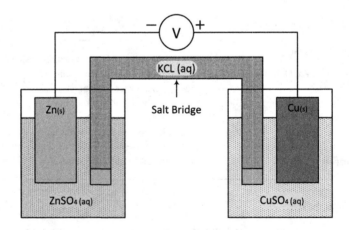

FIGURE C.1.1 Daniell cell in equilibrium mode.

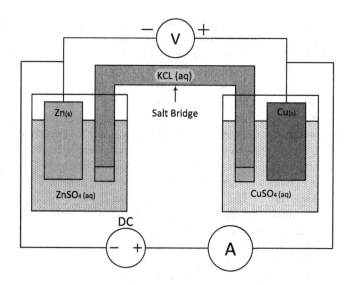

FIGURE C.1.2   Daniell cell in electrolytic mode.

- Power up the ammeter and set it to DCI-1 mode.
- Power up the DC power supply and set it to the "Current Control" (cc) mode: turn the current control knob all the way to the left, and then turn the voltage control knob 8–10 full turns to the right. The power supply is now in current control mode and is reading zero current.
2. Perform the current controlled electrolysis experiment.
    - Get the stop watch ready (reset to 0).
    - Turn the current control knob on the power supply slowly until you reach a forced current of 25 mA. *Then, immediately start the stopwatch.*
    - Allow the current to flow for 40 minutes. Record the current from the ammeter and voltage from the high-impedance voltmeter every 5 minutes. Make sure the power supply is in current control mode (cc indicator lit), and the current value is stable within 0.2 mA of the initial set point.
    - After 40 minutes, disconnect the electrodes from the circuit and record the exact time of the experiment.
3. Measure the weight change of the electrodes.
    - Remove the Cu electrode from solution, holding it by the wire; be very careful not to touch the active area of the electrode.
    - Hold the Cu electrode above a waste beaker and rinse it by spraying distilled water *above the affected area* on both sides; but do not spray the active area directly.
    - Dry the electrode by touching a corner very lightly to a Kimwipe just enough to wick off the bulk of the water, and then hang the electrode on a clamp inside the fume hood.

- Repeat the removal procedure with the Zn electrode. Be extra careful not to touch or brush off the affected surface.
- Once the electrodes are dry, weigh each of the electrodes using an analytical balance with an accuracy of at least 1 mg. Take five repeated readings after each 30 seconds, and calculate the average values and standard deviation of the data.

C. Electrolytic mode—Polarization curve (voltage–current relationship

1. Re-assemble the Daniell cell the same way as for the experiment in part B (Figure C.1.2). You do not need to re-polish the electrodes.
   - Attach the negative terminal of the power supply to the Zn electrode and positive terminal (through the ammeter) to the Cu electrode
   - Power up the DC power supply and set it to the "voltage control" (vc) mode: turn the voltage control knob all the way to the left, and then, turn the current control knob 3–4 full turns to the right. The power supply is now in voltage control mode and is reading zero current and zero voltage.
2. Perform the electrolytic polarization experiment.
   - Apply successively 2, 3, 4, 5 .... 15 V on to the cell slowly turning the voltage control knob to the right and record current readings at each voltage value.
   - Allow ~1 minute for the cell to reach a steady state at each voltage.
   - Repeat the measurements by stepping down the voltage from 15 to 2 V.

D. Galvanic mode—Polarization curve (voltage-current relationship).

1. Assemble the Daniell cell as shown in Figure C.1.3.
   - Tip: simply replace the power supply with Agilent electronic load (the same cables can be used).

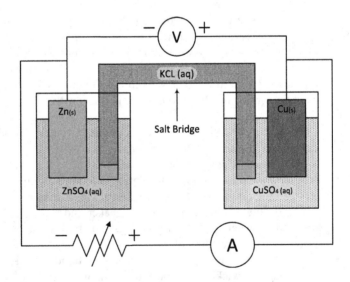

FIGURE C.1.3   Daniell cell in galvanic mode.

- Connect the negative terminal of the electronic load to the Zn electrode and positive terminal (through ammeter) to Cu electrode.
2. Perform galvanic polarization
   - Power up the electronic load.
   - Press "Mode," then "Res," and then "Enter" to set the load to "Resistance" mode.
   - Press "Res" then type "1000", and press "Enter" to set resistance to 1000 Ω.
   - Record the voltage from the high-impedance voltmeter and current from ammeter (check: the voltage should be very close to the OCP.)
   - Repeat the measurement for 200, 100, 50, 25, 10, and 1 Ω.
   - Finally, bypass the electronic load completely giving an effectively 0 Ω external resistance.

CLEAN-UP

1. Power off and disconnect all electronics.
2. Rinse each electrode into the waste beaker.
3. Rinse the salt bridge into the waste beaker and return it to storage container.
4. Pour the solutions back to stock containers.
5. Rinse each 150-mL beaker into the waste beaker.
6. Dispose off the waste in the labeled storage container under the sink.
7. Wash all glassware with tap and distilled water and clean the lab station.

DATA TREATMENT

1. Explain the main difference between electrolytic and galvanic cells.
2. Draw a schematic of the experimental setup indicating all connections and polarities.
3. Provide the electrochemical scheme of the Daniell cell, using "|" to separate phases, indicate all concentrations.
4. Write the reduction half-reactions that take place in the Daniell cell.
5. Calculate the theoretical electrode potentials for the zinc and copper electrode reactions (using Nernst equation) and then find the theoretical OCP value as their difference. Compare this value to the OCP measured in Part A of the lab. Discuss possible reasons for a discrepancy between the observed and calculated values. Make conclusions on the directions of the electrode half-reactions. Write the total spontaneous reaction.
6. Plot current values (in A) vs. time (in s), as observed in Part B of the lab. Integrate the function $i(t)$ to find the charge, $Q(C)$, passed over the duration of the experimental time:

$$Q = \int_0^t i(t)\,dt,$$

where $i(t)$ is the electric current, Q is the charge, and t is time. [Mathematica is the suggested program to use for the integration. Hint: use the Interpolation and NIntegrate functions].

7. Using Faraday's law of electrolysis, determine the theoretical change in mass of both the zinc and copper electrodes. The Faraday's law is a relationship between the quantity of charge, Q, passed through an electrode and the amount of the ith chemical, m_i, consumed or produced at the electrode:

$$m_i = v_i M_i Q / (Fn),$$

where v_i is stoichiometric coefficient of the chemical in the half-reaction, M_i is the molar mass of the chemical, and n is the electron number in half-reaction.

8. Compare the theoretical mass gain/loss for the copper and zinc electrodes found in Step 7 to the respective experimental mass changes measured in Part B of the lab. Discuss the reasons for possible discrepancies between the experimental and calculated values.

9. Plot the electrolytic polarization curve (voltage vs. current) based on data from Part C of the lab. Estimate the decomposition potential (DP) and internal resistance from these results. Compare the measured decomposition potential with the theoretical value and discuss possible reasons for the discrepancy.

10. Plot the galvanic polarization curve (voltage *vs.* current) based on data from Part D. of the lab. Estimate the internal resistance of the cell as the slope of the linear part of the polarization curve and the open-circuit potential (OCP) as well.

LABORATORY: ELECTRONIC AND IONIC CONDUCTIVITY

Video URL: http://youtu.be/GpdErVzWaQU
Video-Based Assignment 3 (Appendix B)

OBJECTIVES

- Exercise basic electrical wiring and metering.
- Experimentally determine the electrical conductivity of a copper wire.
- Experimentally determine the ionic conductivity of aqueous electrolyte solutions.
- Experimentally determine the cell constant of a conductivity cell.
- Learn how basic impedance spectroscopy techniques can determine the ionic conductivity.
- Understand the dissociation behavior of strong and weak electrolytes.
- Demonstrate proficiency in solution preparation (purity and accuracy are important.)
- Follow the Safety Guidelines in carrying out this lab.

MATERIALS

- Conductivity cell
- LCR Digibridge
- Ammeter
- DC power supply
- High-impedance voltmeter
- Standard conductivity solution
- Thermometer

- 0.01 mol L^{-1}(M) CH$_3$COOH(aq) stock solution
- 0.01 mol L^{-1} (M) NaCl(aq) stock solution
- Cu wire (on circuit board)
- 150 mL beaker
- Eight 100 mL volumetric flasks
- Graduated cylinder and pipettes
- Distilled H$_2$O

EXPERIMENTAL SETUP

Solution Preparation

1. Prior to the beginning of the lab session, calculate the amounts of stock solutions necessary to make 100 mL of each of the test solutions listed in the table (Table C.1):
2. Use a graduated cylinder or pipettes to carefully measure the portion of a stock solution (#1 in Table) and put it into a 100 mL volumetric flask. Add distilled water to the mark using wash-bottle. Note that solutions #2 and #3 can be prepared directly from the stock solution, and solutions #4 and #5 can be prepared from ten time dilution of solutions #2 and #3, respectively.
3. Make sure to label the flasks indicating # or type of solution and concentration.

Setup of LCR Digibridge

- The electrometer to be used in this experiment is an LCR Digibridge, which allows for impedance magnitude (Z_s) measurements at four different frequencies: 100, 120, 1000, and 10,000 Hz (Figure C.2.1).
- Prior to testing, connect the Digibridge to the conductivity cell. First, connect the black wire from the cell to the negative lead from the Digibridge, and the white wire from the cell to the positive red lead from the electrometer. The conductivity cell has three unused wires (red and green for temperature measurement and silver for ground; it is not necessary to ground the conductivity cell for this experiment).

TABLE C.1
Concentrations of Test Solutions

#	CH$_3$COOH(aq) (mol L^{-1})	NaCl(aq) (mol L^{-1})
1	0.01	0.01
2	0.005	0.005
3	0.001	0.001
4	0.0005	0.0005
5	0.0001	0.0001

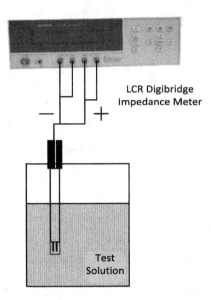

FIGURE C.2.1   Wiring diagram for measuring ionic conductivity of electrolyte solutions.

- Next, power up the system and give 1 minute for boot up.
- Submerge the conductivity cell in the standard solution (1 mol L^{-1} KCl), be sure that the Pt plates are completely submerged and no bubbles are on their surfaces, and allow the cell to sit in solution for 1 minute before taking measurements.
- Set the electrometer to measure impedance. F-2 is set to use the default voltage of 1 V as the amplitude of AC. Press F-3, then press left, and select Z_s by pressing down until the display shows Z_s. Repeat it for the right-side output to read θ, phase angle. These are the impedance measurements we want.
- To cycle through the four frequency levels, press F-1 to choose one of the values—100, 120, 1000, and 10,000 Hz. For each frequency, the electrometer will display the impedance magnitude and phase angle.

EXPERIMENTAL PROCEDURE

Ionic Conductivity Measurements

1. Determine the cell constant of the conductivity cell using 0.01 mol L^{-1} KCl standard solution.
   - Make sure to use clean glassware for all measurements. Also rinse conductivity cell with distilled water. Even small amounts of contamination can corrupt the measurements.
   - Place the conductivity probe into the beaker with the standard 0.01 mol L^{-1} KCl solution.
   - Place the thermometer into the beaker and take a temperature reading. Record and report the temperature of the solution in the final report. Note that conductivity significantly depends on temperature.

- Cycle through the four frequencies values on the Digibridge and record the impedance magnitude (Z_s) and the phase angle (θ) at each set of parameters.
- These data and the known conductivity of 0.01 mol L^{-1} KCl will be used further to calculate the cell constant of the conductivity cell. The cell constant should be on the order of 0.1 cm^{-1}.
- Thoroughly rinse and shake the electrode dry. Thoroughly rinse the beaker with distilled water.

2. Measure the ionic conductivity of strong and weak electrolytes as a function of concentration.
   - Pour some of the 0.0001 mol L^{-1} NaCl(aq) solution from the flask into the beaker, and dip the conductivity probe into the solution. Let it equilibrate for 1 minute.
   - Put the thermometer into the test solution and take a temperature reading.
   - Cycle through the four frequency values on the Digibridge and record the impedance magnitude (Z_s) and the phase angle (θ) at each set of parameters.
   - Repeat the same measurements for all other concentrations of NaCl(aq) from the most dilute to the most concentrated as listed in Part A of the lab. Make sure to triple-rinse the conductivity cell and the beaker after each solution (detergents are not necessary).
   - Repeat this process for the CH$_3$COOH(aq) solutions listed in Part A of the lab. Make sure to triple rinse the conductivity cell and the beaker after each solution (detergents are not necessary).

Electronic Conductivity Measurements

1. Set up equipment and connections.
   - For measuring the electrical conductivity of a wire, we will use a DC power supply, a high-impedance electrometer, and an ammeter.
   - Measure and record the length of the 0.5 mm diameter copper wire.
   - Connect the electrometers and electrical connections exactly as shown in Figure C.2.2.

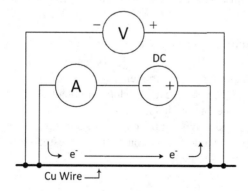

FIGURE C.2.2   Wiring diagram for measuring conductivity of a copper wire.

- Set the DC power supply to the current control (cc) mode: turn the current control knob all the way to the left, and then, turn the voltage control knob 3–4 full turns to the right.
2. Measure the electronic conductivity of the copper wire.
   - Apply successively DC currents of 25, 50, 75, 100, 125, and 150 mA to the wire by turning the current control knob slowly to the right. **Do not exceed 150 mA!**
   - At each setting, obtain a voltage reading from the high-impedance voltmeter and a current reading from the ammeter.

CLEAN-UP

1. Power off and disconnect all electronics.
2. Disconnect and rinse the electrode in the conductivity probe into the waste beaker.
3. Store the solutions in capped flask stock containers.
4. Dispose of the waste in the labeled storage container under the sink.
5. Wash all glassware with tap and distilled water and clean the lab station.

DATA TREATMENT

Ionic Conductivity Measurements

1. Create a plot of impedance magnitude (Z_s) vs. frequency$^{(-0.5)}$ (similar to Figure 3.15) for each electrolyte solution tested during the ionic conductivity measurements.
2. Use the plot to determine the solution resistance and conductance of each solution, including the conductivity standard solution.
3. Calculate the cell constant for the conductivity probe using measured solution resistance and known conductivity of 0.01 mol L^{-1} KCl.
4. Calculate the conductivity of each NaCl(aq) and CH_3COOH(aq) solution.
5. Calculate the molar conductivities for all solutions.
6. Plot the molar conductivity vs. concentration$^{(0.5)}$ (similar to Figure 3.20) of all solutions on a single plot and include a fitting curve for each type of solutions.

Electronic Conductivity Measurements

1. Plot the potential difference (voltage) vs. current relationship obtained in electronic conductivity measurements.
2. Estimate the wire resistance as the slope of potential difference vs. current, approximating the data with a linear fit.
3. Calculate the conductivity of the copper wire using its length and diameter.
4. Use the reference value for Cu conductivity to estimate your experimental error for each measurement.

LABORATORY: ELECTRODE POTENTIAL OF Cu(S)|Cu²⁺(aq) ELECTRODE

Video URL: http://youtu.be/BvMIhH2tiiw
Video-Based Assignment 4 (Appendix B)

OBJECTIVES

- Experimentally observe the dependence of the Cu/Cu^{2+}(aq) electrode potential on the concentration of $CuSO_4$ solution using galvanic cell Cu'(s) | Ag(s) | AgCl(s)| KCl(aq) || $CuSO_4$(aq)| Cu(s) | Cu'(s).
- Learn how the Nernst equation can be used to theoretically predict the electrode potential values.
- Compare experimental data with theoretical calculations.
- Follow the Safety Guidelines in carrying out this lab.

MATERIALS

• Glass flask with lid for electrochemical experiments	• Ninety-eight percent H_2SO_4
	• Distilled water for rinsing
• Copper electrode (plate)	• Sandpaper
• Copper wire	• 1 L Glass beaker for solution disposal
• Commercial Ag/AgCl reference electrode containing 4.0 mol L^{-1} KCl saturated with AgCl	• Thermometer
	• Stopwatch
	• High-impedance electrometer
• Five 1 L solution flasks	• Stirring plate and stirring bar
• 1 L of each of the following $CuSO_4$(aq) solutions: 0.001, 0.003, 0.005, 0.01, and 0.03 mol L^{-1}	• Stand and clamp holder for the flask
	• Kimwipes

EXPERIMENTAL SETUP AND PROCEDURE

The following half-reactions take place in the cell:

Left: $2AgCl(s) + 2e^- \rightarrow 2Ag(s) + 2Cl^{-1}(aq)$,
Right: $Cu^{2+}(aq) + 2e^- \rightarrow Cu(s)$.

Solutions of 0.001, 0.003, 0.005, 0.01, and 0.05 mol L^{-1} $CuSO_4$ (aq) should be prepared in advance, and 5 mL of 98% H_2SO_4 should be added into each 1000 mL solution in order to create a pH less than 4. According to the Pourbaix Diagram for Cu, a pH of 4 or less ensures that Cu^{2+} ions are the dominant species (rather than Cu_2O or CuO), and the desired cathodic reaction is as specified above.

1. If the 4.0 mol L^{-1} KCl solution in the Ag/AgCl reference electrode has not been changed within the past few weeks, replace the solution with fresh filling solution to ensure the correct concentration.
2. Clean the Cu electrode with sandpaper, removing oxidized layer. Rinse electrode with distilled water and dry electrode with Kimwipes, making certain all copper particles are removed.
3. Add approximately 750 mL of 0.001 mol L^{-1} CuSO$_4$ aqueous solution into the flask, place the stirring bullet in the solution, and put on the glass lid.
4. Construct the electrochemical cell as shown in Figure C.3.1. Place the copper plate electrode and the commercial Ag/AgCl reference electrode and in their slots in the glass lid (normally the central slot is used for Cu electrode). Open the blowhole of the commercial Ag/AgCl reference electrode. Make sure that the tip of the Ag/AgCl reference electrode is fully immersed in the solution.
5. Insert the thermometer and turn on the stirrer to "2." Turn on the stop watch.

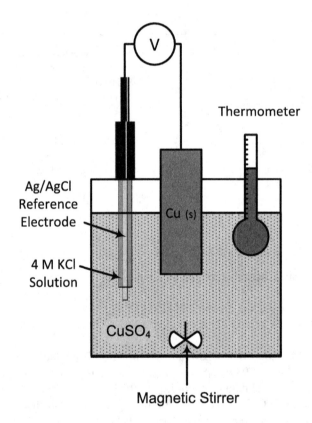

FIGURE C.3.1 Potentiometric cell for studying the concentration dependence of the Cu electrode potential.

6. Connect the electrodes to the high-impedance electrometer: Cu electrode—to the positive (red), and reference electrode – to the negative (black) clips.
7. Wait for 3 minutes to allow the system to reach a stable potential. Record the cell voltage and temperature five times with 30-second intervals. Normally, variation of the cell voltage should be within 1 mV. Calculate the average value and the standard deviation of the measurements.
8. Turn off the stirrer and disconnect the cables. Rinse the copper electrode and reference Ag/AgCl electrode by distilled water, and dry them with Chimwipes. Pour the used $CuSO_4$ solution back into the storage container (via 1 L beaker). Do not rinse the flask to avoid dilution of the next solution with droplets of water.
9. Repeat the measurements (Steps 3–8) with the $CuSO_4$ concentrations 0.003, 0.005, 0.01, and 0.05 mol L^{-1}, in that order.
10. Record all experimental data in the form of the following table:

$CuSO_4$ Concentration, mol kg^{-1}	Temperature, °C	Time Exp. min:sec	Cell Voltage, mV	Cell Voltage Average (At Each Concentration)

11. In the end of the experiment, turn off the electronics and rinse and store the reference electrode in its storage solution. Also rinse the beaker, flask, thermometer, and stirrer.

DATA TREATMENT

1. Using the experimentally collected voltage data (E_{cell}) for the galvanic cell

$$Cu(s)|Ag(s)|AgCl(s)|KCl(aq)4\,M\|CuSO_4(aq)|Cu(s),$$

determine the electrode potential of the copper electrode, taking into account that

$$E_{cell} = E_R - E_L,$$

where E_R is the potential of the right-side electrode (half-reaction), and E_L is the potential of the left-side electrode (half-reaction). Note that the potential of the Ag/AgCl reference electrode (E_L) is constant in all our measurements (see Lab 3); hence, the observed variation in cell voltage is solely due to the concentration dependence of the copper electrode (E_R).

Plot the experimental copper electrode potential values as a function of log $m(CuSO_4)$. Use the standard deviation values for the repeated measurements in experimental step (7) as error bars for the data points.

2. Theoretically calculate the electrode potential of the copper electrode over the range of $CuSO_4$ concentrations using the corresponding Nernst equation to theoretically calculate the electrode potential. The following information (a)–(f) can be helpful in this calculation.

   a. Half-reaction: $Cu^{2+}(aq) + 2e^- \rightarrow Cu(s)$.

   b. Nernst equation: $E_{Cu/Cu^{2+}} = E^0_{Cu/Cu^{2+}} + \dfrac{RT}{nF} \ln a_{Cu^{2+}}$.

   c. Define all parameters in the Nernst equation:

      $E^0_{Cu/Cu^{2+}} = 0.342\,V$ [Standard electrode potential],

      $R = 8.314\,J\,mol^{-1}K^{-1}$ [Gas constant],

      T = experimental temperature $+ 273.15$ K [absolute temperature],

      $n = 2$ [number of electrons transferred in the electrode process],

      $F = 96,485\,C\,mol^{-1}$ [Faraday constant].

   d. Define the activity of Cu^{2+} ion: $a_{Cu2+} = m_{Cu2+} \times \gamma_{Cu2+}$.

   e. Use the Debye–Hückel theory to obtain the activity coefficient for Cu^{2+} ion:

$$\log \gamma_i = -\frac{A z_i^2 \sqrt{I}}{1 + B\mathring{a}\sqrt{I}},$$

where γ_i is the activity coefficient of species i, z_i is the charge of species i, I is the ionic strength of the solution, and $\mathring{a}$ is a parameter dependent on solute radius, with $\mathring{a} = 4.5$ Å. A and B are Debye–Hückel parameters dependent upon the solvent and temperature. For a solvent of water at 25°C, $A = 0.5098$ (kg mol^{-1})$^{1/2}$ and $B = 0.3284$ (kg mol^{-1})$^{1/2}$ (Å)$^{-1}$.

   f. Ionic strength (I) in the Debye–Hückel equation can be obtained as follows:

$$I = 1/2 \sum \left(b_i \times z_i^2 \right),$$

where b_i are the molal concentrations of each ion in solution and z_i are the charges of each ion in solution.

   g. Plot the $E_{Cu/Cu2+}$ values calculated by the Nernst equation as a function of log b_{CuSO4} and make linear approximation.

3. Compare the concentration dependence of the copper electrode potential ($E_{Cu/Cu2+}$) obtained from your experiment with the one predicted by the Nernst equation on the same graph. Comment on the slope of the dependence, and discuss possible reasons for the deviation of actual electrode potential from the theoretical values.

LABORATORY: Ag/AgCl REFERENCE ELECTRODE AND DIFFUSION POTENTIAL

Video URL: http://youtu.be/LzjL2jkaUuQ
Video-Based Assignment 5 (Appendix B)

OBJECTIVES

- Learn how to make and evaluate a standard Ag/AgCl reference electrode.
- Learn how to use a standard Ag/AgCl reference electrode.
- Learn how to calculate a potential difference in a concentration cell.
- Learn how to calculate a liquid junction (diffusion) potential.
- Learn how to convert the molar concentration to molal.
- Follow the Safety Guidelines in carrying out this lab.

MATERIALS

• Silver wire	• 0.1 M HCl solution Pliers
• Wire cutters or scissors	• Heat gun
• Sandpaper	• Propane torch and flint striker
• Silver powder	• Commercial Ag/AgCl electrodes containing
• AgCl powder	fresh 1 and 4 M KCl saturated with AgCl
• Shrinkable Teflon for 0.5 mm wire	• Glass beakers
• 1 M (mol L⁻¹) KCl solution*	• High-impedance electrometer
• 4.0 M KCl solution*	
• 1 M Nitric acid**	

* It is suggested that the KCl solution must be saturated with AgCl so that the AgCl coating is not stripped from the silver wire. The solubility of AgCl in pure water is approximately 10^{-5} M at 25°C. The solubility of AgCl in saturated KCl solution is approximately 6×10^{-3} M. The increased solubility is due to the formation of aqueous complexes such as $AgCl_2^-$. The "classical" Ag/AgCl electrode contains saturated KCl, but in most of the cases, other concentrations of KCl are used.

** 3 M nitric acid is a recommended solution for cleaning the silver before applying AgCl coating.

EXPERIMENTAL SETUP

Preparation of Ag/AgCl Electrode

1. Cut two pieces of silver wire approximately 6 inches long with scissors or wire cutters.
2. Clean the ends of the silver wires with a sandpaper, especially if wire was oxidized.
3. Clean the end of the sliver wires, dipping them in 1 M nitric acid for 5 minutes, and rinse with DI water and wipe dry.
4. Put a small amount of silver powder and AgCl powder into mortars.
5. Light the propane torch.
6. Hold one wire with pliers or hold at the furthest end of the wire while wearing gloves. Heat the silver wire in the yellow tip of the blue flame. Do not allow the wire to be heated too quickly, or the tip of the wire will melt and fall off.

7. While the wire is red-hot, dip it <u>quickly</u> into the silver powder (if you wait too long, the powder will not stick). You may have to repeat it several times until there is a silver bulb at the end of the wire. Heat the powder in the tip of the flame until all of the white powder turns silver. The silver powder will melt very quickly in the flame, so be careful not to let it drop off.

8. Once the bulb of silver is formed on the tip of the wire, dip it into AgCl powder. If you wait too long, the silver will be too cool and AgCl will not stick to the silver bulb.

9. With AgCl on the silver bulb, hold the wire high above the flame, gradually bringing it lower until the AgCl starts to melt. Tilt the electrode in different directions so that the AgCl melt does not roll up the wire. Hold AgCl above the flame until it melts, but do NOT touch the AgCl to the flame. Repeat this step a few times until the silver bulb is thinly coated with AgCl. Make sure silver wire is not exposed on the tip and fully covered by AgCl.

10. Repeat Steps (6)–(9) with a second silver wire.

11. Cut two pieces of shrinkable Teflon tube and cover the lengths of the Ag wires that are not coated with AgCl. Leave about a half inch on the other end of the wire uncoated for electrical connection.

12. Holding the wire with pliers, use a heat gun at over 1150°F (621°C) to shrink Teflon onto each of the electrodes.

13. Place AgCl-coated tips of the prepared electrodes in 0.1 M HCl solution in a dark place, and let the electrode core age for 1–2 days. Note the 0.1 M HCl can be reused, so do not dispose of it. After 2 days, place the electrodes in 1 M KCl solution (saturated with AgCl) for 2 more days of ageing. After that, the electrodes are ready for evaluation and measurements.

EXPERIMENTAL PROCEDURE

Electrode Evaluation and Diffusion Potential Measurement

1. To test a homemade silver–silver chloride electrode, place it with the AgCl tip in a beaker of 1 M KCl together with the commercial Ag/AgCl reference electrode filled with 1 M KCl and connect both electrodes to high-impedance electrometer: home-made to [+ red] and commercial to [− black], as shown in Figure C.4.1. [Note: If the commercial reference electrode has not been used recently, it is advisable to replace the KCl solution in the electrode with fresh 1 M KCl].

2. Turn on the electrometer, and let the potential stabilize within 3 minutes. The measured potential difference should be less than ± 3 mV for a well-functioning electrode. Test each electrode of the two self-made electrodes three separate times, and record the average potential values.

3. If any of the electrodes exhibit either a significantly higher potential or an unstable reading, put it away. Use only the electrode that passes the check in Step 2, and choose the best of the two for further measurements.

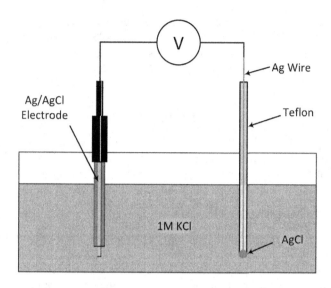

FIGURE C.4.1   Setup of the concentration cell used to evaluate the performance of the self-made Ag/AgCl reference electrode versus a commercial Ag/AgCl reference electrode.

4. Replace the commercial Ag/AgCl reference electrode filled with 1 M KCl with the one filled with 4 M KCl. This creates a concentration cell and difference in electrode potentials.
5. After waiting for 3 minutes, record the potential difference from the electrometer five times with 1-minute intervals and take an average value.
6. After the measurements, put the 1 M KCl solution from the beaker back in its storage container. It can be reused.

DATA TREATMENT

Part I: Calculation of the Ag/AgCl Electrode Potential Using Nernst Equation

1. Which one of the home-made electrodes did you choose for further testing and why?
2. Write the reduction reaction for the Ag/AgCl electrode, and write a cell schematic for the experimental setup given in Figure C.4.1 (i.e., a schematic that gives the species and phases divided by | or other appropriate notations).
3. Use the Nernst equation to find the electrode potential (half-cell potential) of the Ag/AgCl reference electrode in the 1 M (mol L^{-1}) and 4 M KCl solutions, using the appropriate activity coefficients. Since the activity chart is in terms of molality (mol kg^{-1}), use the instructions below to convert from molarity (mol L^{-1}) to molality [mol/(kg of H$_2$O)]. Then, once you have the

solution concentration in terms of molality, use the activity coefficient from the chart that can be obtained using an interpolation procedure. In order to convert from molarity to molality, solution density is needed. Use the following equation to determine the KCl solution's density, ρ, at a given molarity, M:

$$\rho = A + B * M,$$

where ρ is the KCl solution density in units of kg m^{-3}, and M is the KCl solution molarity in units of mol L^{-1}, A=999.83 kg m^{-3}, B=43.315 kg L m^{-3} mol^{-1}.

CONVERT FROM MOLARITY TO MOLALITY

Be certain to watch that the units you use properly cancel and are converted when necessary.

    a. Assume that you have 1 L of solution (to make the number of moles of solute easier to work with and so that you have a given amount that you are working with when performing conversions).

    b. Determine the number of moles of solute that you have (this is the same as the molarity since you have 1 L of solution).

    c. Determine the mass of the solute by multiplying the number of moles of solute by the molar mass of the solute.

    d. Determine the mass of the solution by multiplying the volume of the solution (1 L) times the solution density.

    e. Determine the mass of solvent (water) in 1 L of solution. Do this by subtracting the mass of the solute from the mass of the solution since $g_{solvent} = g_{solution} - g_{solute}$.

    f. Determine the molality (mol kg^{-1}) of the solution by dividing the number of moles of solute by the mass of the solvent.

Part II: Calculation of the Diffusion Potential

    4. The experimentally measured potential in the concentration cell is the result of the potential difference between the Ag/AgCl electrodes in 1 and 4 M KCl and the diffusion potential.

    5. Use the following equation as a guide to estimate the diffusion potential, $\Delta\phi_{diff}$:

$$\Delta E = E_R - E_L + \Delta\phi_{diff},$$

where ΔE is the measured potential difference, E_R is the Nernst potential for the Ag/AgCl electrode in contact with 1 M KCl, and E_L is the Nernst potential for the Ag/AgCl electrode in contact with 4 M KCl.

6. Calculate the theoretical value of diffusion potential, which contains both the equation for the diffusion potential of a concentration cell and the values of the transport number for K^+ and Cl^- ions.

7. Compare the calculated and experimental values and discuss possible reasons for discrepancies.

LABORATORY: LINEAR SWEEP VOLTAMMETRY AND ROTATING DISK ELECTRODE

Video URL: http://youtu.be/LYUEFuAM62c
Video-Based Assignment 6 (Appendix B)

ADDITIONAL SAFETY GUIDELINES

- Both $KNO_3(aq)$ and $K_3Fe(CN)_6(aq)$ are toxic solutions. Use proper safety methods to avoid exposure, and do not, under any circumstance, dispose of the chemicals down the drain.
- **DO NOT** mix $K_3Fe(CN)_6(aq)$ with acids, it is possible to form $HCN(g)$ which is highly toxic, flammable and can be fatal.

OBJECTIVES

- Use linear sweep voltammetry (LSV) to understand electrochemical kinetics.
- Use the rotating disc electrode (RDE) to understand mass transfer effects.
- Apply the Butler–Volmer equation (via Tafel analysis) to the potassium ferricyanide system.
- Apply the limiting current relation to the RDE in potassium ferricyanide system.
- Follow the safety guidelines when carrying out this lab.

MATERIALS

• Safety goggles and gloves	• Three-electrode cell for electrochemistry
• Distilled water for rinsing	• Platinum rotating disk electrode
• Paper towels	• AFCTR1 platinum auxiliary electrode
• 1 M $KNO_3(aq)$, electrolyte solution	• Ag/AgCl (reference) electrode
• 6.4 mM $K_3Fe(CN)_6(aq)$ in 1 M $KNO_3(aq)$, analyte solution	• 0.05 micron alumina polishing solution
	• Polishing microcloth
• Pine AFCBP1 Bipotentiostat	
• Computer with *PineChem* software	
• Pine AFMSRCE analytical rotator	

EXPERIMENTAL SETUP:

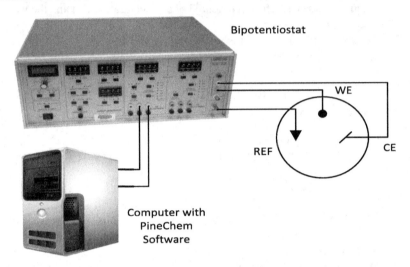

FIGURE C.6.1 Connections of the electrochemical cell, bipotentiostat, and computer for the linear sweep and rotating disk electrode measurements.

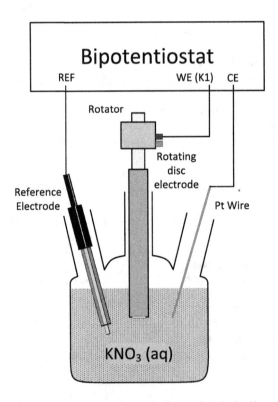

FIGURE C.6.2 Schematic of the three-electrode electrochemical cell used for linear sweep and rotating disk electrode measurements.

EXPERIMENTAL PROCEDURE:

A. Solution preparation
  1. Background Electrolyte Solution (KNO_3): The electrolyte solution is 1.0 M potassium nitrate (KNO_3) in water. This solution provides the electric conductivity for correct electrochemical measurements. To prepare 1.0 M KNO_3 solution in a 250 mL volumetric flask, add 25.30 g of potassium nitrate to the distilled water. Mix the solution well, and adjust water to the hair-mark. Make sure to rinse the flask at least three times with distilled water before use.
  2. Analyte Solution [$K_3Fe(CN)_6$]: The analyte solution is a 6.4 mM solution of potassium ferricyanide $K_3Fe(CN)_6$ in the electrolyte solution. To prepare 6.4 mM solution in a 100 mL flask, add exactly 210.7 mg of potassium ferricyanide [$K_3Fe(CN)_6$] to the KNO_3 electrolyte solution prepared in (1). Mix and adjust the solution level to the hair-mark. The electrochemical half-reaction for this system is $Fe(CN)_6^{3-} + e^- \Leftrightarrow Fe(CN)_6^{4-}$.
B. Electrode preparation
  1. Working Electrode: Polish the Pt disk working electrode using the polishing kit and following the instructions inside. Use the "routine cleaning" method with the 0.05 micron alumina solution and the Buehler microcloth. Rinse electrode with distilled water and wipe clean. Check that the electrode is mirror bright and free of defects. **Remember** to record the working electrode's surface area. Once finished, screw the Pt disk electrode onto the rotator shaft while holding the shaft still. The rotator should be in the raised position on its bar.
  2. Reference Electrode: Rinse the (reference) electrode with distilled water and wipe clean. Open the middle section of the cap by pushing down on the cap [It should have been closed (pushed up) for storage]. This is done to equalize the pressure inside the electrode with the ambient pressure in the cell.
C. Background Scan
     Pure electrolyte solution (1 M KNO_3) will be used in a background voltammetric scan to check the purity of the solution, cleanliness of on the glassware, and functionality of the working Pt electrode. Any impurities in the solution will show unexplained peaks in the background scan, and a fouled or improperly polished electrode surface will cause a large background current.
  1. Rinse the glass electrochemical cell.
  2. Fill the cell with electrolyte solution (1 M KNO_3).
  3. Lower the Pt rotating disk working electrode into the cell. Place the reference electrode and the platinum auxiliary (counter) electrode in the cell. Do not connect the electrodes to the bipotentiostat yet. Make certain all electrodes are in contact with the solution. Note that the Pt working electrode should not be rotating during the background scan. (Rotator speed should be set at approximately 0 RPM or left "Off.")
  4. Plug in and turn on the bipotentiostat and the speed control of the rotator. Press in and hold the control source button to change from "panel" to "external."

5. Open the PineChem software on the computer connected to the bipotentiostat. Choose device "Pine" and click <u>done</u>. On the "Instrument
Status" tab, place the bipotentiostat in "Dummy" mode. Properly connect the electrochemical cell to the bipotentiostat. Note that the connector for the Pt working disk electrode is the top (red) connector on the
rotator. (See Figures C.6.1 and C.6.2.)

6. On the "Instrument Status" tab, adjust the "Idle Conditions" to look as
shown in Figure C.6.3. Note that in addition to changing the K1 electrode
to read 800 mV, you must also change from "Dummy" to "Normal" mode.

7. Check the bipotentiostat settings and readings. The bipotentiostat
should be in "Normal" mode. The front panel voltammeter should display the working electrode potential (E1), which should confirm that the
K1 electrode is idling near +800 mV.

8. Choose the "Experiment" menu option, and select "Analog Sweep
Voltammetry." Adjust settings to match those shown in Figure C.6.4.
Note that these are only the beginning settings and may need to be
adjusted (wider or narrower) to obtain a nice voltammogram. "Electrode
Sensitivity" for K1 Current may particularly need to be adjusted.

9. Click on "Perform" to start the measurement. The cyclic voltammogram should be relatively flat with no significant peaks. Choose the
"Plot" tab, and click on "I1 vs. E1" to eliminate the line from the inactive K2 electrode. The plot should be a somewhat jagged, thick fairly
straight line. Some current may be expected in the cathodic region due
to the reduction of hydronium ions. If the plot is not fairly straight, the
electrode should be polished again and the background scan should be
repeated. If the problem is not solved, the cell or electrolyte solution
may be the problem and should be cleaned or replaced.

10. Save an image of the background scan for the lab report by copying the
image (with "Copy" button) and pasting it onto a Word document.

11. Unhook the potentiostat from the electrochemical cell by unhooking the
cable to the working electrode and counter electrode and removing the
reference electrode from the cell and placing it in a beaker with a small
amount of distilled water at the bottom. Lift the shaft of the Pt working
electrode out of the solution.

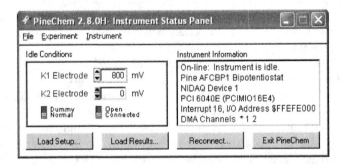

FIGURE C.6.3   PineChem Instrumental Status Panel.

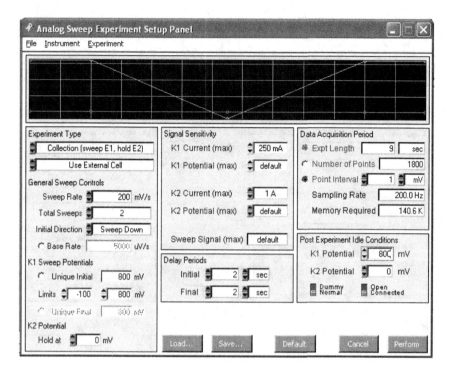

FIGURE C.6.4 PineChem Setup Panel for Background Scan.

12. Pour the used electrolyte solution back in its storage container unless it was determined to be contaminated. If contaminated, dispose of it properly.

D. Linear Sweep Voltammetry

In this section, we are going to use LSV to perform a kinetic analysis of our system. Two sweeps will be performed over the same potential range: one starting at the more positive potential and sweeping downwards, and the second starting at the more negative potential and sweeping upwards.

1. Pour the analyte solution [6.4 mM $K_3Fe(CN)_6$(aq) in 1 M KNO_3(aq)] into the glass cell, and put counter electrode and reference electrode in place. Lower Pt working electrode into solution. Make certain that all electrodes are fully in the analyte solution, then connect to the bipotentiostat.

2. Choose the "Experiment" menu option and select "Analog Sweep Voltammetry." Adjust the settings to match those shown in Figure C.6.5.

This will allow for a slow sweep of 10 mV s^{-1} with the potential range from +750 to −500 mV. Note that these are only the beginning settings and may need to be adjusted (wider or narrower) to obtain a nice voltammogram. "Electrode Sensitivity" for K1 Current, initially 500 µA, may particularly need to be adjusted.

3. Click on "Perform" to start the measurement. The resulting curve should show both cathodic and anodic peaks. Note that PineChem software is

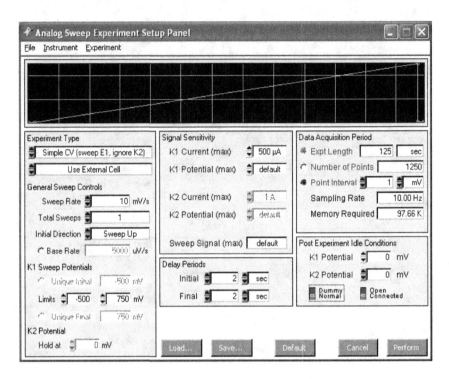

FIGURE C.6.5   PineChem Setup Panel for LSV.

set up to plot positive potentials to the right and cathodic currents are plotted toward the bottom.

4. Once the data have been obtained, export the data to a text file (Figure C.6.6). Under the "File" menu, click "Export." Only the E1 and I1 data are needed, all other boxes can be un-checked. Save these data as a text file on a removable storage drive.

5. Repeat the previous steps, only this time sweep in the upwards direction (from −500 to +750 mV).

E. Rotating Disc Electrode

Now that the kinetic analysis has been completed, the RDE will be used to analyze the mass transfer effects on the system. Note that the RDE should never be set higher than 2500 PRM, as this could damage or destroy the electrode.

1. Using the same solution and setup as the previous tests, turn on the RPM controller for the RDE. Set the rotation speed to 400 RPM.

2. Choose the "Experiment" menu option and select "Analog Sweep Voltammetry." Adjust the settings to match those shown in Figure C.6.7. This will allow for a slow sweep of 10 mV s^{-1} with the potential range from 500 to −100 mV and back. Note that in this test, we are only interested in the reduction reaction, and as such the potential range has been limited to the cathodic region (Figure C.6.7).

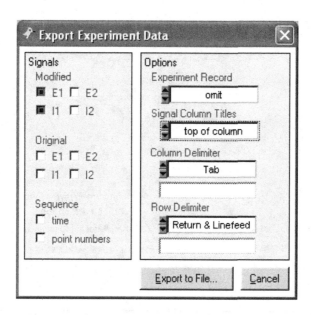

FIGURE C.6.6   Export Experimental Data Panel.

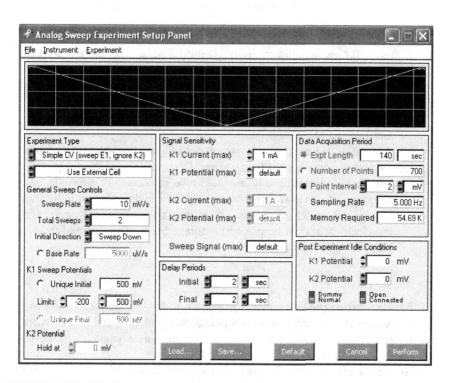

FIGURE C.6.7   PineChem Setup Panel for testing RDE.

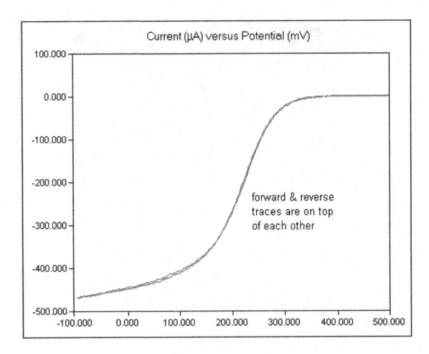

FIGURE C.6.8 A typical sigmoidal voltammogram.

3. Once the data have been obtained, export the data to a text file. Under the "File" menu, click "Export." Only the E1 and I1 data are needed, and all other boxes can be un-checked. Save these data as a text file on a removable storage drive. A typical sigmoidal voltammogram can be seen in Figure C.6.8.

4. Once the data have been saved, repeat the test at rotation speeds of 900 and 1600 RPM.

5. After all tests have been completed, pour the analyte solution back into its storage container for future usage. Clean the rest of the setup properly, including rinsing cell and glass tube of counter Pt electrode, and disposing of rinsed water appropriately. The middle section of the cap of the reference electrode must be pushed up (closed) for storage, and the electrode should be placed in storage bottle. Turn off and unplug bipotentiostat and Rotator speed control.

DATA TREATMENT: LSV

1. For each sweep, produce a plot of Log[j] vs. E. Use this plot to identify linear regions at least 25 mV away from the open circuit potential (this is where the anodic and cathodic regions come together in an asymptote). Note that the diameter of the Pt working electrode is 5 mm.

2. Isolate the anodic and cathodic linear regions, and make a linear fit. Set these two lines equal and solve for the open circuit potential E_{oc} and the exchange current density j_o.

3. Using the overpotential found above, replot the linear regions with respect to the overpotential η. Add a linear fit and use the Tafel relations to calculate j_o and the symmetry parameter β for the anodic and cathodic regions.

4. Make a plot of j versus η for the anodic and cathodic regions within 20 mV of E_{oc}. Make a linear fit of these data, and analyze to calculate j_o.

DATA TREATMENT: RDE

1. Give the graphs of current density vs. potential for each of your experimental results. The diameter of the Pt working electrode is 5 mm.

2. Read the limiting current density on sigmoidal voltammograms at each rotational rate.

3. Use the formula $\omega = 2\pi f$ where f is the rotation rate in revolutions per second to convert all rotation rates in RPMs to angular rotation rates (ω) in radians second^{-1}.

4. Plot the limiting current density vs. $\omega^{1/2}$. Show the graph and discuss the effect of reducing the rotational rate. Explain why this change occurs.

5. Use the Levich equation to find the diffusion coefficient of analyte solution. Be careful with unit conversions. Note that the Levich current/limiting cathodic currents were negative because a cathodic polarization was applied. Therefore, use the absolute value of the experimental current density.

    a. The Levich equation is $j_L = (0.620)nFAD^{2/3}\omega^{1/2}\upsilon^{-1/6}C$, where C is the analyte concentration, υ is the kinematic viscosity, A is the surface area of the electrode, F is Faraday's constant, D is the analyte's diffusion coefficient, and n is the number of electrons involved in the half-reaction (see the "Solution Preparation" section for the half-reaction).

    b. The kinematic viscosity, υ, of 1.0 M KNO$_3$ is 0.00916 cm^{2}s^{-1}.

    c. Use cm for units of length and volume.

6. Repeat the calculations with the Levich equation, and only this time use the value of the diffusion coefficient to calculate the concentration of ferricyanide in the test solution. Literature values for the diffusion constant are 8.96×10^{-6} and 7.35×10^{-6} cm^{2}s^{-1} for the Fe(CN)$^{3-}$ and Fe(CN)$^{4-}$ ions, respectively.

LABORATORY: CYCLIC VOLTAMMETRY

Video URL: http://youtu.be/4fjVCuZEMz0
Video-Based Assignment 7 (Appendix B)

ADDITIONAL SAFETY GUIDELINES

- Both KNO$_3$(aq) and K$_3$Fe(CN)$_6$(aq) are toxic solutions. Use proper safety methods to avoid exposure and do not, under any circumstance, dispose of the chemicals down the drain.

- **DO NOT** mix K$_3$Fe(CN)$_6$(aq) with acids, because it could form HCN(g) which is highly toxic and flammable, and can be fatal.

OBJECTIVES

- Learn how to test for electrochemical reversibility from simple diagnostic tests using the cyclic voltammogram (CV) peak currents and potentials.
- Study the effect of scan rate on the voltammograms.
- Observe the effect of Pt electrode and glassy carbon (GC) electrode on electrochemical behavior.
- Follow the safety guidelines carrying out in this lab (Figure C.7.1).

MATERIALS

- Safety goggles and gloves
- Distilled water for rinsing
- Paper towels
- 1 M KNO_3(aq), electrolyte solution
- 6.4 mM $K_3Fe(CN)_6$(aq) in 1 M KNO_3(aq), analyte solution
- Pine AFCBP1 Bipotentiostat
- Computer with *PineChem* software

- Three-electrode cell for electrochemistry
- Platinum disk electrode
- Cu electrode
- GC electrode
- AFCTR1 platinum counter electrode
- Saturated calomel (reference) electrode (SCE)
- 0.05 Micron alumina polishing solution
- Polishing microcloth

EXPERIMENTAL SETUP

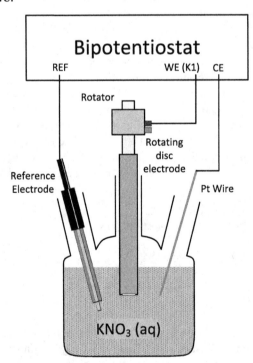

FIGURE C.7.1   Schematic of the three-electrode electrochemical cell used for Cyclic Voltammetry.

EXPERIMENTAL PROCEDURE

Solution Preparation

1. Background Solution (KNO_3)

   The electrolyte solution is 1.0 M potassium nitrate (KNO_3) in water. This solution provides the electrical conduction for correct electrochemical measurements. To prepare 1.0 M KNO_3 solution in a 250 mL volumetric flask, add 25.30 g of potassium nitrate to the distilled water. Mix the solution well and adjust water to the hair-mark. Make sure to rinse the flask at least three times with distilled water before use.

2. Analyte Solution [$K_3Fe(CN)_6$]

   The analyte solution is a 6.4 mM solution of potassium ferricyanide $K_3Fe(CN)_6$ in the electrolyte solution. To prepare 6.4 mM solution in a 100 mL flask, and add exactly 210.7 mg of potassium ferricyanide [$K_3Fe(CN)_6$] to the KNO_3 electrolyte solution prepared in (1). Mix and adjust the solution level to the hair-mark. The electrochemical half-reaction for this system is $Fe(CN)_6^{3-} + e^- \Leftrightarrow Fe(CN)_6^{4-}$.

Electrode Preparation

1. Working Electrode
   - Polish the Pt disk working electrode using the polishing kit and following the instructions inside. Use the "routine cleaning" method with the 0.05 micron alumina solution and the Buehler microcloth. Rinse the electrode with distilled water and wipe clean. Check that the electrode is mirror bright and free of defects. Once finished, screw the Pt disk electrode onto the rotator shaft while holding the shaft still. The rotator should be in the raised position on its bar.
   - Polish the glassy carbon (GC) disk working electrode using the polishing kit and following the instructions inside. Use the "routine cleaning" method with the 0.05 micron alumina solution and the Buehler microcloth. Rinse the electrode with distilled water and wipe clean. Check that the electrode is mirror bright and free of defects. Once finished with the Pt electrode CV testing, screw the GC disk electrode onto the rotator shaft while holding the shaft still. The rotator should be in the raised position on its bar.

2. Reference Electrode

   Rinse the saturated calomel (reference) electrode (SCE) with distilled water and wipe clean. Open the middle section of the cap by pushing down the cap [It should have been closed (pushed up) for storage]. This is done to equalize the pressure inside the electrode with the ambient pressure in the cell.

BACKGROUND SCAN

Pure electrolyte solution (1 M KNO_3) will be used in a background voltammetric scan to check the purity of the solution, cleanliness of the glassware, and functionality of the working Pt electrode. Any impurities in the solution will show unexplained peaks in the background scan, and a fouled or improperly polished electrode surface will cause a large background current.

1. Rinse the glass electrochemical cell.
2. Fill the cell with electrolyte solution (1 M KNO₃).
3. Lower the Pt rotating disk working electrode into the cell. Place the SCE reference electrode and the platinum auxiliary (counter) electrode in the cell. Do not connect the electrodes to the bipotentiostat yet. Make certain all electrodes are in contact with the solution. Note that the Pt working electrode should not be rotating during the background scan. (Rotator speed should be set at approximately 0 RMP or left "Off.")
4. Plug in and turn on the bipotentiostat and the speed control of the rotator. Press in and hold the control source button to change from "panel" to "external."
5. Open the PineChem software on the computer connected to the bipotentiostat. Choose device "Pine" and click done. On the "Instrument Status" tab, place the bipotentiostat in "Dummy" mode. Properly connect the electrochemical cell to the bipotentiostat. Note that the connector for the Pt working disk electrode is the top (red) connector on the Rotator.
6. Check the bipotentiostat settings and readings. The bipotentiostat should be in "Normal" mode. The front panel on the bipotentiostat should display the working electrode potential (E1), which should confirm that the K1 electrode is idling near 0 V.
7. Choose the "Experiment" menu option and select "Analog Sweep Voltammetry." Adjust settings to match those shown in Figure C.7.2. Note that these are only the beginning settings and may need to be adjusted

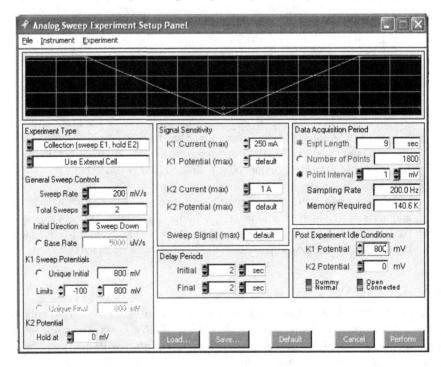

FIGURE C.7.2 PineChem software Experimental Setup Panel for performing the background scan.

(wider or narrower) to obtain a nice voltammogram. "Electrode Sensitivity" for K1 Current may particularly need to be adjusted.

8. Click on "Perform" to start measurement. The cyclic voltammogram should be relatively flat with no significant peaks. Choose the "Plot" tab and click on "I1 vs. E1" to eliminate the line from the inactive K2 electrode. The plot should be a somewhat jagged, thick fairly straight line. Some current may be expected in the cathodic region due to the reduction of hydronium ions. If the plot is not fairly straight, the electrode should be polished again and the background scan should be repeated. If the problem is not solved, the cell or electrolyte solution may be the problem and should be cleaned or replaced.

9. Save an image of the background scan for the lab report by copying the image (with "Copy" button) and pasting it onto a Word document.

10. Unhook the potentiostat from the electrochemical cell by unhooking the cable to the working electrode and counter electrode and removing the SCE reference electrode from the cell and placing it in a beaker with a small amount of distilled water at the bottom. Lift the shaft of the Pt working electrode out of the solution.

11. Pour the used electrolyte solution back in its storage container unless it was determined to be contaminated. If contaminated, dispose of it properly.

CYCLIC VOLTAMMETRY

A. Platinum Electrode
1. Pour the analyte solution [6.4 mM $K_3Fe(CN)_6$(aq) in 1 M KNO_3 (aq)] into the glass cell, and put the counter electrode and reference electrode in place. Lower the Pt working electrode into the solution. Make certain that all electrodes are fully in the analyte solution, then connect to the bipotentiostat.

2. Choose the "Experiment" menu option, and select "Analog Sweep Voltammetry." Adjust the settings to match those shown in Figure C.7.3.

3. Click on "Perform" to start the measurement. The resulting curve should show both cathodic and anodic peaks. Note that the PineChem software is set up to plot positive potentials to the right and cathodic currents are plotted toward the bottom.

4. Once the data have been obtained, export the data to a text file. Under the "File" menu, click "Export." Only the E1 and I1 data sets are needed, all other boxes can be un-checked. Save these data as a text file on a removable storage drive (Figure C.7.4).

5. It is quite common that the initial data points taken during the first sweep deviate significantly from the rest of the sweeps. The starting behavior in sweep one should be deleted from the data set before analysis. Aside from the initial data points, all other sweeps for the same scan rate should be fairly repeatable.

6. Repeat steps 2–5 for a scan rate of 200 and 300 mV s^{-1}, and export the data to a flash drive. The resulting data should look similar to Figure C.7.5 for the Pt electrode.

7. Check that the default applied potential for the bipotentiostat is set to zero, and disconnect the working electrode from the bipotentiostat. Raise the working electrode from the glass vessel, then clean and remove the Pt working electrode.

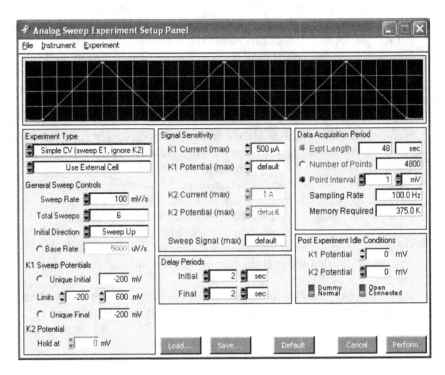

FIGURE C.7.3 PineChem software Experimental Setup Panel for performing cyclic voltammetry.

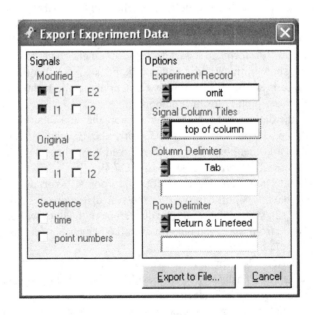

FIGURE C.7.4 PineChem software Panel for exporting experimental data.

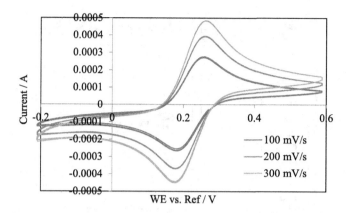

FIGURE C.7.5   Example cyclic voltammetry data at different scans rates.

B. Glassy Carbon Electrode

1. Connect the polished GC working electrode and lower it into the solution. Make certain that all electrodes are fully in the analyte solution, then reconnect to the bipotentiostat.

2. Choose the "Experiment" menu option and select "Analog Sweep Voltammetry." Adjust the settings to match those shown in the Pt electrode settings EXCEPT changing the potential sweep limits from −200 mV: 600 mV to the values of −400 mV: 800 mV.

3. Click on "Perform" to start the measurement. The resulting curve should show both a cathodic and anodic peak. Note that PineChem software is set up to plot positive potentials to the right and cathodic currents are plotted toward the bottom.

4. Once the data have been obtained, export the data to a text file. Under the "File" menu, click "Export." Only the E1 and I1 data sets are needed, all other boxes can be un-checked. Save these data as a text file on a removable storage drive (Figure C.7.6).

5. It is quite common that the initial data points taken during the first sweep deviate significantly from the rest of the sweeps. The starting behavior in sweep one should be deleted from the data set before analysis. Aside from the initial data points, all other sweeps for the same scan rate should be fairly repeatable.

6. Repeat steps 2–5 for a scan rate of 200 and 300 mV s⁻¹, and export the data to a flash drive.

7. Check that the default applied potential for the bipotentiostat is set to zero and disconnect the working electrode from the bipotentiostat. Raise the working electrode from the glass vessel, then clean and remove the Pt working electrode.

8. After all tests have been completed, pour the analyte solution back into its storage container for future usage. Clean the rest of setup properly, including rinsing the cell and glass tube of the counter Pt electrode, and disposing of rinsed water appropriately. The middle section of the cap

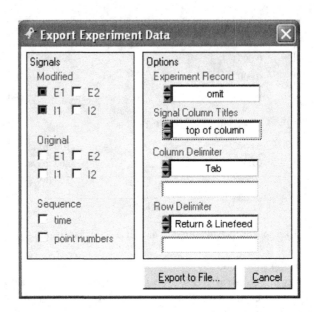

FIGURE C.7.6   PineChem software Panel for exporting experimental data.

of the SCE electrode must be pushed up (closed) for storage, and the electrode should be placed in storage bottle of saturated KCl storage solution. Turn off and unplug the bipotentiostat.

DATA TREATMENT

1. Plot current–potential curves with excel for the Pt and GC electrodes for three different scan rates.
2. Identify the peak currents and potentials at which the peak currents occur for the anodic and cathodic sweeps (remember anodic currents are positive, and cathodic currents are negative).
3. Calculate the peak potential separation for each scan rate.
4. Calculate the ratio of the maximum and minimum current peaks
5. Determine if the scan rate is proportional to the peak current values.
6. Estimate the diffusion coefficient using the Randles–Sevcik equation assuming the electrode area is $0.196\,cm^2$ and the symmetry parameter is 0.5.
7. Comment on your findings. What are sources of potential error?

LABORATORY: PEM FUEL AND ELECTROLYSIS CELLS

Video URL: http://youtu.be/EVM04HEc8u0
Video-Based Assignment 8 (Appendix B)

OBJECTIVES

- Study water electrolysis as a method of producing hydrogen and oxygen.
- Apply Faraday's law to determine the theoretical production of hydrogen and oxygen.

- Calculate the efficiency of the electrolysis cell.
- Study the components and operation of a Proton Exchange Membrane (PEM) H_2/O_2 fuel cell.
- Obtain the polarization (current-voltage) curve of the fuel cell.
- Compare the theoretical open circuit potential (OCP) to the experimentally measured value.
- Follow the safety guidelines while carrying out this lab.

MATERIALS AND TOOLS

- PEM fuel cell/electrolysis system
- Power supply
- Cables
- Distilled water
- Waste beaker (100–500 mL)
- Paper towels

EXPERIMENTAL SETUP

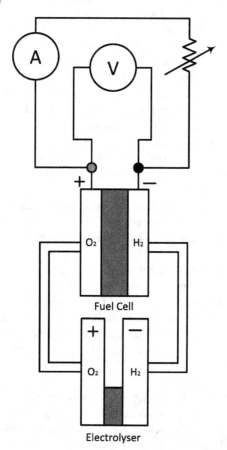

FIGURE C.8.1 Diagram showing connections between the electrolysis cell and the fuel cell, and the fuel cell to ammeter (A), voltmeter (V), and load (R).

System Connections and Setup

- Before any electrical connections are made, connect the tubing from the electrolyzer to the fuel cell stack. The exhaust tubes from the fuel cell should be clamped shut.
- The smaller, free-standing electrolysis cell is used for this experiment. Place the electrolyzer on a paper towel. Add distilled water to the top anode and cathode chambers of the electrolysis system. To prevent overflow, do not completely fill the top chambers; the gas evolving during electrolysis will fill the bottom chamber and will displace the water upward.
- Connect the power supply to the electrolyzer by connecting the positive lead to the anode and the negative lead to the cathode (Figure 8.1.1).

EXPERIMENTAL PROCEDURE

Electrolysis of Water

- Turn on the power supply and set the voltage to 1.7 V. [**Important:** Do not allow the system to exceed 2.0 V. The maximum current allowed for the electrolysis cell is 0.5 A. Monitor the current and reduce the potential if the current across the cell goes above 0.5 A limit]. Then, the gas evolution can be seen inside the O_2 and H_2 gas chambers.
- In order to force the air out of the system, vent the anode and cathode chambers of the electrolytic cell by opening the exhaust lines from the fuel cells. Do NOT allow the gas chambers to be completely empty, as this prevents the water from blocking the tubing. Repeat this venting process three or four times over 10–20 minutes to vent the air out of the oxygen and hydrogen lines; then, close the exhaust line clamps.
- For the electrolytic cell study, disconnect the positive lead from the power supply and add the ammeter in series. Connect the voltmeter in parallel across the electrolyzer.
- Break the circuit by disconnecting one lead from the power supply. Vent both anode and cathode chambers until the water level is at the 0.0 mL level. Use binder clips on the tubes from the electrolyzer to the fuel cell to stop gas flow.
- Reconnect the electrical circuit and turn on the stopwatch simultaneously.
- The system is now generating hydrogen and oxygen. Observe the gas volumes growing as the process goes. After running the electrolysis cell for 2 minutes, break the circuit or turn off the power supply and simultaneously stop the watch to record the precise total time of electrolysis.
- Record the volumes of each gas (H_2 and O_2) produced in the electrolysis cell.

Fuel Cell Operation

- After disconnecting the ammeter and voltmeter from the electrolytic circuit, reconnect the electrolysis cell directly to the power supply. And the voltage is still at 1.7 V as before. Make sure that the produced gases in the electrolysis cell can flow through the fuel cells.

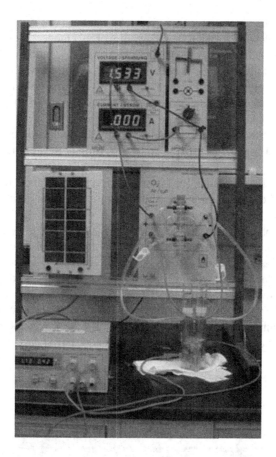

FIGURE C.8.2 View of the PEM fuel cell system in the lab: connections between the power supply, electrolyzer, and fuel cell.

- For the setup of fuel cell study, first, connect the two fuel cells to be stacked in series (see Figure C.8.2). Note that connecting the two fuel cells in series will cause the current to remain equal to the current produced by one fuel cell, but the potential will be twice the potential produced by one fuel cell.
- Second, connect the fuel cell stack in series to the variable electronic load, and connect the load in series with the ammeter, as shown in the video. Make sure the connections are through proper wiring of terminal polarities.
- Finally, connect the voltmeter in parallel across the fuel cell stack. **Remember to always attach the ammeter in series and the voltmeter in parallel.** Now the setup has been completed as shown in Figure C.8.1. The electrolyzer still produces O_2 and H_2 and provides to the fuel cell stack.
- Make sure you are following the steps in the same sequence as shown in the video to avoid any confusion.
- Set the variable electronic load to "Open Circuit," which is a very high resistance. Record the voltage when it gets steady, which is the open circuit

potential. The current should be very close to zero because ideally there is no current flowing through the system under this mode.

• Rotate the load resistance controller clockwise, beginning with 100 Ω and proceeding to low values. At each load value, record the voltage and current generated by the fuel cell system, waiting 30–60 seconds each time before taking the measurement. Continue the procedure down to the resistance of 1 Ω.

• Short circuit the variable electronic load by disconnecting one terminal of the load and connecting it to the opposite terminal. Record the potential and current.

• Disconnect the resistance load from the circuit and connect the fan into its place, as shown in Figure C.8.2. Observe the fan spinning and record the voltage and current across it.

• After completion of the above, shut down the system by turning off the power supply and removing the leads from the fuel cell. Use the clamps to close off the tubing leading to and from the fuel cell. Disconnect the tubing from the electrolysis cell. Empty the water from the electrolysis cell and place it to the storage container.

• Look at the small, dismantled fuel cell and identify the main components of the membrane electrode assembly. Identify the anode and cathode flow plate/current collector, diffusion layers, and catalyst layers. Identify the polymer electrolyte membrane. Note the importance of seals (Figure C.8.3).

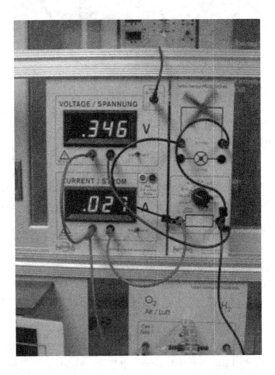

FIGURE C.8.3   Connection of the fan to the fuel cell system.

DATA TREATMENT

Fuel Cell Test

1. Start with drawing a schematic of a PEM fuel cell, labeling the key components and inputs.

   Plot the measured cell voltage vs. current. Identify and label the region dominated by activation polarization losses, the region dominated by ohmic losses, and the region dominated by concentration polarization losses. The concentration polarization region may not be identifiable from the experimental data—if not, indicate the region in which the concentration polarization losses would dominate.

2. Calculate the theoretical open circuit potential, $E_{theoretical}$, using the Nernst equation. Assume the fugacity (activity) of the hydrogen and oxygen is approximately equal to 1 bar and consider the behavior of the gases to be close to ideal. Assume the pressure at the anode (p_a) and the cathode (p_c) is ≈ 1 bar, and that $p_{O2} = p_{H2O} = p_c$ and $p_{H2} = p_a$.

   Helpful Info:

Electrode processes presented as reduction half-reactions	**Right: 0.5 O_2(g) + 2 H^+(aq) + 2 e^- = H_2O(l)** **Left: 2 H^+(aq) + 2 e^- = H_2(g)**
Cell open circuit potential:	$E = E_R - E_L$

3. Compare the calculated equilibrium potential to the experimental open circuit voltage value, taking into account that the experimental system consists of two fuel cells connected in series. Discuss possible reasons for discrepancy.

4. Calculate and plot the cell power vs. current.

Electrolysis Test

1. Calculate the voltage efficiency, ξ_V, of the electrolysis using the equation below and the experimental circuit voltage, E_{EC}, applied to the electrolysis cell:

$$\xi_V = E_D / E_{EC},$$

   where E_D (decomposition potential) is the theoretically determined minimum voltage that must be applied in order to carry out the production of hydrogen and oxygen.

2. Use Faraday's law to calculate the maximum (theoretical) mass (m_i) of hydrogen and oxygen produced in the water electrolysis cell:

$$m_i = It v_i M_i / (nF),$$

   where I is the current passed through the cell, t is time of electrolysis, F is Faraday constant, M_i is the molar mass of the ith chemical, v_i is the stoichiometric number of the ith chemical, and n is the number of electrons in the electrode reaction.

3. Use the ideal gas law

$$pV = (m/M)RT,$$

to determine the theoretical change in volume for each gas. In the ideal gas law, V is the volume of the produced gas, T is the absolute temperature (in K), and R is the molar gas constant ($8.314\,J\,mol^{-1}\,K^{-1}$). Assume that the ambient pressure $p = 1$ bar.

4. Calculate the current (Faradaic) efficiency using the equation below for both hydrogen and oxygen. Use the calculated values of gas volume (V_{calc}) from Steps 2 and 3 and the experimentally measured volumes (V_{exp}):

$$\xi_i = i/i_F = V_{exp}/V_{calc}.$$

5. Calculate the standard thermodynamic efficiency $\xi^\circ_{th,EC}$ of the electrolysis cell (the overall reaction) using the following equation:

$$\xi^\circ_{th,EC} = \Delta_r H^\circ / \Delta_r G^\circ,$$

where $\Delta_r H^\circ$ and $\Delta_r G^\circ$ are, respectively, the standard enthalpy and Gibbs energy of the electrolysis reaction.

6. Calculate the total efficiency ξ_{EC} of the electrolysis cell through the following equation, assuming that

$$\xi_{EC} = \xi_{th,\,EC}\xi_V\xi_i.$$

7. Explain why an applied voltage that is used for water electrolysis is significantly higher than the decomposition potential. Discuss possible reasons for discrepancies between the theoretical and experimental volumes of gas production. Discuss all contributions to the efficiency of this process.

LABORATORY: ELECTROCHEMICAL CORROSION MEASUREMENTS

Video URL: https://www.youtube.com/watch?v=kQTdqmFx_KQ
Video-Based Assignment 9 (Appendix B)

OBJECTIVES

- Understand the corrosion phenomenon as a galvanic cell and the concepts of corrosion potential E_{corr} and corrosion current I_{corr}.
- Obtain E_{corr} by measuring the open circuit potential OCP.

- Calculate the in-situ corrosion current and corrosion rate based on the Stern–Geary equation and Tafel Analysis from linear polarization resistance (LPR) and linear sweep voltammetry (LSV) data.
- Follow the Safety Guidelines in carrying out this lab.

EQUIPMENT AND MATERIALS

- Safety goggles and gloves
- Distilled water
- Paper towels
- 1 M H_2SO_4(aq), electrolyte solution
- *Gamry* Potentiostat
- Computer with *Gamry Framework*
- Rotating speed controller

- Three-electrode cell for electrochemical measurement
- Steel working electrode sample
- Platinum counter electrode
- Ag/AgCl reference electrode (4 mol L^{-1} KCl)
- 1 and 0.05 μm alumina polishing solutions
- Polishing microcloth

EXPERIMENTAL SETUP

Preparing the Cell

1. Polish the steel sample

    Place and polish the steel sample in the PTFE sample holder using the same method of cleaning and polishing that was used for the platinum working electrode in the RDE lab. Check that the steel electrode is mirror bright and free of defects. Remember to note down the working electrode's surface area (0.196 cm² with the diameter of 0.5 cm).

2. Assemble the RDE working electrode

    Carefully remove the polished steel sample from the sample holder. Place the steel sample into the sample holder section of the RDE tip. Use the press to position the steel sample within the holder so that the steel surface is flush with the PTFE around the sample. Attach the newly assembled tip to the threaded part of the RDE sample tip. Carefully screw the RDE sample tip onto the RDE shaft with the spring-loaded electric connection rod in place. Pay particular attention to the steel sample while screwing in the tip. Overtightening will result in the sample being pushed out of the sample holder, leaving the steel no longer flush with the PTFE or worse, ejecting the sample completely out of the holder. Under tightening will result in a poor electric connection.

3. Test Solution

    Rinse and clean the multiple-port glass cell. Fill with approximately 100 mL of 1 mol L^{-1} H_2SO_4(aq) test solution.

4. Reference Electrode

    Rinse the Ag/AgCl reference electrode with the distilled water and wipe clean. Place the reference electrode in the glass cell.

5. Counter Electrode

    Rinse the platinum coil counter electrode with the distilled water and wipe clean. Place the electrode in the glass cell.

6. Carefully bring down the rotor assembly of the RDE working electrode so that roughly 1 cm of the RDE tip is submerged. All other electrodes must also be in contact with the test solution for the test to work correctly.

Connecting the Circuit

1. Connect the Gamry potentiostat to a computer with the Gamry Framework installed via the USB connector.
2. Connect the blue and green colored leads from Gamry to the blue and red ports at the side of the RDE, respectively (Figure C.9.1).
3. Connect the red colored lead from Gamry with an alligator clip to the exposed tip of the platinum counter electrode outside the glass vessel (Figure C.9.1).
4. Connect the white colored lead from Gamry to the reference electrode using an alligator clip (Figure C.9.1).

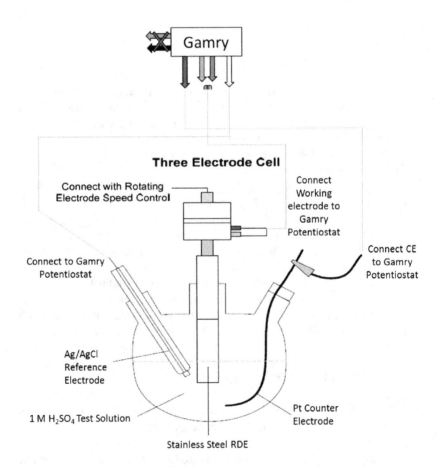

FIGURE C.9.1   Schematic of the experimental system.

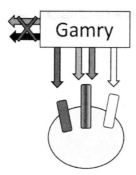

FIGURE C.9.2 Electric connections between Gamry and the electrochemical cell: **White**: Reference; **Red**: Counter; **Green** (with blue stripe): Working.

5. Leave the black and orange colored leads from Gamry disconnected. Make sure they are NOT in contact with each other or with any other electrically conductive surface (Figure C.9.2).
6. Make sure no connections are touching each other, with the exception of the blue and green leads as shown in Figure C.9.2.

Experimental Procedure

A. Measuring Corrosion Potential using OCP
- After setting up the system as instructed above, power on the RDE speed control and set the speed to 500 RPM. Ensure that roughly 1 cm of the RDE is immersed in the solution.
- On the computer, open the Gamry Framework software and click on "Experiment" in the Menus tab and navigate as shown in Figure C.9.3.

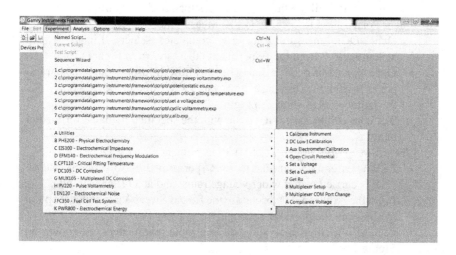

FIGURE C.9.3 Choosing the OCP setup using Gamry Framework.

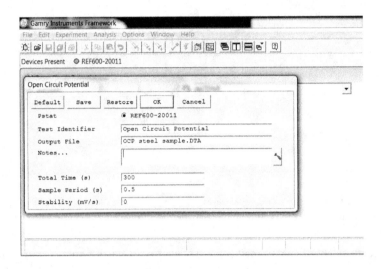

FIGURE C.9.4   Settings for the open circuit potential test.

- Upon clicking on the Open Circuit Potential, a new dialog box opens allowing the user to input the settings.
- Make sure to clearly name the data output file, preferably with your initials and the date and set the other parameters as shown in Figure C.9.4.
- The potentiostat should now measure/record OCP, i.e., E_{corr} values until/ if the steady state is reached. Here, we will assume steady state is when the E_{corr} does not change by more than 3 mV over a 5-minute period.
- Observe the trend of the recorded data and check to see whether the observed potential drift by more than 3 mV in 5 minutes. If it does, re-run the test until the fluctuation is less than 3 mV in 5 minutes.

B. Linear Polarization Resistance (LPR) Measurement
- Once OCP measurements are complete, select "Linear Sweep Voltammetry" from the Experiment tab as shown in Figure C.9.5.
- Once again, this opens a dialog box to allow the user to input the parameters. You will be running a downward sweep around the E_{corr} or E_{oc}. Set the parameters as shown in Figure C.9.6.
- The data will be saved in the file with the specified filename for further analysis.

C. Linear Sweep Voltammetry (LSV) Measurement
- LSV is essentially similar to LPR (performed above) except that an LSV is carried over a wider voltage range and at a faster scan rate.
- From the Experiment tab, choose the Linear Sweep Voltammetry option as shown in the previous section.
- Change the filename, and set the input parameters as shown in Figure C.9.7.
- Note that in LSV, we start from anodic polarization and sweep downward.

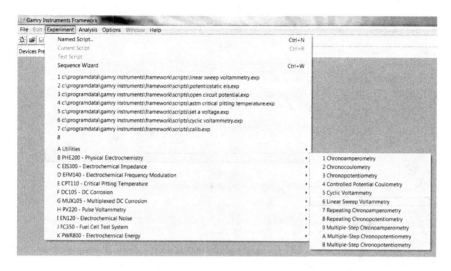

FIGURE C.9.5   Choosing the LPR option in Gamry Framework.

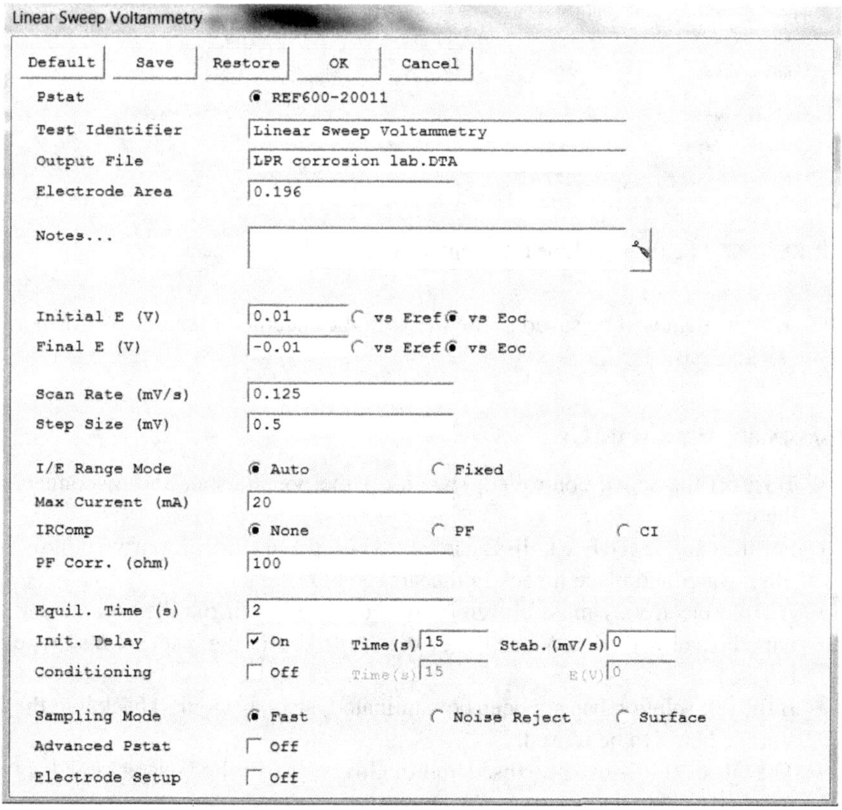

Linear Sweep Voltammetry

| Default | Save | Restore | OK | Cancel |

Pstat	⦿ REF600-20011
Test Identifier	Linear Sweep Voltammetry
Output File	LPR corrosion lab.DTA
Electrode Area	0.196

Notes...

| Initial E (V) | 0.01 | ○ vs Eref ⦿ vs Eoc |
| Final E (V) | -0.01 | ○ vs Eref ⦿ vs Eoc |

| Scan Rate (mV/s) | 0.125 |
| Step Size (mV) | 0.5 |

I/E Range Mode	⦿ Auto	○ Fixed
Max Current (mA)	20
IRComp	⦿ None	○ PF	○ CI
PF Corr. (ohm)	100

Equil. Time (s)	2
Init. Delay	☑ On	Time(s) 15	Stab. (mV/s) 0
Conditioning	☐ Off	Time(s) 15	E(V) 0

Sampling Mode	⦿ Fast	○ Noise Reject	○ Surface
Advanced Pstat	☐ Off
Electrode Setup	☐ Off

FIGURE C.9.6   Settings for Linear Polarization Resistance Measurement.

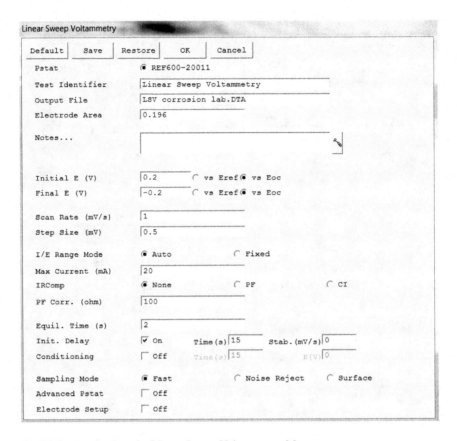

FIGURE C.9.7 Settings for Linear Sweep Voltammetry Measurement.

- The data will be saved in the file with the specified filename for further analysis.

DISASSEMBLY AND CLEAN-UP

- Turn off the speed controlled, switch off the potentiostat, and disconnect the wires.
- For the sample RDE kit, disassemble the setup and rinse all parts with distilled water and place it back in the storage box.
- All the electrodes must be removed and rinsed with distilled water and wiped clean with KimWipes. Place the electrodes in the respective storage boxes.
- If the test solution has not been contaminated, it can be poured back into the storage bottle to be reused.
- The glass cell must be rinsed thoroughly with distilled water and kept inverted to dry.
- If any of the test solutions must be disposed, it must be done in a manner suitable for hazardous solutions.

REPORT GUIDELINES

1. Introduction

    Address the following questions:

    • Describe the corrosion of a metal from the electrochemical perspective. Explain in simple sentences the processes involved.

    • Brief description of the statement of objectives.

2. Experimental Setup

    • Provide a brief description of the experimental procedures.

    • Show a schematic of the experimental setup indicating all connections and polarities.

    • Provide the electrochemical scheme of the corrosion cell, using "|" to separate phases, and indicate all concentrations.

3. Theory

    • Show a schematic demonstrating the process of electrochemical corrosion.

    • Write the half-reactions (as reduction reaction) that take place in the corrosion cell. Show the standard electrode potentials at 25°C for each half-reaction. Write the total spontaneous reaction.

    • Provide the Stern–Geary equation with the applicable assumption. Provide the equation to calculate the corrosion rate as the penetration depth per time from I_{corr}.

4. Data Treatment

    A. OCP Measurements

        – Show the plot of measured OCP values over time.

        – Does the plot indicate stability of the potential measurements?

        – Explain how this measured potential is a mixed potential.

    B. LPR Measurements

        – Show the current vs. potential plot of the LPR sweep.

        – Calculate the slope of linear section of the plot, which is usually within $\pm 5\,mV$ around E_{corr}. The slope gives the value of linear polarization resistance, R_{pol}.

        – Assuming the values of the symmetry factor $\beta_a = \beta_b = 0.5$, calculate the Tafel slopes and use them to calculate the corrosion current according to the Stern–Geary equation.

        – Using the corrosion current, calculate the corrosion rate in mm year^{-1}.

    C. LSV Measurements

        – Show a plot of the log current vs. potential from the LSV.

        – Estimate the Tafel slopes B_a' and B_c' from the linear regions of log$I$-E plot at anodic and cathodic polarizations. Calculate the corrosion current I_{corr} and corrosion rate using the measured B_a', B_c', and R_{pol} from LPR measurement according to the Stern–Geary equation.

        – As a second part of the calculations, find the point of intersection of the linear approximation from the anodic and cathodic regions of the log I-E plot, noted as (E_{corr}, I_{corr}). Calculate the corrosion rate (CR) taking the current value at the point of intersection as the corrosion current I_{corr}. Compare E_{corr} with the measured E_{corr} from OCP measurement.

- Compare the three I_{corr} and CR in a table. Describe the difference among these values. Determine the corrosion resistance level according to the textbook.
5. Conclusion
   • Summarize the findings in the parts A, B, and C of the measurements.
   • Discuss the advantage and disadvantage of the electrochemical method for calculating the corrosion rate.
   • How would you further improve the accuracy of the measurement of I_{corr}?
   • Was the internal resistance accounted for? How is this beneficial and what type of error does it lead to?

LAB REPORT TEMPLATE

Enter Title Here
   Name:
   Date:

STATEMENT OF OBJECTIVES

Summarize in a few short sentences the objective of this lab session.

BRIEF DESCRIPTION OF LAB PROCEDURES

Based on the lab instruction, summarize the procedures that you followed to successfully carry out the lab. Describe the procedure in a manner fit to conduct the lab exercise in person.

THEORY

Summarize the theory and background for the lab. Make sure to include electrochemical schematics, electrochemical reactions, and relevant equations where applicable. All equations must be numbered and the terms described.

DATA TREATMENT

List each question from the "Data Treatment" section of the instructions sheet followed by your response. You can refer to equations by their number if they have already been defined in the "Theory" section. All the relevant data must be presented in tables. Keep the formatting of the tables consistent. Show the relevant calculations and steps for each question. For Lab 6 (LSV-RDE), Lab 7 (CV), and Lab 9 (Corrosion), you do not need to show the raw data.

CONCLUSION

Summarize your conclusions from the data treatment section. List any trends and observations. If your answers vary from the theoretically expected trend of value, speculate on possible reasons for the discrepancy.

REFERENCES

Provide all used references in the correct format.

GENERAL NOTES

Please make sure your font is consistent and use only black colored text. All figures and images must be centered, numbered, and labeled. Make sure you specify the units while describing any terms or constants. Please cite anything that is not your own, including figures, data values and paraphrased text material. You may refer to (and cite) the equations and equation numbers from S.N. Lvov, *Introduction to Electrochemical Science and Engineering*, 2nd Ed., CRC Press, Boca Raton, 2021. Use all quantities, their names, units and symbols as suggested in this book. The detailed guideline for preparation of your lab reports is below.

STYLE GUIDELINE FOR PREPARATION OF LAB REPORTS

INTRODUCTION

In your future profession you will be preparing a lot of technical papers using some guidelines. The lab reports in this course should be prepared with the guideline given below without any deviations. A lack of time is not a suitable excuse. In preparing your reports, you should use only the recent MS Office software available from Penn State.

TEXT FORMATTING

- All texts should be written in Times New Roman 12-point font.
- Underline, **Bold**, *Italic*, ^{Superscript} and $_{\text{Subscript}}$ are acceptable (when appropriate).
- Avoid indentation of text. Paragraphs should be separated by a single space.
- Do not use multiple spaces, or any other form of in-text formatting until it is specially required.
- Always use "Align left" for text alignment.
- Use only one space after each sentence in a paragraph.
- Use page setup with 1-inch margins on all sides.
- Paginate your text in the bottom right corner as this document.

EQUATIONS AND SYMBOLS

- Do not use in-text formatting to align equations (e.g., indenting the numerator to align with denominator).
- When possible, type equations in single line textual format, e.g.,

$$E = E^* + \left[RT/(nF) \right] \ln\left(m_{H+} \gamma_{H+} \right). \tag{C.1}$$

Avoid using numbers in equations such as

$$E = E^* + \left[8.314T/(n96{,}485) \right] \ln\left(m_{H+} \gamma_{H+} \right). \tag{C.2}$$

- All coefficients and variables must be defined in the text as soon as they appear, usually just after equations. If an equation contains a number in its standard form (such as the Levich equation), you may replace that number with a variable as long as it is defined immediately afterward in the text.
- When defining the coefficients and variables, use examples from S.N. Lvov, *Introduction to Electrochemical Science and Engineering*, 2nd Ed., CRC Press, Boca Raton, 2022.
- Use one space before and after each equation.
- Always number the equations. Align each equation left, and numerate them on the right.
- Avoid using x, *, •, etc. for multiplication (use a space to ensure clarity).
- If the equation is too complex to be presented clearly in a single line textual format, use Microsoft Word Equation Editor (via "Insert" > "Equation"), e.g.,

$$ E = E^0 + \frac{RT}{nF} \ln\left(\frac{m_{H^+}\gamma_{H^+}}{m_{H^+}^0} \right). \tag{C.3} $$

- Never cut and paste equation images from PDF files or online resources, as this can drastically reduce the appearance quality and increase the file size.
- Use italic symbols for all variables, but non-italic for chemicals, numbers, and logs.
- Refer to S.N. Lvov, *Introduction to Electrochemical Science and Engineering*, 2nd Ed., CRC Press, Boca Raton, 2022 as the main source on quantities, units, symbols, fundamental constants, etc.
- Use a necessary number of digits for fundamental constants. For example, in most cases we use $96,485 \, C \, mol^{-1}$ for Faraday's constant. Do not use 9.6 $10^4 C \, mol^{-1}$ or $96,485.3399 \, C \, mol^{-1}$.
- Use a common practice to keep 5–6 digits in all calculations, and report the number of digits that is meaningful according to an estimated error of the values. Make sure you know the error of the reported values.
- Always spell out abbreviations when they first appear in the text (showing abbreviated form in parenthesis).
- In sentences, units follow the number they describe with one space between them (e.g., 5 cm, 2 A, 25°C).
- Phases for a chemical should follow the chemical abbreviation without a space [e.g., $NaCl(s)$, $H_2(g)$, and $CO_2(aq)$].
- S.N. Lvov, *Introduction to Electrochemical Science and Engineering*, 2nd Ed., CRC Press, Boca Raton, 2022 is a good reference to be used when writing down new equations. Find the same or similar equation and use it as an example.

TABLES

- Font in a table should be the same as in the paper text.
- Use the quotient of physical quantity and a unit in such a form that the values to be tabulated are pure numerical values (see IUPAC Manual for details).
- Use a reasonable number of digits for all numbers given in tables.

Figures

- The figures should always be kept in an "active" (editable) format in all working drafts. The acceptable formats are Excel or Visio.
- Color could be used in figures for better presentation, but try to use only black and white figures with following suggestion. Use marker symbols and solid/dashed lines to differentiate data and trends instead of different colors.
- Symbols, legend, and axis labels should be approximately of the same size as text font.
- Use the quotient of physical quantity and a unit in such a form that the values to be shown in the graph are pure numerical values (see IUPAC Manual for details).
- Scale axes such that plot fills as much of figure as possible (when appropriate).
- If convention requires scales to remain in unit ratio (such as Nyquist plot), do so regardless of image.
- Avoid any labels on the plot, but instead include them in the caption. This generally applies to legends and chart titles, not to the axes.
- Captions should precisely and concisely describe the image: specifying all experimental parameters. Additional sentences can be used to describe the different data series or conditions. Ideally, the captioned figure should be self-descriptive and need no external references to the text.
- It is preferred to use the caption to identify data sets if multiple sets are used. Example: "x–25°C; o–100°C; +- 200°C…" Try to use markers that can be easily identified in the caption.

IUPAC Conventions

Please adhere strictly to the conventions set forth by IUPAC Manual. A complete electronic copy of the third edition is now available online via Penn State Libraries. You should spend some time to read the manual. The particular chapters of interest are

- Section 1 (all subsections, i.e., from 1.1 to 1.6)
- Section 2 (any of the subsection might be important but particularly 2.10, 2.11, 2.12, and 2.13). Subsection 2.13 should be studied in all details.
- Section 3 should be used as needed
- Section 4 should always be used for recommended mathematical symbols
- A copy of Section 5 should always on your desk.
- Section 7 should be used to deal with units properly. Use the quantity calculus described in Subsection 7.1.
- Section 8 is one of the most important for us, but has not been properly used thus far. Learn this section, and implement the IUPAC recommendations in defining that data uncertainty.
- Section 9 should be used as needed.
- Use Section 11 to properly use Greek characters and know their pronunciations.
- Section 12 can help you to correctly use the physico-chemical symbols.
- Section 13 is a convenient subject index.
- In pages 233 and 234, you can find the pressure and energy conversion factors.

REFERENCE DATA

- You may refer to (and cite) the equations and equation numbers from S.N. Lvov, *Introduction to Electrochemical Science and Engineering*, 2nd Ed., CRC Press, Boca Raton, 2022.
- You may refer to (and cite) the reference data from S.N. Lvov, *Introduction to Electrochemical Science and Engineering*, 2nd Ed., CRC Press, Boca Raton, 2022.
- Use *CRC Handbook of Chemistry and Physics*, *CRC Press*, (latest edition) to find the necessary data for your calculations.
- Use IUPAC Manual - 1. E.R. Cohen et al., *Quantities, Units, and Symbols in Physical Chemistry*, 3rd ed., RSC Publishing, Cambridge, UK, 2007.
- Use other reference books and original papers as soon as *CRC Handbook of Chemistry and Physics* is not sufficient.

Index

Printed in the United States
by Baker & Taylor Publisher Services